# Lecture Notes in Computer Science 4475

*Commenced Publication in 1973*
Founding and Former Series Editors:
Gerhard Goos, Juris Hartmanis, and Jan van Leeuwen

Lecture Notes in Computer Science 4475

Pierluigi Crescenzi   Giuseppe Prencipe
Geppino Pucci (Eds.)

# Fun with Algorithms

4th International Conference, FUN 2007
Castiglioncello, Italy, June 3-5, 2007
Proceedings

 Springer

Volume Editors

Pierluigi Crescenzi
Università degli Studi di Firenze, Dipartimento di Sistemi e Informatica
Viale Morgagni 65, 50134 Firenze, Italy
E-mail: piluc@dsi.unifi.it

Giuseppe Prencipe
Università degli Studi di Pisa, Dipartimento di Informatica
Largo Bruno Pontecorvo 3, 56127 Pisa, Italy
E-mail: prencipe@di.unipi.it

Geppino Pucci
Università degli Studi di Padova, Dipartimento di Ingegneria dell'Informazione
Via Gradenigo 6/B, 35131 Padova, Italy
E-mail: geppo@dei.unipd.it

The logo is the work of Hadrien Dussoix.

Library of Congress Control Number: 2007927412

CR Subject Classification (1998): F.2, F.1, E.1, I.3.5, G.1-4

LNCS Sublibrary: SL 1 – Theoretical Computer Science and General Issues

ISSN      0302-9743
ISBN-10   3-540-72913-5 Springer Berlin Heidelberg New York
ISBN-13   978-3-540-72913-6 Springer Berlin Heidelberg New York

Springer is a part of Springer Science+Business Media

springer.com

© Springer-Verlag Berlin Heidelberg 2007
Printed in Germany

Typesetting: Camera-ready by author, data conversion by Scientific Publishing Services, Chennai, India
Printed on acid-free paper      SPIN: 12072804      06/3180      5 4 3 2 1 0

# Preface

This volume contains the papers presented at the Fourth International Conference on Fun with Algorithms (FUN 2007), held June 3–5, 2007 in the beautiful Tuscanian coastal town of Castiglioncello, Italy.

FUN is a three-yearly conference dedicated to the use, design, and analysis of algorithms and data structures, focusing on results that provide amusing, witty but nonetheless original and scientifically profound contributions to the area. The previous three meetings were held on Elba Island, Italy, and special issues of the journals *Theoretical Computer Science* (FUN 1998), *Discrete Applied Mathematics* (FUN 2001), and *Theory of Computing Systems* (FUN 2004) feature extended versions of selected papers from the three conference programs.

In response to the Call for Papers for FUN 2007, we received 41 submissions from 25 countries. Each submission was reviewed by at least three Program Committee members. At the end of the selection process, the committee decided to accept 20 papers. The program also includes three invited talks by Giuseppe Di Battista (U. Rome III, Italy), Nicola Santoro (Carleton U., Canada), and Luca Trevisan (U.C. Berkeley, USA).

We wish to thank all the authors who submitted their papers to FUN 2007 and thus contributed to the creation of a high-quality program and entertaining meeting, as well as the colleagues who accepted to serve on the Program Committee and provided invaluable help with the reviewing process. We also wish to thank the external reviewers (listed on the following pages) including those who completed urgent reviews during the discussion phase. Paper submission, selection, and generation of the proceedings was greatly eased by the use of the public-domain *EasyChair* Conference System (http://www.easychair.org). We wish to thank the EasyChair creators and maintainers for their selfless committment to the scientific community. Finally, special thanks go to Vincenzo Gervasi, whose constant help and dedication was crucial in making FUN 2007 a successful event.

April 2007

Pierluigi Crescenzi
Giuseppe Prencipe
Geppino Pucci

# Conference Organization

## Program Chairs

Pierluigi Crescenzi (University of Firenze, Italy)
Geppino Pucci (University of Padua, Italy)

## Program Committee

Nancy Amato (Texas A & M University, USA)
Nina Amenta (University of California at Davis, USA)
Marcella Anselmo (University of Salerno, Italy)
Anna Bernasconi (University of Pisa, Italy)
Paolo Boldi (University of Milano, Italy)
Irene Finocchi (University of Roma "La Sapienza", Italy)
Luisa Gargano (University of Salerno, Italy)
Sandy Irani (University of California at Irvine, USA)
Christos Kaklamanis (University of Patras, Greece)
Shay Kutten (Technion, Haifa, Israel)
Fabrizio Luccio (University of Pisa, Italy)
Bernard Mans (Macquarie University, Australia)
Paolo Penna (University of Salerno, Italy)
Andrea Richa (Arizona State University, Tempe, USA)
Iain Stewart (University of Durham, UK)
Erkki Sutinen (University of Joensuu, Finland)
Denis Trystram (ID-IMAG Grenoble, France)
Peter Widmayer (ETH Zurich, Switzerland)

## Local Organization

Vincenzo Gervasi (University of Pisa, Italy)
Giuseppe Prencipe (University of Pisa, Italy)

## External Reviewers

Luca Becchetti
Hajo Broersma
Valentina Ciriani
David Coudert
Stefan Dantchev
Annalisa De Bonis
Gianluca De Marco

Miriam Di Ianni
Paola Flocchini
Tom Friedetzky
Giulia Galbiati
Goran Konjevod
Zvi Lotker
Ornella Menchi
Filippo Mignosi
Manal Mohammed
Melih Onus
Linda Pagli
Fanny Pascual
Andrea Pietracaprina
Srinivasa Rao
Adele Rescigno
Andrea Richa
Gianluca Rossi
Massimo Santini
Erik Saule
Marinella Sciortino
Riccardo Silvestri
Corinne Touati
Denis Trystram
Sebastiano Vigna
Ivan Visconti
Donglin Xia
Michele Zito
Rosalba Zizza

# Table of Contents

# On Embedding a Graph in the Grid with the Maximum Number of Bends and Other Bad Features

Giuseppe Di Battista, Fabrizio Frati, and Maurizio Patrignani

Dipartimento di Informatica e Automazione – Università di Roma Tre
{gdb,frati,patrigna}@dia.uniroma3.it

**Abstract.** Graph Drawing is (usually) concerned with the production of readable representations of graphs. In this paper, instead of investigating how to produce "good" drawings, we tackle the opposite problem of producing "bad" drawings. In particular, we study how to construct orthogonal drawings with many bends along the edges and with large area. Our results show surprising contact points, in Graph Drawing, between the computational cost of niceness and the one of ugliness.

## 1 Breaking the Graph Drawing Rules

Up to now, bad diagrams have been produced manually or with the aid of a graphic editor; in both cases placement of symbols and routing of connections are under responsibility of the designer. The goal of this work is to investigate how poor readability of diagrams can be achieved by means of automatic tools.

Indeed, although the opposite problem of automatically producing good quality drawings of graphs has been studied since, at least, three decades by a large research community, called Graph Drawing community, the problem of obtaining drawings where the main quality is unreadability has been, as far as we know, neglected.

One of the most important reference points for the Graph Drawing community is the seminal paper of Tamassia [13] devoted to the minimization of the number of bends in orthogonal drawings. Such a paper can be considered as the milestone of the topology-shape-metric approach (see also, [6,2,5]), in which the process of producing an orthogonal drawing is organized in three steps: in the Planarization step the topology of the drawing, is determined. Such a topology is described by a planar embedding, i.e., the order of the edges around each vertex. In this step the purpose is to minimize the number of crossings. The Orthogonalization step determines the drawing shape, in which vertices do not have coordinates and each edge is equipped with a list of angles, describing the bends featured by the orthogonal line representing the edge in the final drawing. The purpose of this step is the minimization of the total number of bends. The Compaction step determines the final coordinates of the vertices and bends. The target is to minimize the area and/or the total length of the edges.

We look at the topology-shape-metric approach from the opposite perspective. Namely, our purpose is to study how a bad orthogonal drawing of a graph can be constructed by interpreting on the negative side the three mentioned steps. More precisely, we concentrate on Orthogonalization and Compaction, leaving to further studies contributions on the Planarization step.

P. Crescenzi, G. Prencipe, and G. Pucci (Eds.): FUN 2007, LNCS 4475, pp. 1–13, 2007.

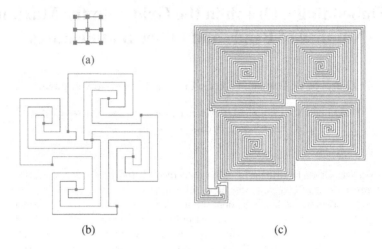

(a)

(b)                                    (c)

**Fig. 1.** A drawing of the 3 × 3 grid with 0 bends per edge (a), 5 bends per edge (b), and 30 bends per edge (c)

In a *planar orthogonal drawing* $\Gamma$ of a plane graph $G$ each edge is drawn as a polygonal chain of alternating horizontal and vertical segments. There are two types of angles in $\Gamma$ [5]. Angles formed by edges incident on a common vertex, called *vertex-angles* and angles formed by bends (formed by consecutive segments on the same edge), called *bend-angles*. The sum of the measures of the vertex-angles around a vertex is $2\pi$. Let $f$ be an internal face. The sum of the measures of the vertex-angles and bend-angles inside $f$ is $\pi(p-2)$, where $p$ is the total number of such angles. If $f$ is the external face, then the above sum is $\pi(p+2)$. A plane graph has a planar orthogonal drawing iff the degree of its vertices is at most 4.

Consider any edge $e$ of $\Gamma$ and an arbitrary direction for $e$. Try to add along $e$ two more bends: a $\pi/2$ bend to the left and a consecutive $\pi/2$ bend to the right (or vice-versa). It is easy to see that there always exists an orthogonal drawing $\Gamma'$ of $G$ with the same shape of $\Gamma$ plus the two mentioned extra bends. Hence, we can arbitrarily increase the ugliness of a drawing inserting consecutive pairs of left and right bends on each edge. However, the aesthetic effect of those bends is not "that bad", in the sense that the human eye can easily "virtually stretch" such two consecutive bends still being able to read the drawing. The effect of sequences of bends all to the left (right) is much worse. Hence, in the following we do not consider drawings that have an edge with two consecutive left-right or right-left bends.

Consider again edge $e$ and try to add to $e$ an arbitrary number of bends all to the left (right). Even in this case it is easy to see that there always exists an orthogonal drawing $\Gamma'$ of $G$ with the prescribed angles on $e$. This implies that, even if we neglect consecutive left-right and right-left bends, it is possible to draw $G$ with a number of bends that is arbitrarily high. However, consider again $\Gamma'$ from the aesthetic perspective. Even if $e$ has now a large number of bends, we do not know anything on the remaining part of the drawing, that, maybe, has in $\Gamma'$ still a nice sub-drawing. At this point it would be easy for the human eye to neglect the bad shape of $e$, concentrating on the remaining part of the drawing and preserving a "side view" of the adjacency expressed by $e$. Hence, to capture the notion of ugly drawing we need a more sophisticated model.

A $k$-bend drawing $\Gamma$ of $G$ is an orthogonal drawing where each edge $e$ has exactly $k$ bends. Traveling on $e$ in any direction such bends are either all to the left or all to the right. We think that the notion of $k$-bend drawing captures very well the notion of bad drawing. Of course the highest is $k$ the worst is the drawing. Examples of 0-bend, 5-bend, and 30-bend drawings are in Figure 1.

In Section 3 we study if it is possible to construct $k$-bend drawings. We show that, unfortunately, there are important classes of graphs that cannot be arbitrarily unpleasant from this perspective. On the other hand there are classes that have this interesting feature. Our results show surprising contact points between the computational cost of niceness and the one of ugliness.

Once the shape has been determined, the topology-shape-metric approach computes the final drawing. The *area* of a *grid drawing* $\Gamma$, where vertices and bends have integer coordinates, is the number of grid points of a minimum size rectangle with sides parallel to the axes that covers the drawing. Of course, a nice drawing is a drawing with limited area. Conversely, a bad drawing is a drawing with large area. Even in this case, to capture the idea of bad drawing it is not enough to simply maximize instead of minimize. In fact, it is easy to see that any grid drawing can be scaled-up to an arbitrarily large value of area. However, the aesthetic effect of this is negligible, since for the human eye is quite easy to re-scale down and to read the drawing. Hence, we adopt a different model. We consider only drawings that do not have "empty strips". Namely, in our drawings if $x_m$ and $x_M$ are the minimum and maximum $x$-coordinate of a vertex or of a bend of $\Gamma$, for each integer $x_i$ with $x_m \leq x_i \leq x_M$ there is either a vertex or a bend in $\Gamma$ with $x$-coordinate equal to $x_i$. The same holds for $y$-coordinates.

In Section 4 we study the problem of maximizing the area in an orthogonal drawing of a graph. Since in the topology-shape-metric approach the final coordinates are computed after the Orthogonalization step, we will assume that the orthogonal shape to draw has been already fixed. In this setting we will also consider the problem of maximizing the total edge length of an orthogonal drawing of a given shape.

Finally, in Section 5 we propose alternative models that can be studied in order to construct bad drawings of graphs and we suggest several open problems that we believe are worth of interest in a hypothetical Bad Graph Drawing community.

## 2   Orthogonal Representations and Flow Networks

To continue our discussion we need some definitions from Graph Drawing.

Let $f$ be a face of a plane graph $G$ of maximum degree four, and let $\Gamma$ be an orthogonal drawing of $G$. Each pair of consecutive (possibly coinciding) segments of $f$ can be associated with a value $\alpha$, where $1 \leq \alpha \leq 4$, such that $\alpha \cdot \pi/2$ is the angle formed by the two segments into $f$.

An *orthogonal representation* or *orthogonal shape* $H$ of $G$ is the equivalence class of planar orthogonal drawings of $G$ with the "same shape", that is, with the same $\alpha$ values associated with the angles of its faces.

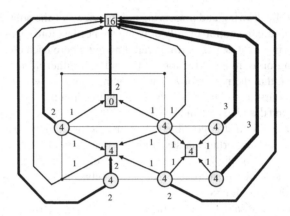

**Fig. 2.** An example of orthogonal shape and the corresponding flow network, where only non-null flows are represented. Vertices (circles) are labeled with the flow they produce. Faces (rectangles) are labeled with the flow they consume.

In [17,13] it is shown that an orthogonal representation of $G$ corresponds to an assignment of $\alpha$ values to the angles such that $1 \leq \alpha \leq 4$ and the sum of $\alpha$ values around an internal (external) face $f$ is 4 ($-4$, respectively).

In [13,5] it is shown that any orthogonal shape $H$ of a degree four plane graph $G$ is associated with a flow into a suitable flow network $N$ defined as follows. $N$ has a node $n_v$ for each vertex $v$ of $G$ and a node $n_f$ for each face $f$. Also, $N$ has a directed arc $(n_v, n_f)$ for each vertex $v$ incident to a face $f$. Finally, for any pair of adjacent faces $f$ and $g$, $N$ has two arcs $(f, g)$ and $(g, f)$. In $N$ each unit of flow is meant to represent a $\pi/2$ angle. Hence, each vertex is a producer of four units of flow and each face $f$ of degree $a(f)$ consumes $2a(f) - 4$ units of flow, if $f$ is internal, or $2a(f) + 4$ units of flow, if $f$ is external. Each bend in $H$ corresponds to one unit of flow across its incident faces. Therefore, by giving unit cost to the flow exchanged between adjacent faces, we have that a drawing with the minimum number of bends corresponds to a flow of minimum cost. This yields a polynomial-time algorithm for bend minimization. This technique was first presented in [13], with variations, refinements, and extensions given in [6,12,14,16]. Linear-time algorithms for constructing planar orthogonal drawings with $O(1)$ bends per edge, but that do not guarantee the minimum number of bends, are given in [15]. Note that it is NP-hard to minimize bends over all possible embeddings of a planar graph [7]. Polynomial-time algorithms exist only for special classes of planar graphs [1,9,11].

## 3   Maximizing the Number of Bends

In this section we deal with the maximization of the number of bends in orthogonal drawings. First, we show that, for all bipartite graphs that admit a straight-line orthogonal drawing, arbitrarily bad drawings can be constructed.

**Theorem 1.** *A bipartite graph admitting a 0-bend drawing admits a $k$-bend drawing, for any positive integer $k$.*

*Proof:* Let $V$ be the vertex set of $G$, with $V = V_1 \cup V_2$ such that $G$ contains only edges from vertices in $V_1$ to vertices in $V_2$. Suppose $G = (V, E)$ admits a 0-bend orthogonal representation $H$. Consider $|V_1|$ cuts such that each cut $c_i$ consists of the edges incident to a distinct vertex in $V_1$. Since $V$ is bipartite each edge belongs to exactly one cut. Cut $c_i$ corresponds, in the flow network associated with $H$, to a cycle $C_i$, which can be assumed arbitrarily oriented. Increase the flow in each $C_i$ of $k$ units. It's easy to see that the obtained flow corresponds to an orthogonal representation $H'$ with exactly $k$ bends on each edge.                                                                              $\square$

For example, Figure 1 shows a $3 \times 3$ grid drawn with 0, 5, and 30 bends per edge.

Next, we show that when dealing with general planar graphs, the ugliness of the drawings cannot be arbitrarily high:

**Theorem 2.** *Let $G$ be a non-bipartite plane graph. There exists an integer $k_0 > 0$ such that, for every integer $k \geq k_0$, $G$ does not admit a $k$-bend drawing.*

*Proof:* Suppose that $G$ is biconnected: the proof for the connected case is analogous. If $G$ is not bipartite, then $G$ has at least one face of odd degree. Consider the odd-degree face $f$ that has the smallest number $2m + 1$ of vertices. Let $k_0 = 2m + 6$. Suppose, as a contradiction, that $G$ admits a $k_0$-bend orthogonal representation $H$. Consider the network flow associated with $H$ and node $n_f$ associated with $f$, which is a sink of $2(2m + 1) - 4 = 4m - 2$ units of flow if $f$ is an internal face, or is a sink of $2(2m + 1) + 4 = 4m + 6$ units of flow if $f$ is the external face. Each edge $e$ of $f$ corresponds to one arc $a_e^+$ entering $n_f$ and one arc $a_e^-$ exiting $n_f$. Since $H$ is a $k_0$-bend orthogonal representation of a $k$-bend drawing, one between $a_e^+$ and $a_e^-$ carries $k_0$ units of flow, while the other carries none. Since $f$ has an odd number of edges, the sum of such flows yields at least $k_0$ units either entering (Case 1) or exiting (Case 2) $n_f$. Also, the $2m + 1$ vertices of $G$ incident to $f$ inject into $n_f$ at least $2m + 1$ and at most $6m + 3$ units of flow. In Case 1 we have at least $(2m + 6) + (2m + 1) = 4m + 7$ units entering $n_f$ that needs at most $4m + 6$ units of flow. In Case 2 we have at most $6m + 3$ units injected by the vertices of $f$, while we need at least $(2m + 6) + (4m - 2) = 6m + 4$ units of flow to balance the flow in $n_f$. Since there is not a network flow associated with $H$ that satisfies the above constraints, we have a contradiction.                      $\square$

In the next theorem we show that for all planar bipartite graphs the possibility of obtaining bad drawings determines also the possibility of obtaining good drawings.

**Theorem 3.** *Let $G$ be a bipartite plane graph. There exists an integer $k(G) > 0$ such that if $G$ admits a $k(G)$-bend drawing, then $G$ admits a 0-bend drawing.*

*Proof:* Suppose that $G$ is biconnected: the proof for the connected case is analogous. Let $k(G) = M + 3$, where $2M$ is the greatest number of vertices incident to a face of $G$. Suppose $G$ admits a $k(G)$-bend orthogonal representation $H$. Consider the network flow $N$ associated with $H$ and the node $n_f$ associated with a face $f$ with $2m \leq 2M$ edges, which, hence, is a sink of $2(2m) - 4 = 4m - 4$ units of flow if $f$ is an internal face, or is a sink of $2(2m)+4 = 4m+4$ units of flow if $f$ is the external face. Each edge $e$ of $f$ corresponds to one arc $a_e^+$ entering $n_f$ and one arc $a_e^-$ exiting $n_f$. Since $H$ is a $k(G)$-bend orthogonal representation of a $k(G)$-bend drawing, one between $a_e^+$ and $a_e^-$ carries $k(G)$ units of flow, while the other carries none. Since $f$ has an even number of edges, the sum of such flows yields at least $2k(G) = 2M+6$ units entering $n_f$ (Case 1), at least $2k(G) = 2M + 6$ units exiting $n_f$ (Case 2), or exactly zero units entering $n_f$ (Case 3). Also, the $2m$ vertices of $G$ incident to $f$ inject into $n_f$ at least $2m$ and at most $6m$ units of flow. In Case 1 we have at least $(2M + 6) + (2m) \geq 4m + 6$ units entering $n_f$ that needs at most $4m + 4$ units of flow. In Case 2 we have at most $6m$ units injected by the vertices of $f$, while we need at least $(2M + 6) + (4m - 4) \geq 6m + 2$ units of flow to balance the flow in $n_f$. Hence, Case 3 is the only possible for each face of $H$, which implies that the $4m - 4$ units of flow needed by each internal face and the $4m+4$ units of flow needed by the external face are balanced by the flow coming from their incident vertices. Therefore, we can obtain a network flow $N'$ from $N$ where, for each edge $e$, the flow on the arcs $a_e^+$ and $a_e^-$ is equal to zero. The orthogonal representation associated with $N'$ has zero bends. $\qquad\square$

Notice that if the integer $k(G)$ of the above theorem exists such that $G$ admits a $k(G)$-bend drawing, then Theorem 1 applies and $G$ admits a $k$-bend drawing, for every $k \geq 0$.

## 4   Maximizing the Area of an Orthogonal Shape

In this section we deal with the problem of obtaining orthogonal drawings of a shape with maximum area. First, we show that both for biconnected and for simply-connected orthogonal shapes the area requirement cannot be arbitrarily high.

**Theorem 4.** *The maximum area of an orthogonal drawing of a connected graph with $n$ vertices and $b$ bends such that every vertex has degree at least 2 is $\lfloor \frac{n+b}{2} \rfloor \cdot \lceil \frac{n+b}{2} \rceil$.*

*Proof:* Consider any orthogonal drawing $\Gamma$ of a graph $G$. Replace each bend with a dummy vertex, obtaining an orthogonal drawing $\Gamma'$ with $n' = n + b$ vertices and no bend. For every vertex $u$ that has only two incident edges $(u, u_1)$ and $(u, u_2)$ that are both vertical, remove $u$, insert an edge $(u_1, u_2)$, and, if there is no other vertex on the same horizontal grid line $R$ of $u$, delete $R$ (all the edges cutting $R$ will be shortened consequently). Analogously, for every vertex $u$ that has only two incident edges $(u, u_1)$ and $(u, u_2)$ that are both horizontal, remove $u$, insert an edge $(u_1, u_2)$, and, if there is no other vertex on the same vertical grid line $C$ of $u$, then delete $C$ (all the edges cutting $C$ will be shortened consequently). Let $r$ and $c$ be the number of horizontal and vertical deleted grid lines, respectively. The resulting $n''$-vertex orthogonal drawing $\Gamma''$, with $n'' \leq n'$, is still such that every vertex has degree at least 2. Moreover, there are at least two vertices for each horizontal and for each vertical grid line of the drawing. Hence, the maximum area of $\Gamma''$ is $(\lfloor n''/2 \rfloor) \times (\lfloor n''/2 \rfloor)$. Observe that the area of $\Gamma'$ is at most $(c + \lfloor n''/2 \rfloor) \times (r + \lfloor n''/2 \rfloor) = (rc + (r + c)\lfloor n''/2 \rfloor + (\lfloor n''/2 \rfloor)^2)$ and recall that $n' = n'' + r + c$. For every $n''$ the area of $\Gamma'$ is maximized when $rc$ is maximal, that is: (i) when $r = c = \frac{n'-n''}{2}$, in the case in which $r + c$ is even; in this case the maximum area of $\Gamma'$ is $(\frac{n'-n''}{2} + \lfloor \frac{n''}{2} \rfloor)^2$, that is equal to $(\frac{n'}{2})^2$ if $n''$ and $n'$ are even and is equal to $(\frac{n'-1}{2})^2$ if $n''$ and $n'$ are odd; (ii) when $r = \frac{n'-n''-1}{2}$ and $c = \frac{n'-n''+1}{2}$ or vice versa in the case in which $r + c$ is odd; in this case the maximum area of $\Gamma'$ is $(\frac{n'-n''-1}{2} + \lfloor \frac{n''}{2} \rfloor)(\frac{n'-n''+1}{2} + \lfloor \frac{n''}{2} \rfloor)$, that is equal to $\frac{n'-2}{2}\frac{n'}{2}$ if $n''$ is odd and $n'$ is even and is equal to $\frac{n'-1}{2}\frac{n'+1}{2}$ if $n''$ is even and $n'$ is odd. From these bounds the claimed lower bound follows by replacing $n'$ with $n+b$. Finally, observe that the bound $\lfloor \frac{n+b}{2} \rfloor \cdot \lceil \frac{n+b}{2} \rceil$ is tight, since there exist shapes that have such an area drawing (see e.g. the shapes in Figs. 3.a and 3.b). $\qquad\square$

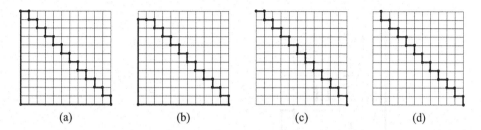

(a)            (b)            (c)            (d)

**Fig. 3.** The bounds in Theorem 4 and 5 are tight: (a) If $n + b$ is even, then there exist connected shapes with all the vertices of degree at least two that admit orthogonal drawings in $(\frac{n+b}{2})^2$ area. (b) If $n+b$ is odd, then there exist connected shapes with all the vertices of degree at least two that admit orthogonal drawings in $\frac{n+b-1}{2}\frac{n+b+1}{2}$ area. (c) If $n + b$ is odd, then there exist connected shapes that admit orthogonal drawings in $(\frac{n+b+1}{2})^2$ area. (d) If $n + b$ is even, then there exist connected shapes that admit orthogonal drawings in $\frac{n+b}{2}\frac{n+b+2}{2}$ area.

Observe that Theorem 4 offers a bound for the maximum area covered by a biconnected graph.

**Theorem 5.** *The maximum area of an orthogonal drawing of a connected graph with $n$ vertices and $b$ bends is $\lfloor \frac{n+b+1}{2} \rfloor \times \lceil \frac{n+b+1}{2} \rceil$.*

*Proof:* Consider any orthogonal drawing $\Gamma$ of a graph $G$. Replace each bend with a dummy vertex, obtaining an orthogonal drawing $\Gamma'$ with $n' = n + b$ vertices and no bend. Till there are vertices of degree 1, remove one of them, say $u$, and its incident edge. If there is no other vertex on the same horizontal grid line $R$ of $u$, delete $R$ (all the edges cutting $R$ will be shortened consequently). Analogously, if there is no other vertex on the vertical grid line $C$ of $u$, delete $C$ (all the edges cutting $C$ will be shortened consequently). Let $r$ and $c$ be the number of deleted horizontal and vertical grid lines, respectively. After the removal of all the vertices of degree 1 two situations are possible. In the first case a graph with all vertices of degree at least 2 is left. By means of a proof similar to the one of Theorem 4, bounds similar to the ones in such a theorem can be proved. In the second case ($G$ is a tree) only one vertex without incident edges is left. In this case every drawing has a total number of rows and columns that is at most $r + 1$ and $c + 1$, respectively. The total area of the drawing is hence at most $(r + 1)(c + 1)$, that is maximized when $rc$ is maximized. Since $r + c = n' - 1$, $rc$ is maximized when $r = c = \frac{n'-1}{2}$ if $n'$ is odd, or when $r = \frac{n'-2}{2}$ and $r = \frac{n'}{2}$ or vice versa if $n'$ is even. In the former case we obtain a maximum area of $(\frac{n'+1}{2})^2$, in the latter one $\frac{n'}{2}\frac{n'+2}{2}$. From these bounds the claimed lower bound follows by replacing $n'$ with $n + b$. Finally, observe that the bound $\lfloor\frac{n+b+1}{2}\rfloor \times \lceil\frac{n+b+1}{2}\rceil$ is tight, since there exist shapes that have such an area drawing (see e.g. the shapes in Figs. 3.c and 3.d). $\square$

Changing the geometry can lead to different area exploitations, as shown in the following:

**Lemma 1.** *There exist shapes that admit an $O(n + b)$ minimum area drawing and an $O((n + b)^2)$ maximum area drawing.*

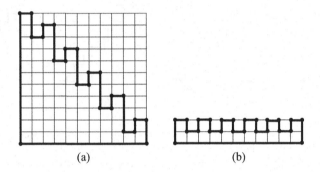

(a)                                        (b)

**Fig. 4.** (a) A quadratic area drawing of the orthogonal shape $H$. (b) A linear area drawing of the orthogonal shape $H$.

*Proof:* Consider the shape $H$ drawn in Fig. 4.a. Let $n$ and $b$ be the vertices and bends of $H$, respectively (identically represented in Fig. 4.a by circles). It's easy to see that $H$ covers an area that is quadratic in $n + b$. The same shape requires linear width and constant height when drawn as in Fig. 4.b. $\square$

While the problem of minimizing the area of an orthogonal drawing with a given orthogonal shape is NP-complete [10], the problem of maximizing the area of a given orthogonal shape can be solved in linear-time, as described in the following.

**Theorem 6.** *Computing a maximum-area drawing of an orthogonal shape has linear-time complexity in the number of vertices of the graph.*

*Proof:* Let $H$ be an orthogonal shape. Obtain an orthogonal shape $H'$ with no bend by replacing each bend of $H$ with a dummy vertex. Orient each horizontal edge of $H'$ from left to right and each vertical edge from the bottom to the top. Consider two maximal sequences $s'$ and $s''$ of vertices connected by horizontal edges. We say that $s'$ *precedes* $s''$ if there is a sequence of vertical edges from $s'$ to $s''$. The set of maximal horizontal sequences and their precedence relation is a partially ordered set. Assign a $y$ coordinate to each sequence according to its position in a linear extension of such a poset. An $x$ coordinate for each maximal sequence of vertices connected by vertical edges can be found in an analogous way. Since each horizontal (vertical) sequence must lie on the same grid line, the area of the produced drawing is maximal. All the steps involved in the computation can be performed in linear time. $\square$

The maximization of the total edge length appears to be a more difficult problem. However, for special families of orthogonal shapes we are able to find a polynomial-time algorithm.

**Lemma 2.** *Let $H$ be an orthogonal shape such that all its faces are rectangular. Computing a drawing of $H$ with maximum total edge length has polynomial-time complexity.*

*Proof:* First, we compute the maximum area drawing $\Gamma$ of $H$ by using the technique described in the proof of Theorem 6. Let $w$ and $h$ be the width and height of $\Gamma$, respectively. We use a technique analogous to that described in [5] for the minimization of the area of an orthogonal drawing with rectangular faces. Namely, we build two flow networks $N_h$ (see Fig. 5.a) and $N_v$ (see Fig. 5.b) where each internal face $f$ of $H$ corresponds to a node $n_f^h$ in $N_h$ and to a node $n_f^v$ in $N_v$. Also, $N_h$ ($N_v$) has two special nodes $s^h$ and $t^h$ ($s^v$ and $t^v$, respectively) representing the left and right region (the lower and the upper region, respectively) of the external face. Two nodes of $N_h$ ($N_v$) are connected by an arc if the corresponding faces share a vertical (respectively, horizontal) segment. The arcs of $N_h$ ($N_v$) are oriented from $s^h$ to $t^h$ ($s^v$ to $t^v$, respectively). Each arc has a minimum flow of 1 unit and cost 1. In [5] it is shown that two admissible flows in $N_h$ and $N_v$ correspond to a drawing $\Gamma$ of $H$. Hence, the maximum-cost flow of $N_h$, when $s^h$ ($t^h$) is a source (sink) of $h$ units of flow and analogous maximum-cost flow of $N_v$, when $s^v$ ($t^v$) is a source (sink) of $w$ units of flow provide a drawing of $H$ with the maximum total edge length. □

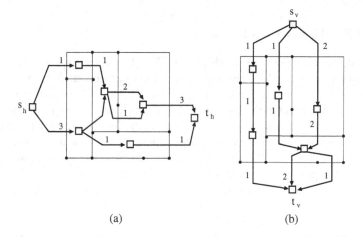

**Fig. 5.** The two flow networks $N_h$ (a) and $N_v$ (b) in the proof of Lemma 2

Observe that Lemma 2 can be extended to turn-regular [3] orthogonal shapes, by suitably transforming them into rectangular orthogonal shapes.

## 5  Other Models for Bad Drawings and Open Problems

In this paper we have considered the problem of producing bad drawings of graphs, mostly focusing on the maximization of the number of bends. Namely, we have shown that, for some classes of planar graphs, drawings with an arbitrarily high number of bends can be produced, while for other classes of planar graphs it's not possible to obtain drawings with the same arbitrarily high number of bends on all the edges. However, there are still some intriguing open questions left:

*Problem 1.* Given a maximum degree 4 plane graph $G$ and an integer $k$, which is the time complexity of deciding whether $G$ admits a $k$-bend drawing?

Notice that for $k = 0$ polynomial time suffices (simple variation of the technique in [13]).

There are, of course, other models for studying bad drawings. For example one can study the maximization of the number of bends in an elementary setting, where a rectangle is given and the drawing of a graph should be squeezed into the rectangle. We consider this problem in the case in which the input graph is a single edge or a path. In this setting we do not use the model for bends introduced before and allow an edge to have bends in any direction.

Let $A$ be a finite rectangular grid with width $X$ and height $Y$.

**Lemma 3.** *An edge $e = (u, v)$ admits a drawing in the grid $A$ with $b$ bends, where $b$ is as follows:*

- *if* $\min\{X, Y\}$ *is even, then* $b = XY - \min\{X, Y\}$
- *if* $\min\{X, Y\}$ *is odd:*
  - *if* $X \neq Y$, *then* $b = XY - \max\{X, Y\}$
  - *if* $X = Y$, *then* $b = XY - \max\{X, Y\} - 1$

*Proof:* First, consider the case in which $\min\{X, Y\}$ is even. Suppose that $X$ is the minimum between $X$ and $Y$. Draw vertex $u$ in the top-left corner of $A$, and fill the first two columns of $A$ with a sequence of segments right-down-left-down-right-down... (see Fig. 6.a), till a point in the last row is drawn. Then draw segments to the right till the third column is reached. Now fill the third and the fourth column with a sequence of segments right-up-left-up-right-up..., till a point in the first row is drawn. Then draw segments to the right till the fifth column is reached. The whole drawing can be constructed by repeating these two steps till there are pairs of columns to fill. The last point inserted is vertex $v$. It's easy to see that for each two columns only two grid points don't contain bends. If $\min\{X, Y\} = Y$ an analogous construction fills $A$ so that for each two rows only two grid points don't contain bends.

Now consider the case in which $\min\{X, Y\}$ is odd. Suppose that $X$ is the minimum between $X$ and $Y$. Again start by placing vertex $u$ in the top-left corner of $A$, and fill the first two rows by a sequence of segments down-right-up-right-down-right-up..., till a point in the last column is reached. Then draw segments towards the bottom part of $A$ till the third row is reached. Fill the last two columns by a sequence of segments left-down-right-down-left-down-right-down..., till a point in the last row is reached. Then draw segments to the left till the third column from the right is reached. Fill the last two rows by a sequence of segments up-left-down-left-up-left-down-left..., till a point in the first column is reached. Then draw segments towards the top part of $A$ till the third row from the bottom is reached. Fill the first two columns by a sequence of segments up-right-up-left-up-right-up-left..., till a point in the third row is reached. The whole drawing can be constructed by repeating these four steps till there are pairs of columns to fill. When only one column is missing, and a point on it has been drawn, then one bend can still be introduced if $X \neq Y$ (see Figs. 6.b and 6.c). In this case it can be noticed that each row has only one grid point that doesn't contain a bend, otherwise one row has two grid points that don't contain bends.                                        □

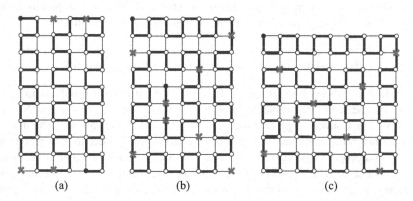

**Fig. 6.** Drawing an edge in a rectangular grid with the maximum number of bends. Black circles represent end-vertices of the edge, white circles represent bends, and red crosses represent grid points in which there is neither a vertex nor a bend. (a) $\min\{X,Y\}$ is even. (b) $\min\{X,Y\}$ is odd and $X \neq Y$. (c) $\min\{X,Y\}$ is odd and $X = Y$.

**Lemma 4.** *An n-vertex path p admits a drawing in the grid A with b bends, where b is as follows:*

- *if* $\min\{X,Y\}$ *is even, then* $b = XY - \max\{n, \min\{X,Y\}\}$
- *if* $\min\{X,Y\}$ *is odd:*
  - *if* $X \neq Y$, *then* $b = XY - \max\{n, \max\{X,Y\}\}$
  - *if* $X = Y$, *then* $b = XY - \max\{n, \max\{X,Y\} + 1\}$

*Proof:* The construction described in the proof of Lemma 3 can be slightly modified so that each grid point that doesn't contain a bend or a vertex is internal to a segment of the drawing. Then we can draw an edge and insert the $n - 2$ vertices that turn the edge in an $n$-vertex path exactly in those grid points in which there aren't bends. When the number of points that don't contain bends is less than $n - 2$, vertices must be introduced also instead of bends. The claimed bounds directly follow from this observation.        □

Since the previous two lemmas provide upper bounds on the maximum number of bends that can be introduced when drawing an edge or a path inside a rectangle, one can ask the following:

*Problem 2.* Are the bounds in Lemmas 3 and 4 tight?

Of course the problem considered here, when extended to graphs richer than paths, is much more interesting:

*Problem 3.* Given a maximum degree 4 plane graph $G$ and an $X \times Y$ rectangular grid $A$, which is the maximum number of bends that a planar orthogonal drawing of $G$ inside $A$ can have?

Once the orthogonal shape has been fixed, one can ask for the maximization of several other graph features, like the area or the total edge length of the drawing. In the case of the area we have shown that a simple linear-time algorithm allows to maximize the area of the drawing of a given orthogonal shape. However, the time complexity for the maximization of the total edge length of an orthogonal shape is still unknown.

*Problem 4.* Given a maximum degree 4 plane graph $G$, which is the time needed to compute an orthogonal drawing of $G$ with maximum total edge length?

We conclude by observing that one of the oldest and still open problems in Graph Drawing can be seen as a problem of obtaining a "bad" drawing of a graph. Namely, in the first 60's Conway suggested the following graph representation: each edge is a simple Jordan curve, each pair of edges cross exactly once, either in a common end-vertex or in a proper crossing. Conway asked for the maximum number of edges that a graph that can be represented in such a way can have. Moreover, he conjectured that any graph that can be drawn in this way doesn't have more edges than vertices. Despite a big research effort, the problem of proving or disproving Conway's conjecture is still open [8,4].

# References

1. Di Battista, G., Liotta, G., Vargiu, F.: Spirality and optimal orthogonal drawings. SIAM J. Comput. 27(6), 1764–1811 (1998)
2. Bertolazzi, P., Di Battista, G., Didimo, W.: Computing orthogonal drawings with the minimum number of bends. In: Dehne, F., Rau-Chaplin, A., Sack, J.-R., Tamassia, R. (eds.) WADS 1997. LNCS, vol. 1272, pp. 331–344. Springer, Heidelberg (1997)
3. Bridgeman, S.S., Di Battista, G., Didimo, W., Liotta, G., Tamassia, R., Vismara, L.: Turn-regularity and optimal area drawings of orthogonal representations. Computational Geometry 16, 53–93 (2000)
4. Cairns, G., Nikolayevsky, Y.: Bounds for generalized thrackles. Discrete & Computational Geometry 23(2), 191–206 (2000)
5. Di Battista, G., Eades, P., Tamassia, R., Tollis, I.G.: Graph Drawing, Upper Saddle River, NJ. Prentice-Hall, Englewood Cliffs (1999)
6. Fößmeier, U., Kaufmann, M.: Drawing high degree graphs with low bend numbers. In: Brandenburg, F.J. (ed.) GD 1995. LNCS, vol. 1027, pp. 254–266. Springer, Heidelberg (1996)
7. Garg, A., Tamassia, R.: On the computational complexity of upward and rectilinear planarity testing. SIAM J. Comput. 31(2), 601–625 (2001)
8. Lovász, L., Pach, J., Szegedy, M.: On Conway's thrackle conjecture. In: Symposium on Computational Geometry, pp. 147–151 (1995)
9. Nakano, S.-I., Yoshikawa, M.: A linear-time algorithm for bend-optimal orthogonal drawings of biconnected cubic plane graphs. In: Marks, J. (ed.) Graph Drawing 2000. LNCS, vol. 1984, pp. 296–307. Springer, Heidelberg (2000)
10. Patrignani, M.: On the complexity of orthogonal compaction. Comput. Geom. 19(1), 47–67 (2001)

11. Rahman, M.S., Nakano, S.-I., Nishizeki, T.: A linear algorithm for bend-optimal orthogonal drawings of triconnected cubic plane graphs. J. Graph Algorithms Appl. 3(4), 31–62 (1999)
12. Tamassia, R.: New layout techniques for entity-relationship diagrams. In: Proc. 4th Internat. Conf. on Entity-Relationship Approach, pp. 304–311 (1985)
13. Tamassia, R.: On embedding a graph in the grid with the minimum number of bends. SIAM J. Comput. 16(3), 421–444 (1987)
14. Tamassia, R., Di Battista, G., Batini, C.: Automatic graph drawing and readability of diagrams. IEEE Trans. Syst. Man. Cybern. SMC-18(1), 61–79 (1988)
15. Tamassia, R., Tollis, I.G.: Planar grid embedding in linear time. IEEE Trans. Circuits Syst. CAS-36(9), 1230–1234 (1989)
16. Tamassia, R., Tollis, I.G., Vitter, J.S.: Lower bounds and parallel algorithms for planar orthogonal grid drawings. In: Proc. IEEE Symposium on Parallel and Distributed Processing, pp. 386–393 (1991)
17. Vijayan, G., Wigderson, A.: Rectilinear graphs and their embeddings. SIAM J. Comput. 14, 355–372 (1985)

# Close Encounters with
# a Black Hole
# or
# Explorations and Gatherings in
# Dangerous Graphs

Nicola Santoro*

School of Computer Science, Carleton University, Canada
santoro@scs.carleton.ca

**Abstract.** Consider a netscape inhabited by mobile computational entities (e.g., robots, agents, sensors). In the algorithmic literature, these environments are usually assumed to be safe for the entities. Outside of the literature, this is hardly the case: highly harmful objects can operate in the netscape rendering the environment dangerous for the entities. A particular example is the presence of a *black hole*: a network site (node, host) that disposes of any incoming robot/agent, leaving no observable trace of such a destruction. The reasons why a node becomes a black hole are varied; for example, the presence at a node of a harmful static process (e.g., a virus) that destroys incoming code and messages transforms that node into a black hole; the undetectable crash failure of a host renders that host a black hole; "receive-omission" failures in the communication software of a site makes that site act as a black hole. Indeed, this type of danger is not rare.

Clearly the presence of a black hole renders computations in the net dangerous to be performed, and some tasks become impossible to be carried out. We will examine two classic problems for mobile entities, *Exploration* and *Gathering* (or *Rendezvous*), and discuss how they are affected by the presence of a black hole. In particular, we will view them with respect to a new task that, in this context, is even more basic and essential: *Black Hole Search*, the problem of a team of mobile entities locating the black hole. Obviously, any entity entering the black hole is destroyed; the black hole location problem is solved if at least one agent survives, and all surviving agents know the location of the black hole.

Not satisfied with correctness, our focus is on efficiency. The basic cost measures are the number of entities (and of casualties), and the number of moves.

**Keywords:** Harmful Host, Exploration, Rendezvous, Gathering, Mobile Agents, Robots, Asynchronous, Anonymous Networks, Anonymous Agents, Whiteboards, Tokens.

---

* This work is supported in part by the Natural Sciences and Engineering Research Council of Canada.

P. Crescenzi, G. Prencipe, and G. Pucci (Eds.): FUN 2007, LNCS 4475, p. 14, 2007.

# Fun with Sub-linear Time Algorithms

Luca Trevisan[*]

Computer Science Division, U.C. Berkeley
679 Soda Hall, Berkeley, CA 94720
luca@cs.berkeley.edu

**Abstract.** Provided that one is willing to use randomness and to toler-
ate an approximate answer, many computational problems admit ultra-
fast algorithms that run in less than linear time in the length of the
input. In many interesting cases, even algorithms that run in *constant*
time are known, whose efficiency depends only on the accuracy of the
approximation and not on the length of the inputs.

Algorithms for graph problems on *dense graphs* are especially efficient
and simple. I will describe an algorithm that estimates the size of the
maximum cut in a dense graph, and its specialization to the task of
distinguishing bipartite dense graphs from dense graphs that are "far
from bipartite." Results "explaining" the simplicity of such algorithms
will also be discussed.

Some sublinear-time algorithms are also known for graph problems
in *sparse graphs*, but they are typically more elaborate. I will describe
a simple but very clever algorithm that approximates the number of
connected components of a given graph, and its generalization to the
problem of approximating the weight of the minimum spanning tree of
a given weighted graph. The algorithm runs in time dependent only on
the maximum degree, the required quality of approximation, and the
range of weights, but the running time is independent of the number of
vertices.

[*] This material is based upon work supported by the National Science Foundation
under grant CCF 0515231 and by the US-Israel Binational Science Foundation Grant
2002246.

# Wooden Geometric Puzzles: Design and Hardness Proofs

Helmut Alt[1], Hans Bodlaender[2], Marc van Kreveld[2], Günter Rote[1], and Gerard Tel[2]

[1] Department of Computer Science, Freie Universität Berlin
{alt,rote}@inf.fu-berlin.de
[2] Department of Information and Computing Sciences, Utrecht University
{hansb,marc,gerard}@cs.uu.nl

**Abstract.** We discuss some new geometric puzzles and the complexity of their extension to arbitrary sizes. For gate puzzles and two-layer puzzles we prove NP-completeness of solving them. Not only the solution of puzzles leads to interesting questions, but also puzzle design gives rise to interesting theoretical questions. This leads to the search for instances of partition that use only integers and are uniquely solvable. We show that instances of polynomial size exist with this property. This result also holds for partition into $k$ subsets with the same sum: We construct instances of $n$ integers with subset sum $O(n^{k+1})$, for fixed $k$.

## 1  Introduction

Many good puzzles are instances of problems that are in general NP-complete. Conversely, NP-complete problems may be the inspiration for the design of nice puzzles. This is true for puzzles based on combinatorics, graphs, and geometry.

A puzzler's classification system of geometric puzzles exists that includes the classes Put-Together, Take Apart, Sequential Movement, and various others [1]. Although instances of puzzles in these classes have constant size, the natural generalization of many of them to sizes based on some parameter are NP-complete. For example, Instant Insanity is NP-complete [7,10], sliding block puzzles like the 15-puzzle, Sokoban, and Rush Hour are NP-complete or PSPACE-complete [2,6,9], and puzzles related to packing like Tetris are NP-complete [5]. Some overviews are given by Demaine [3] and Demaine and Demaine [4].

In this paper we discuss some new geometric puzzles of the Put-Together type and analyze their complexity. We also discuss the creation of good instances of certain geometric puzzles based on set partition.

*Gate puzzles.* Gate puzzles consist of a board that is a regular square grid of holes and a number of pieces called *gates*. Gates consist of three rods, two vertical and one horizontal, connecting the tops of the vertical rods. The vertical rods are called *legs* and have a certain *leg distance* that allows the gate to be placed on the board. A gate has a leg distance of 1 if the two legs are in adjacent holes. Furthermore, gates have a height, taken from a small set of values. To solve a

P. Crescenzi, G. Prencipe, and G. Pucci (Eds.): FUN 2007, LNCS 4475, pp. 16–29, 2007.

**Fig. 1.** Gate puzzles. Left with two-legged gates, right also with three-legged gates.

gate puzzle, a given set of gates must be placed in the board. Every hole of the board must contain exactly one of the legs, and two gates can only intersect in the vertical projection if they have a different height, and the intersection is not at the vertical rods of the higher gate. Figure 1 shows an example. On the left, a $5 \times 5$ grid is shown with eleven normal gates of heights 2, 3, and 4, and three loose pegs (one-legged gates) of height 1. On the right, a variation is shown where many gates have an extra leg: Two gates have two legs and seven gates have three legs. Most puzzlers take half an hour to a full hour to solve one of these puzzles. Gate puzzles were first described by the third author in [12]. In this paper we show that solving gate puzzles is NP-complete, which we prove by reduction from the strongly NP-complete problem 3-PARTITION (see for instance [7]).

*Two-layer puzzles.* Two-layer puzzles consist of a set of pieces that must be arranged in two layers, where touching pieces from opposing layers must fit. The simplest type of such a puzzle consists of $2k$ pieces of base $k \times 1$, and every $1 \times 1$ unit has a height 1 or 2. The pieces must be arranged to make a solid $k \times k \times 3$ block. To this end, $k$ of the pieces must be arranged as rows, and the other $k$ pieces must be arranged upside down and as columns. Other two-layer puzzles can have pieces that use more than two heights, or pieces that do not have different heights, but use slanted tops in one of the four orientations [11]. See Figure 2 for two examples.

A different realization of simplest type of two-layer puzzle is also known as the 16-holes puzzle. It consists of eight flat pieces of $4 \times 1$, with one, two or three holes. The objective is to cover the 16 holes of a $4 \times 4$ grid by placing the pieces on the grid in two layers, see Figure 3.

Two-layer puzzles are NP-complete to solve, which we prove by reduction from Hamiltonian Circuit on graphs of degree three.

**Fig. 2.** Examples of two-layer puzzles

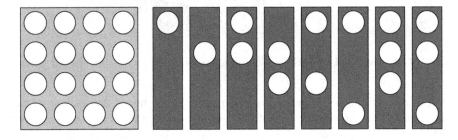

**Fig. 3.** The 16-holes puzzle by Wim Zwaan

*Partition puzzles.* Partition puzzles are puzzles that are based on the well-known PARTITION problem: Given a set of positive integers $v_1, \ldots, v_n$, partition them in two subsets of equal total value. This problem is NP-complete [7]. The easiest realization as a geometric puzzle is to consider each integer value $v_i$ as a $1 \times 1 \times v_i$ block and the puzzle is to pack the blocks in a (very long) box of dimensions $1 \times 2 \times V/2$, where $V = \sum_{i=1}^{n} v_i$.

Another partition problem that is NP-complete is 3-PARTITION, which involves partitioning a set of $3n$ positive integers into $n$ sets of three elements each and with the same subset sum. One puzzle that appears to be directly based on 3-PARTITION is Kunio Saeki's *Pipes in Pipe*, designed for the 18th International Puzzle Party in 1998. It has 21 little cylinders of different lengths that must fit in seven holes of equal length, see Figure 4.

Obviously, partitioning a set of integers into three or four subsets of the same total sum is also NP-complete. A realization of a partition puzzle that uses three subsets is shown in Figure 5. In this puzzle, the slant of $\pi/3$ and the different ways to deal with the corners make it a variation on a 3-partition puzzle.

Not only solving puzzles based on partition problems is difficult, the creation of geometrically good instances of such partition puzzles is also challenging. A good geometric puzzle has the property that it is clear whether a particular solution is the correct solution. Furthermore, it should not be too large, physically. Finally, most good puzzles have only few pieces but are still very hard. The last property can be interpreted for partition puzzles that there should be only

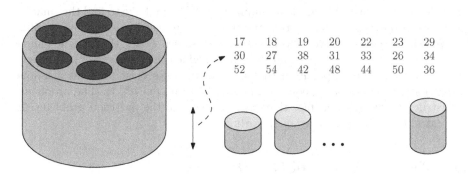

|    |    |    |    |    |    |    |
|----|----|----|----|----|----|----|
| 17 | 18 | 19 | 20 | 22 | 23 | 29 |
| 30 | 27 | 38 | 31 | 33 | 26 | 34 |
| 52 | 54 | 42 | 48 | 44 | 50 | 36 |

**Fig. 4.** Partition puzzle by Kunio Saeki

**Fig. 5.** Partition puzzle based on covering an equilateral triangle with nine pieces of different lengths and shapes

one solution. The presence of equal pieces tends to make the solution easier, since it reduces the number of different potential solutions. Therefore, we require that all pieces are distinct. We thus restrict our attention to *sets* of numbers instead of multisets.

The discussion on clearness of the correct solution can be interpreted as follows: if a set of reals has a solution with two sums of value $V$, then there should not be a small $\varepsilon > 0$ such that a different partition into two sets has sums of values $V + \varepsilon$ and $V - \varepsilon$. Here the ratio of $V$ and $\varepsilon$ is important. We will only consider the partition problem for integers. This automatically gives a difference in the subset sums between a correct partition and non-correct partition of 2. Since a difference of length of 2 mm is clearly visible, we could take millimeters as units of measurement. But then the sum of a subset that gives a correct solution is the size of the puzzle in millimeters. We would like to find the smallest instance of partition, meaning that the sum of all integers is as small as possible.

We show that for PARTITION, a set of $n$ values exists that has a unique partition into two subsets of equal sum, and of which the sum is $O(n^3)$. Similarly, we show for $k$-subsets partition that a set of $n$ values exist that has a unique partition into $k$ subsets of equal sum, and the sum is $O(n^{k+1})$. The proofs are constructive: we give schemes that give instances of the partition problems. In all cases, the $k$ subsets have equal cardinality.

## 2   The Complexity of Gate Puzzles

In this section we show that solving gate puzzles is NP-complete. We consider the simplest form where only two-legged gates occur, and only two heights are used.

**Theorem 1.** *Given a grid of $n \times m$, and $nm/2$ gates of height $1$ or $2$, it is NP-complete to decide if they can be placed on the grid.*

*Proof.* Clearly the problem is in NP. To prove NP-hardness we make a reduction from 3-PARTITION, which is NP-complete in the strong sense [7]. An instance of 3-PARTITION consists of $3N$ positive integers $v_1, \ldots, v_{3N}$, where each integer is between $B/4$ and $B/2$ for some given $B$, and $\sum_{i=1}^{3N} v_i = NB$. The problem is to decide whether a partition of the $3N$ integers into $N$ subsets exist such that each of these subsets has sum $B$. We transform an integer $v_i$ into one gate with leg distance $2v_i N^2 - 1$ and height 2, and $v_i N^2 - 1$ gates of leg distance 1 and height 1. We ask if all gates fit on a grid of size $(2N^2 B - 1) \times N$, see Fig. 6.

We first show that gates of height 2 only fit horizontally. It is obvious that they do not fit vertically, but they might fit as the diagonal of a Pythagorean triangle. Note that any gate of height 2 has leg distance $L > N^2$. It can easily be seen that such a gate cannot fit diagonally, since $L - 1 > \sqrt{(L-2)^2 + (N-1)^2}$, see Figure 7.

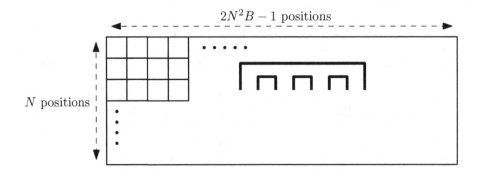

**Fig. 6.** Reduction of 3-PARTITION to gates

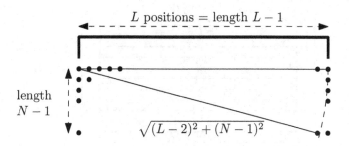

**Fig. 7.** Since gates have large enough leg distance, they cannot be placed diagonally

We showed that there are $3N$ gates that only fit horizontally. There are $N$ rows, and every row will contain three gates of height 2 in any solution. The gates of height 1 are only for making the gate puzzle valid by filling the holes of the whole grid. They fit under the height 2 gate with which they were created.

It is clear that the gate puzzle has a solution if and only if 3-PARTITION has a solution. The reduction is polynomial because 3-PARTITION is NP-complete in the strong sense: even if we write all values in unary notation on the input, the problem is NP-complete. Therefore, the number of gates obtained after the reduction is polynomial in the input size.    □

## 3    The Complexity of Two-Layer Puzzles

For the NP-completeness proof of two-layer puzzles, we choose a version with $2n$ pieces of length $n$. Every piece is a row of elements, each of which has height 1 or 2. We must place $n$ pieces as rows, and the other $n$ pieces upside down as columns on top, such that if a position of the bottom, row layer contains a 1, then the corresponding position of the top, column layer contains a 2, and vice versa.

**Theorem 2.** *Given a set of $2n$ two-layer pieces of length $n$, it is NP-complete to decide if they can be placed to form a solid block of $n \times n \times 3$.*

*Proof.* Clearly, the problem is in NP. To prove NP-hardness, we transform from HAMILTONIAN CIRCUIT FOR CUBIC GRAPHS [8].

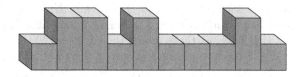

**Fig. 8.** A two-layer piece with two heights, for a $10 \times 10 \times 3$ block

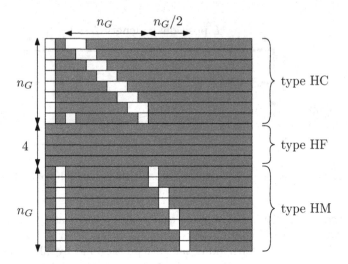

**Fig. 9.** The horizontal pieces for the reduction; grey is height 2 and white is height 1

Let $G = (V, E)$ be a cubic graph, i.e., each vertex in $V$ has exactly three neighbours. Write $n_G = |V|$. Note that $n_G$ is even, as $G$ is cubic. Without loss of generality, assume that $n_G \geq 8$. Assume $V = \{v_1, \ldots, v_{n_G}\}$.

We build a collection of $4n_G + 8$ two-layer pieces of length $n = 2n_G + 4$, and distinguish certain types. The main ones are the H-type and V-type, see Figures 9 and 10. Each of these types has subtypes, and the following pieces per subtype:

- Type HF: four pieces with all positions at height 2. (Horizontal, Full)
- Type HC: $n_G$ pieces with all but three positions at height 2. For $1 \leq i \leq n_G - 1$, we have a piece with positions 1, $i + 2$, and $i + 3$ at height 1 and all other positions at height 2. We also have a piece with positions 1, 3, and $2 + n_G$ at height 1, and all other positions at height 2. (Horizontal, Circuit, as these will be used to model the Hamiltonian circuit. The last piece models the edge that closes the circuit.)
- Type HM: $n_G$ pieces with all but two positions at height 2. For $1 \leq i \leq n_G/2$, we have two pieces with positions 2 and $2 + n_G + i$ at height 1, and all other positions at height 2. (Horizontal, Matching, as these model a matching in $G$.)
- Type V1: two pieces with positions 1 until $n_G$ (inclusive) at height 2, and all other positions at height 1.
- Type VE: one piece for each of the $3n_G/2$ edges in $E$. If $\{v_i, v_j\} \in E$, then we take a piece with positions $i$ and $j$ at height 2, and all other positions at height 1. (Vertical, Edge, as these model the edges of $G$.)
- Type VF: $n_G/2 + 2$ pieces with all positions at height 1. (Vertical, Full)

We claim that this collection of pieces has a solution if and only if $G$ has a Hamiltonian circuit. This claim and the fact that the collection of pieces can be constructed in polynomial time, given $G$, show the NP-hardness.

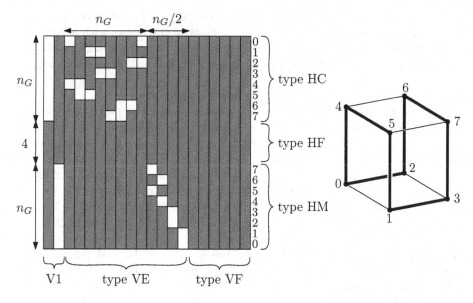

**Fig. 10.** The vertical pieces for the reduction, showing a solution for the graph of the cube (on the right, with the Hamiltonian circuit highlighted). This time, grey is height 1 and white is height 2, complementing the horizontal pieces of Fig. 9 after permuting them within the groups HC and HM. The vertices of the cube are numbered 0–7. Two vertices are adjacent if their difference is a power of 2.

Suppose $v_{j_1}, v_{j_2}, \ldots, v_{j_{n_G}}$ is a Hamiltonian circuit in $G$. Let $M$ be the set of edges in $G$ that do not belong to the circuit. As each vertex in $G$ is incident to two edges on the circuit, $M$ is a matching in $G$. We place the pieces as follows, see Figure 10.

Pieces of H-type will always be placed horizontally, pieces of V-type vertically. If we do not state that a piece is reversed, it is placed like its description above.

- One piece of type V1 is placed in the first column.
- The second piece of type V1 is placed in the second column, but reversed, that is, the height 2 squares are at the intersection with rows $n_G + 5$ until $2n_G + 4$.
- For $1 \leq i < n_G$, the VE-piece which models the edge $\{v_{j_i}, v_{j_{i+1}}\}$ is placed in column $i + 2$.
- The VE-piece that models the edge $\{v_{j_{n_G}}, v_{j_1}\}$ is placed in column $n_G + 2$.
- The $n_G/2$ VE-pieces that model the edges in $M$ are placed reversed in some arbitrary order in the columns $n_G + 3, \ldots, 3n_G/2 + 2$.
- The VF-pieces are placed in columns $3n_G/2 + 3, \ldots, 2n_G + 4$.
- The HF pieces are placed in rows $n_G + 1$, $n_G + 2$, $n_G + 3$, $n_G + 4$.
- The HC-piece with height 1 positions at $1$, $i + 2$, $i + 3$ is placed in row $j_i$. Note that it fits with the VE-pieces!
- The HC-piece with height 1 positions at $1$, $3$, and $2 + n_G$ is placed in row $j_{n_G}$.

– For each $i$, $1 \leq i \leq n_G/2$: consider the edge $\{v_{k_1}, v_{k_2}\} \in M$ whose VE-piece is placed reversed in column $n_G + 2 + i$. The two HM-pieces with height 1 at positions 2 and $n_G + 2 + i$ are placed in rows $2n_G + 4 - k_1$ and $2n_G + 4 - k_2$. Note that this fits with the VE-pieces; here we use that $M$ is a matching.

One can verify that we indeed have a solution for the puzzle.

Suppose the collection of pieces has a solution. Consider an arbitrary piece of type HF. Without loss of generality, suppose it is placed horizontally. Then, all other pieces of type HF must be placed horizontally, otherwise we would have a mismatch at the position where the pieces intersect. Each piece of type HC and HM has at most three positions of height 1, so it cannot be placed vertically. (Otherwise, it would share a position with each of the four HF-type pieces, and at least one of these positions it would also have height 2.) As all $2n_G + 4$ pieces of H-type are placed horizontally, all pieces of V-type are placed vertically.

V1-type pieces have $n_G$ positions of height 2. So, if a V1-type piece is placed in column $i$, then there are $n_G$ H-type pieces with height 1 at position $i$ or $2n_G + 5 - i$. For each $i \in \{3, \ldots, 2n_G + 2\}$, there are at most six H-type pieces with height 1 at positions $i$ or $2n_G + 5 - i$. There are $n_G$ H-type pieces with height 1 at position 1, and $n_G$ H-type pieces with height 1 at position 2. Thus, one V1-type piece must be placed in column 1 or $2n_G + 4$, and one V1-type piece must be placed in column 2 or $2n_G + 3$.

Without loss of generality, we suppose one V1-type piece is placed in column 1, and it is not reversed. Consider the H-type pieces at rows $1, \ldots, n_G$. At their first position, they meet the height 2 position of the V1-type piece, so they must have height 1 at their first position, and hence be a HC-type piece. Also, their orientation cannot be reversed.

For $1 \leq i < n_G$, if the HC-piece with height 1 positions at 1, $i+2$, $i+3$ is in row $j$, set $v_{j_i} = j$. Similarly, if the piece with height 1 positions at 1, 3, $2 + n_G$ is in row $j$, set $v_{j_{n_G}} = j$. This gives a Hamiltonian Circuit. Consider a pair of successive vertices $v_{j_i}$, $v_{j_{i+1}}$. Note that the HC-type pieces in rows $j_i$ and $j_{i+1}$ have height 1 at their position $i + 3$. So the V-type piece in column $i + 3$ must have height 2 at positions $j_i$ and $j_{i+1}$. It cannot be a V1-type piece, see above. So, we have a VE-type piece with height 2 at positions $j_i$ and $j_{i+1}$, and hence $\{v_{j_i}, v_{j_{i+1}}\} \in E$. A similar argument shows that $\{v_{j_{n_G}}, v_{j_1}\} \in E$, and hence we have a Hamiltonian circuit.                                                                $\square$

## 4   Designing Partition Puzzles

In this section we consider partition problems for integers. From the introduction we know that we are mostly interested in instances that are uniquely solvable and have a small total value. We will concentrate on instances of $2n$, or more generally, $kn$ integers that have a unique partition into 2 subsets (or $k$ subsets, respectively). Moreover, we want all subsets of the partition to have the same cardinality $n$. It is easy to adapt the instances to subsets of different cardinalities, by simply combining pieces.

**Lemma 1.** *Let $S$ be a set of integers that has a partition into two subsets $S_1$ and $S_2$, each of total value $\frac{1}{2} \cdot \sum_{v \in S} v$. Then this is the unique partition with this property, if and only if no proper subset of $S_1$ has equal value to any proper subset of $S_2$.*

*Proof.* If two such subsets exist, they can be exchanged and the partition is not unique. Conversely, let $S_1'$, $S_2'$ be a different partition with equal sums. Since $S_1' = (S_1 \setminus (S_1 - S_1')) \cup (S_1' - S_1)$, the sets $S_1 - S_1'$ and $S_1' - S_1$ are nonempty proper subsets of $S_1$ and of $S_2$ with equal sums.     □

For $k \geq 3$, the condition of Lemma 1 applied to pairs of sets, is not sufficient to guarantee uniqueness. For example, no two proper subsets of $S_1 = \{8, 14, 78\}$, $S_2 = \{9, 15, 76\}$, or $S_3 = \{10, 13, 77\}$ have equal sums, but $S_1' = \{9, 13, 78\}$, $S_2' = \{10, 14, 76\}$, $S_3' = \{8, 15, 77\}$ is a different solution.

We present schemes that generate instances of partition with a set $S$ of $2n$ or $kn$ integers that have a strongly unique solution and a polynomial bound on $\sum_{v_i \in S} v_i$.

*A simple scheme for partition.* Suppose we wish to generate a set $S$ of $2n$ integers that has a unique partition into two subsets $S_1$ and $S_2$ of cardinality $n$ each.

$$S_1 = \{ 1, \qquad 2, \qquad \ldots, n-1, \qquad N \qquad\qquad \}$$
$$S_2 = \{ \tfrac{1}{2}n(n-1)+1, \tfrac{1}{2}n(n-1)+2, \ldots, \tfrac{1}{2}n(n-1)+n-1, \tfrac{1}{2}n(n-1)+n \}$$

where $N = \sum_{i=1}^{n}(\frac{1}{2}n(n-1)+i) - \frac{1}{2}n(n-1)$. Since all integers in $S_2$ are larger than the sum of the smallest $n-1$ integers in $S_1$, and the $n$-th integer from $S_1$ is larger than all integers from $S_2$, no proper subset sum from $S_1$ can be equal to any proper subset sum of $S_2$.

**Theorem 3.** *For any $n \geq 2$, an instance of PARTITION exists with $2n$ values which has a unique solution in two subsets. Both subsets have $n$ integers, and the subset sum is $(n^3 + n)/2 = O(n^3)$.*     □

The obvious lower bound corresponding to the theorem is $\sum_{i=1}^{2n} i = \Omega(n^2)$.

*A simple scheme for partition into $k$ subsets.* The scheme for partition can easily be extended to a scheme for partition into $k$ subsets. The scheme gives subset sums that are polynomial in $n$, but exponential in $k$. We write $V_i$ for the sum of the smallest $n-1$ elements in $S_i$.

$$
\begin{aligned}
S_1 &= \{ 1, & 2, & \ldots, n-1, & N_1 & \} \\
S_2 &= \{ V_1+1, & V_1+2, & \ldots, V_1+n-1, & N_2 & \} \\
S_3 &= \{ V_2+1, & V_2+2, & \ldots, V_2+n-1, & N_3 & \} \\
&\cdots & \cdots \\
S_{k-1} &= \{ V_{k-2}+1, V_{k-2}+2, \ldots, V_{k-2}+n-1, N_{k-1} & \} \\
S_k &= \{ V_{k-1}+1, V_{k-1}+2, \ldots, V_{k-1}+n-1, V_{k-1}+n \}
\end{aligned}
$$

The integers $N_1, \ldots, N_{k-1}$ are chosen so that all subsets have the same subset sum as $S_k$. As before we can argue that $N_1$ is such that only $1, \ldots, n-1$ are small enough to be with $N_1$ and give the right subset sum. Since this fixes $S_1$, we can repeat the argument by observing that $N_2$ is such that of the remaining integers, only $V_1 + 1, \ldots, V_1 + n - 1$ are small enough to be with $S_2$ and give the right subset sum.

**Theorem 4.** *For any $k \geq 2$ and $n \geq 2$, an instance of* PARTITION INTO $k$ SUBSETS *exists with $kn$ values which has a unique solution in $k$ subsets. All subsets have $n$ integers, and the subset sum is $O(n^{k+1})$, if $k$ is fixed.* □

Although we presented a scheme that gives uniquely solvable instances of partition of cubic size, the scheme is not satisfactory from the puzzle point of view. It contains an integer that is so large that it is clear which other integers should go in the same subset (which was the argument for uniqueness). So in this case, uniqueness of solution does not imply that a puzzle using this scheme will be difficult. Therefore, we will present another partition scheme and its extension for $k$ subsets that does not have this problem. We will bound the value of the largest integer in the partition problem while obtaining the same bound on the subset sum.

*An improved scheme for partition.* To obtain a scheme that does not have the disadvantage of the simple scheme, choose two sets of integers $1, 2, \ldots, n$ and $1, 2, \ldots, n - 1, n + 1$. Multiply each integer in the first set by $n$. Multiply each integer in the second set by $n$ and subtract 1. This way we get $S_1$ and $S_2$:

$$S_1 = \{\, n, \quad 2n, \quad 3n, \quad \ldots, (n-1)n, \quad n^2 \quad\quad \}$$
$$S_2 = \{\, n-1, 2n-1, 3n-1, \ldots, (n-1)n-1, n^2+n-1 \,\}$$

Every subset sum from $S_1$ is a multiple of $n$. No proper subset sum from $S_2$ is a multiple of $n$, because each integer is $\equiv -1 \bmod n$. Hence, no proper subset sum of $S_1$ can be equal to any proper subset sum of $S_2$. The sum of all integers in $S_1$ is equal to the sum of all integers in $S_2$, and is equal to $\frac{1}{2}n(n+1) \cdot n = (n^3 + n^2)/2$.

**Theorem 5.** *For any $n \geq 2$, an instance of* PARTITION *exists with $2n$ values which has a unique solution in two subsets. Both subsets have $n$ integers, all integers have value $\Omega(n)$ and $O(n^2)$, and the subset sum is $(n^3 + n^2)/2 = O(n^3)$.* □

It is easy to adapt the scheme to yield a partition in subsets of different cardinalities: we let $n$ be the desired cardinality of the larger subset in the scheme, and generate $S_1$ and $S_2$. Then we add $n - m + 1$ values in $S_2$ to get any cardinality $m$ for the smaller subset, and the partition itself remains unique.

For small values of $n$, we have computed the uniquely solvable instances of PARTITION with smallest subset sum with the help of a computer, by an enumeration algorithm. The instances in the following table turned out to be the unique instances with the given sums, where the two subsets have equal cardinality. (For $n = 7$ there are two different smallest instances.) For decompositions into parts of distinct cardinalities, there are smaller solutions. For examplem $\{1, 3, 4, 5, 6, 7\} \cup \{2, 24\}$ is the unique solution for an 8-element set, with sums 26, but clearly, this leads to a very easy puzzle.

One can see in the table that the constructions of Theorems 3 and 5 are not far from the optimum. Also, the instances for $n = 5$, $n = 6$, and $n = 7$ share certain characteristics with the construction of Theorem 5: they contain arithmetic progressions, which tends to reduce the number of different subset sums that can be built from a given set. The two solutions for $n = 7$ seem to be based on arithmetic progressions with step lengths 7 and 6, respectively.

| $n$ | $S_1$ | $S_2$ | minimum subset sum | simple scheme $\frac{1}{2}(n^3 + n)$ | improved scheme $\frac{1}{2}(n^3 + n^2)$ |
|---|---|---|---|---|---|
| 2 | 1, 4 | 2, 3 | 5 | 5 | 6 |
| 3 | 1, 3, 9 | 2, 5, 6 | 13 | 15 | 18 |
| 4 | 2, 7, 10, 12 | 3, 5, 8, 15 | 31 | 34 | 40 |
| 5 | 2, 7, 12, 17, 22 | 3, 5, 10, 15, 27 | 60 | 65 | 75 |
| 6 | 3, 7, 10, 21, 28, 35 | 4, 11, 14, 18, 25, 32 | 104 | 111 | 126 |
| 7 | 2, 9, 16, 23, 30, 37, 44 | 5, 7, 14, 21, 28, 35, 51 | 161 | 175 | 196 |
| 7 | 5, 11, 17, 23, 29, 35, 41 | 1, 6, 12, 18, 24, 47, 53 | 161 | 175 | 196 |

*An improved scheme for partition into $k$ subsets.* We now present a scheme to generate instances of unique partition into 3 subsets, of $n$ integers each, and bounded integers. Below we will generalize it to larger values of $k$. Because we wish to avoid large integers in the instance, we cannot use the inductive argument that was used in the simple scheme for partition into $k$ subsets to obtain uniqueness. Instead, we use the following property to guarantee uniqueness.

*Strong Uniqueness.* If the total sum of the set $S$ is $kN$, there are only $k$ subsets of $S$ whose sum is $N$.

Choose two integers $p = n$ and $q = n+1$, and let $r = p \cdot q$. The sets of integers in $S_1$, $S_2$, and $S_3$ are:

$$
\begin{aligned}
S_1 &= \{\, r+q,\ 2r+q,\ \ldots,\ (n-1)r+q,\ nr+q \,\} \\
S_2 &= \{\, r+p,\ 2r+p,\ \ldots,\ (n-1)r+p,\ nr+2p \,\} \\
S_3 &= \{\, r,\quad 2r,\quad \ldots,\ (n-1)r,\quad nr+r \,\}
\end{aligned}
$$

It is easy to see that the three subset sums are the same. Also, $\sum_{v \in S_3} v = (\frac{1}{2}n(n+1)+1)pq = (\frac{1}{2}n(n+1)+1)n(n+1) = \Theta(n^4)$.

To prove uniqueness of the partition, we show strong uniqueness: $S_1$, $S_2$, and $S_3$ are the only subsets of $S = S_1 \cup S_2 \cup S_3$, with subset sum $(\frac{1}{2}n(n+1)+1)r = \frac{1}{3} \cdot \sum_{v_i \in S} v_i$. Let $S'$ be any subset $S$, and let $S'$ have $h$ elements from $S_1$, $i$ elements from $S_2$, and $j$ elements from $S_3$, where $0 \le h, i, j \le n$. Since elements from $S_1$ are $\equiv q \bmod r$, and $p$ and $q$ are relatively prime, $h > 0$ implies that $h = n$ to obtain a total sum of $S'$ that is $\equiv 0 \bmod r$. This subset is already $S_1$, and $i$ and $j$ have to be 0, otherwise the total sum is too large. Similarly, $i > 0$ implies that $S'$ must contain all elements of $S_2$ to be $\equiv 0 \bmod r$, and $h = 0$ and $j = 0$. Finally, if $h = 0$ and $i = 0$, we need $j = n$ to get a subset $S'$ of large enough total sum. Hence, $S_1$, $S_2$, and $S_3$ are the only subsets of $S$ with sum $(\frac{1}{2}n(n+1)+1)r$.

To extend this scheme to $k$ sets we need $k-1$ integers $p_1, \ldots, p_{k-1}$ that are at least $n$ and pairwise relatively prime. One way to construct such integers is as follows: Let $K$ be the least common multiple of $1, 2, \ldots, k-2$. Select $p_1$ in the interval $n \le p_1 < n + K$, relatively prime to $K$, and set $p_{i+1} = p_i + K$, for $i = 1, \ldots, k-2$. This yields integers $p_i < n+(k-1)!$ that are pairwise relatively prime.

We let $r = \prod_{i=1}^{k-1} p_i$, and construct sets based on $r$ and $p_1, \ldots, p_{k-1}$ as above. For $i = 1, \ldots, k-1$, we define

$$S_i = \left\{ r + a_1 \cdot \frac{r}{p_i}, \ 2r + a_2 \cdot \frac{r}{p_i}, \ \ldots, \ nr + a_n \cdot \frac{r}{p_i} \right\},$$

where $(a_1, a_2, \ldots, a_n)$ is a sequence of small positive integers summing $p_i$. (A different sequence $(a_1, a_2, \ldots, a_n)$ is chosen for every $i$.) As before, the last set is just

$$S_k = \{ \ r, \ 2r, \ldots, \ (n-1)r, \ (n+1)r \ \}$$

The subset sum of each set is $\frac{1}{2}n(n+1) \cdot r + r = \Theta(n^{k+1})$.

**Theorem 6.** *For any $k \ge 2$ and $n \ge 2$, an instance of* PARTITION INTO $k$ SUBSETS *exists with $kn$ values which has a unique solution in $k$ subsets. All subsets have $n$ integers, all integers have value $\Omega(n^{k-1})$ and $O(n^k)$, and the subset sum is $O(n^{k+1})$, if $k$ is fixed.* □

## 5   Conclusions and Open Problems

We showed that two new types of geometric puzzles—gate puzzles and two-layer puzzles—are NP-complete to solve. For puzzles based on partition, we constructed instances with polynomially bounded values that have unique solutions. The sum of the values relates to the physical size of the geometric puzzle. Uniqueness tends to make a puzzle harder, but we saw that a uniquely solvable puzzle may still be easy (for instance, in the simple scheme for partition).

The *strong uniqueness* property for partition in three or more subsets is stronger than necessary for having a unique solution. Also for this reason, it may be possible to improve upon the $O(n^{k+1})$ bound on the summed value of instances with a unique solution. Moreover, the strong uniqueness property makes the puzzle *easier*: if the puzzler finds a subset with the right sum, then this subset is certainly part of the overall solution. So for puzzle design purposes, it is interesting to have instances of partition into three or more subsets that have a unique solution, but many subsets with the right summed value.

## References

1. Cubism For Fun website. `http://cff.helm.lu/`
2. Culberson, J.: SOKOBAN is PSPACE-complete. In: Proceedings in Informatics 4 Int. Conf. FUN with Algorithms 1998, pp. 65–76 (1999)

3. Demaine, E.D.: Playing games with algorithms: Algorithmic combinatorial game theory. In: Proc. of Math. Found. of Comp. Sci. pp. 18–32 (2001)
4. Demaine, E.D., Demaine, M.L.: Puzzles, art, and magic with algorithms. Theory Comput. Syst. 39(3), 473–481 (2006)
5. Demaine, E.D., Hohenberge, S., Liben-Nowell, D.: Tetris is hard, even to approximate. In: Technical Report MIT-LCS-TR-865, MIT, Cambridge (2002)
6. Flake, G.W., Baum, E.B.: Rush Hour is PSPACE-complete, or why you should generously tip parking lot attendants, Manuscript, (2001)
7. Garey, M.R., Johnson, D.S.: Computers and Intractability: A Guide to the Theory of NP-Completeness. W. H. Freeman, New York (1979)
8. Garey, M.R., Johnson, D.S., Tarjan, R.E.: The planar Hamiltonian circuit problem is NP-complete. SIAM J. Comput. 5(4), 704–714 (1976)
9. Ratner, D., Warmuth, M.: Finding a shortest solution for the $N * N$-extension of the 15-puzzle is intractable. J. Symb. Comp. 10, 111–137 (1990)
10. Robertson, E., Munro, I.: NP-completeness, puzzles, and games. Util. Math. 13, 99–116 (1978)
11. van Kreveld, M.: Some tetraform puzzles. Cubism For. Fun. 68, 12–15 (2005)
12. van Kreveld, M.: Gate puzzles. Cubism For. Fun. 71, 28–30 (2006)

# HIROIMONO Is NP-Complete

Daniel Andersson

Department of Computer Science,
University of Aarhus, Denmark
koda@daimi.au.dk

**Abstract.** In a Hiroimono puzzle, one must collect a set of stones from a square grid, moving along grid lines, picking up stones as one encounters them, and changing direction only when one picks up a stone. We show that deciding the solvability of such puzzles is **NP**-complete.

## 1 Introduction

Hiroimono (拾い物, "things picked up") is an ancient Japanese class of tour puzzles. In a Hiroimono puzzle, we are given a square grid with stones placed at some grid points, and our task is to move along the grid lines and collect all the stones, while respecting the following rules:

1. We may start at any stone.
2. When a stone is encountered, we must pick it up.
3. We may change direction only when we pick up a stone.
4. We must not make 180° turns.

Figure 1 shows some small example puzzles.

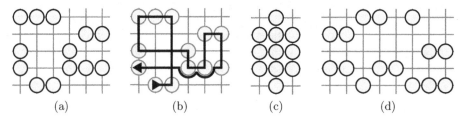

**Fig. 1.** (a) A Hiroimono puzzle. (b) A solution to (a). (c) Unsolvable. (d) Exercise.

Although it is more than half a millennium old [1], Hiroimono, also known as Goishi Hiroi (碁石ひろい), appears in magazines, newspapers, and the World Puzzle Championship. Many other popular games and puzzles have been studied from a complexity-theoretic point of view and proved to give rise to hard computational problems, e.g. Tetris [2], Minesweeper [3], Sokoban [4], and Sudoku (also known as Number Place) [5]. We shall see that this is also the case for Hiroimono.

P. Crescenzi, G. Prencipe, and G. Pucci (Eds.): FUN 2007, LNCS 4475, pp. 30–39, 2007.

We will show that deciding the solvability of a given Hiroimono puzzle is **NP**-complete and that specifying a starting stone (a common variation) and/or allowing 180° turns (surprisingly uncommon) does not change this fact.

**Definition 1.** HIROIMONO *is the problem of deciding for a given nonempty list of distinct integer points representing a set of stones on the Cartesian grid, whether the corresponding Hiroimono puzzle is solvable under rules 1–4. The definition of* START-HIROIMONO *is the same, except that it replaces rule 1 with a rule stating that we must start at the first stone in the given list. Finally,* 180-HIROIMONO *and* 180-START-HIROIMONO *are derived from* HIROIMONO *and* START-HIROIMONO, *respectively, by lifting rule 4.*

**Theorem 1.** *All problems in Definition 1 are* **NP***-complete.*

These problems obviously belong to **NP**. To show their hardness, we will construct a reduction from 3-SAT [6] to all four of them.

## 2    Reduction

Suppose that we are given as input a CNF formula $\phi = C_1 \wedge C_2 \wedge \cdots \wedge C_m$ with variables $x_1, x_2, \ldots, x_n$ and with three literals in each clause. We output the puzzle p defined in Fig. 2–4. Figure 5 shows an example.

$$p :=$$

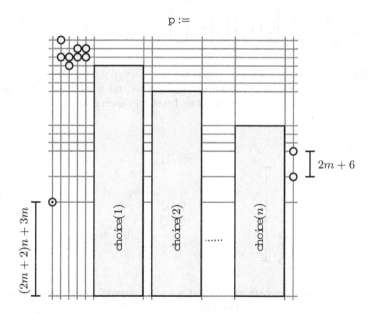

**Fig. 2.** The puzzle p corresponding to the formula $\phi$. Although formally, the problem instances are ordered lists of integer points, we leave out irrelevant details such as orientation, absolute position, and ordering after the first stone ◉.

choice($i$) :=

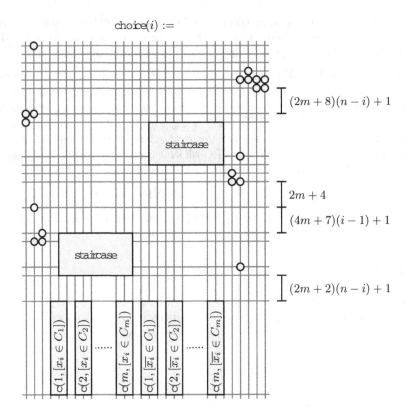

**Fig. 3.** The definition of choice($i$), representing the variable $x_i$. The two staircase-components represent the possible truth values, and the c-components below them indicate the occurrence of the corresponding literals in each clause.

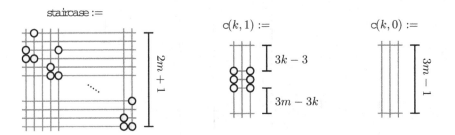

**Fig. 4.** The definition of staircase, consisting of $m$ "steps", and the c-components. Note that for any fixed $k$, all c($k$, 1)-components in p, which together represent $C_k$, are horizontally aligned.

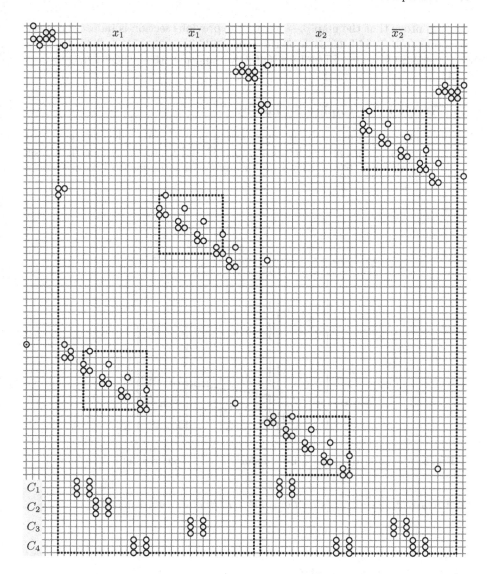

**Fig. 5.** If $\phi = (x_1 \vee x_2 \vee x_2) \wedge (x_1 \vee x_1 \vee x_1) \wedge (\overline{x_1} \vee \overline{x_2} \vee \overline{x_2}) \wedge (x_1 \vee x_2 \vee \overline{x_2})$, this is p. Labels indicate the encoding of clauses, and dotted boxes indicate choice(1), choice(2), and staircase-components. The implementation that generated this example is accessible online [7].

# 3   Correctness

From Definition 1, it follows that

$$\text{START-HIROIMONO} \underset{\subseteq}{\subseteq} \begin{array}{c} \text{HIROIMONO} \\ \subseteq \\ \text{180-START-HIROIMONO} \end{array} \underset{\subseteq}{\subseteq} \text{180-HIROIMONO}.$$

Thus, to prove that the map $\phi \mapsto \mathsf{p}$ from the previous section is indeed a correct reduction from 3-SAT to each of the four problems above, it suffices to show that $\phi \in$ 3-SAT $\Rightarrow \mathsf{p} \in$ START-HIROIMONO and $\mathsf{p} \in$ 180-HIROIMONO $\Rightarrow \phi \in$ 3-SAT.

### 3.1   Satisfiability Implies Solvability

Suppose that $\phi$ has a satisfying truth assignment $t^*$. We will solve $\mathsf{p}$ in two stages. First, we start at the leftmost stone ⊙ and go to the upper rightmost stone along the path $R(t^*)$, where we for any truth assignment $t$, define $R(t)$ as shown in Fig. 6–8.

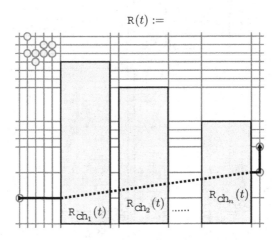

**Fig. 6.** The path $R(t)$, which, if $t$ satisfies $\phi$, is the first stage of a solution to $\mathsf{p}$

**Definition 2.** *Two stones on the same grid line are called* neighbors.

By the construction of $\mathsf{p}$ and $R$, we have the following:

**Lemma 1.** *For any $t$ and $k$, after $R(t)$, there is a stone in a* $\mathsf{c}(k,1)$-*component with a neighbor in a* staircase-*component if and only if $t$ satisfies $C_k$.*

In the second stage, we go back through the choice-components as shown in Fig. 9 and 10. We climb each remaining staircase by performing $R_{SC}$ backwards, but whenever possible, we use the first matching alternative in Fig. 11 to "collect a clause". By Lemma 1, we can collect all clauses. See Fig. 12 for an example.

Since this two-stage solution starts from the first stone ⊙ and does not make 180° turns, it witnesses that $\mathsf{p} \in$ START-HIROIMONO.

### 3.2   Solvability Implies Satisfiability

Suppose that $\mathsf{p} \in$ 180-HIROIMONO, and let $s$ be any solution witnessing this (assuming neither that $s$ starts at the leftmost stone nor that it avoids 180° turns).

$R_{ch_i}(t) :=$

*if* $t(x_i) = \top$

*if* $t(x_i) = \bot$

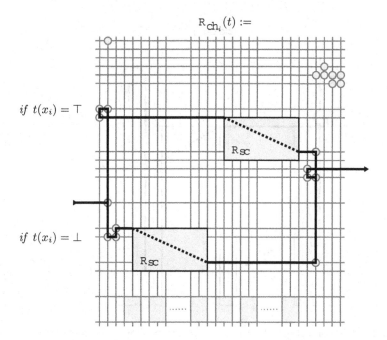

**Fig. 7.** Assigning a truth value by choosing the upper or lower staircase

$R_{SC} :=$

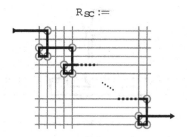

**Fig. 8.** Descending a staircase

p

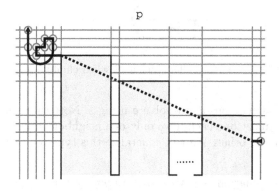

**Fig. 9.** The second stage of solving p

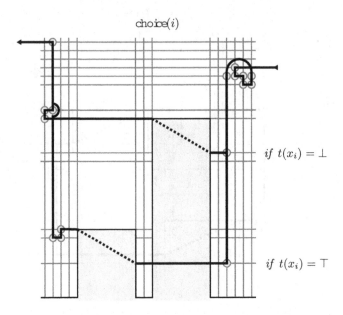

**Fig. 10.** In the second stage, the remaining staircase-component in choice($i$) is collected

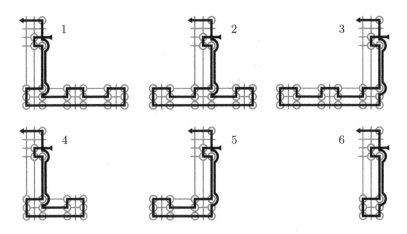

**Fig. 11.** Six different ways to "collect a clause" when climbing a step in a staircase

Now consider what happens as we solve p using $s$. Note that since the topmost stone and the leftmost one each have only one neighbor, $s$ must start at one of these and end at the other. We will generalize this type of reasoning to *sets* of stones.

**Definition 3.** *A* situation *is a set of remaining stones and a current position. A* dead end D *is a nonempty subset of the remaining stones such that:*

— *There is at most one remaining stone outside of* D *that has a neighbor in* D.
— *No stone in* D *is on the same grid line as the current position.*

A hopeless *situation is one with two disjoint dead ends.*

Since the stones in a dead end must be the very last ones picked up, a solution can never create a hopeless situation. If we start at the topmost stone, then we

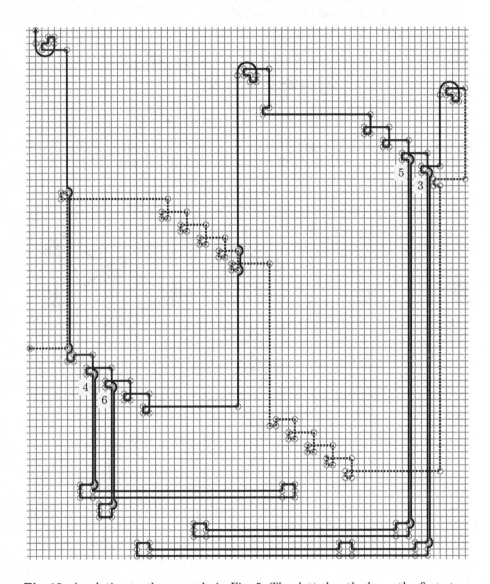

**Fig. 12.** A solution to the example in Fig. 5. The dotted path shows the first stage $R(t^*)$, with $t^*(x_1) = \top$ and $t^*(x_2) = \bot$. The solid path shows the second stage, with numbers indicating the alternative in Fig. 11 used to collect each clause.

38     D. Andersson

will after collecting at most four stones find ourselves in a hopeless situation, as is illustrated in Fig. 13. Therefore, $s$ must start at the leftmost stone and end at the topmost one.

We claim that there is an assignment $t^*$ such that $s$ starts with $R(t^*)$. Figure 14 shows all the ways that one might attempt to deviate from the set of R-paths and the dead ends that would arise. By Lemma 1, we have that if this $t^*$ were to fail to satisfy some clause $C_k$, then after $R(t^*)$, the stones in the $c(k,1)$-components would together form a dead end. We conclude that the assignment $t^*$ satisfies $\phi$.

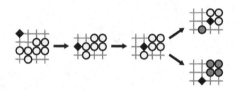

**Fig. 13.** Starting at the topmost stone inevitably leads to a hopeless situation. A ◆ denotes the current position, and a ◉ denotes a stone in a dead end.

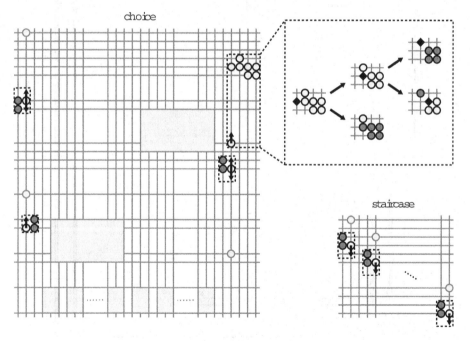

**Fig. 14.** Possible deviations from the R-paths and the resulting dead ends

**Acknowledgements.** I thank Kristoffer Arnsfelt Hansen, who introduced me to Hiroimono and suggested the investigation of its complexity, and my advisor,

Peter Bro Miltersen. I also thank the anonymous reviewers for their comments and suggestions.

# References

1. Costello, M.J.: The greatest puzzles of all time. Prentice-Hall, Englewood Cliffs (1988)
2. Demaine, E.D., Hohenberger, S., Liben-Nowell, D.: Tetris is hard, even to approximate. In: Warnow, T.J., Zhu, B. (eds.) COCOON 2003. LNCS, vol. 2697, pp. 351–363. Springer, Heidelberg (2003)
3. Kaye, R.: Minesweeper is NP-complete. Mathematical Intelligencer 22, 9–15 (2000)
4. Culberson, J.: Sokoban is PSPACE-complete. In: Proceedings of the International Conference on Fun with Algorithms, Carleton Scientific, pp. 65–76 (1998)
5. Yato, T., Seta, T.: Complexity and completeness of finding another solution and its application to puzzles. IEICE Transactions on Fundamentals of Electronics, Communications and Computer Sciences 86, 1052–1060 (2003)
6. Garey, M.R., Johnson, D.S.: Computers and Intractability: A Guide to the Theory of NP-Completeness. W.H. Freeman & Co, New York (1979)
7. Andersson, D.: Reduce 3-SAT to HIROIMONO. http://purl.org/net/koda/s2h.php

# Tablatures for Stringed Instruments and Generating Functions

Davide Baccherini, Donatella Merlini, and Renzo Sprugnoli

Dipartimento di Sistemi e Informatica
viale Morgagni 65, 50134, Firenze, Italia
baccherini,merlini,sprugnoli @dsi.unifi.it
baccherini@gmail.com

**Abstract.** We study some combinatorial properties related to the problem of tablature for stringed instruments. First, we describe the problem in a formal way and prove that it is equivalent to a finite state automaton. We define the concepts of *distance between two chords* and *tablature complexity* in order to study the problem of tablature in terms of music performance. By using the Schützenberger methodology we are then able to find the generating function counting the number of tablatures having a certain complexity and we can study the average complexity for the tablatures of a music score.

## 1 Introduction

The music perfomance process involves several representation levels [13], such as physical, perceptual, operational, symbolic, structural. Consequently, a performance environment should be concerned at least in:

1. getting a score in input (symbolic level);
2. analysing it, like a human performer would do (structural);
3. modelling the constraints posed by body-instrument interaction (operational);
4. manipulating sound parameters (physical).

In particular, in the present paper we focus our attention on a problem present in both structural and operational levels, and very relevant for string instrument, namely the problem of *tablature*. Some instruments, such as the piano, have only one way to produce a given pitch. To play a score of music on a piano, one needs only to read sequentially the notes from the page and depress the corresponding keys in order. Stringed instruments, however, require a great deal of experience and decision making on the part of the performer. A given note on the guitar may have as many as six different positions on the fretboard on which it can be produced. A fretboard position is described by two variables, the string and the fret. To play a piece of music, the performer must decide upon a sequence of fretboard positions that minimize the mechanical difficulty of the piece to at least the point where it is physically possible to be executed. This process is time-consuming and especially difficult for novice and intermediate players and,

P. Crescenzi, G. Prencipe, and G. Pucci (Eds.): FUN 2007, LNCS 4475, pp. 40–52, 2007.

as a result, the task of reading music from a page, as a pianist would, is limited only to very advanced guitar players. To address this problem, a musical notation known as *tablature* was devised. A tablature describes to the performer exactly how a piece of music is to be played by graphically representing the six guitar strings and labeling them with the corresponding frets for each note, in order. *Fingering* is the process that, given a sequence of notes or chords (set of notes to be played simultaneously), yields to assigning to each note one position on the fretboard and one finger of the left hand. Fingering and tablature problem has been studied from many points of view (see, e.g., [12,14,18]).

In this paper, we study some combinatorial properties of tablature problem. This problem can be described by a finite state automaton (or, equivalently, by a regular grammar) to which we can apply the Schützenberger methodology (see [8,9,15,16] for the theory and [6,7,11] for some recent applications). We define the concepts of *distance between two chords* and *tablature complexity* to study the problem of tablature in terms of the music performance. It is then possible to compute the average complexity of a tablature. In particular, we prove the following basic results:

1. every tablature problem is equivalent to a finite state automaton (or to a regular grammar);
2. an algorithm exists that finds the finite state automaton corresponding to a tablature problem;
3. by using the Schützenberger methodology we find the generating function $\Xi(t) = \sum_n \Xi_n t^n$ counting the number $\Xi_n$ of tablatures with complexity $n$.

The concept of complexity introduced in this paper takes into consideration the total movement of the hand on the fretboard during the execution. This quantity is certainly related to the difficulty of playing a score of music but, of course, many other measures could be considered. Moreover, the difficulty depends on the artist who plays the song.

We will show an example taken from *Knocking on Heaven's Door* by Bob Dylan.

## 2  Stringed Instruments, Tablature and Symbolic Method

A *note* is a sign used in music to represent the relative duration and pitch of sound. A note with doubled frequency has another but very similar sound, and is commonly given the same name, called *pitch class*. The span of notes within this doubling is called an *octave*. The complete name of a note consists of its pitch class and the octave it lies in. The pitch class uses the first seven letters of the latin alphabet: A, B, C, D, E, F, and G (in order of rising pitch). The letter names repeat, so that the note above G is A (an octave higher than the first A) and the sequence continues indefinitely. Notes are used together as a musical scale or tone row. In Italian notation, the notes of scales are given in terms of Do - Re - Mi - Fa - Sol - La - Si rather than C - D - E - F - G - A - B. These names follow the original names reputedly given by Guido d'Arezzo, who had taken them from the first syllables of the first six musical phrases of a Gregorian Chant melody *Ut queant laxis*, which began on the appropriate scale degrees.

In this section we define in a formal way the problem of the tablature of a stringed instrument score. In order to do this, we take into consideration the set of the notes defined in the MIDI standard (see [4,5]), which uses the note-octave notation. In fact, this set contains a range of integer numbers $\daleth = \{0, ..., 127\}$, where every element represents a note of an octave.

**Table 1.** A representation of notes in the MIDI standard

| Octave | Note numbers | | | | | | | | | | | |
|---|---|---|---|---|---|---|---|---|---|---|---|---|
| | Do | Do# | Re | Re# | Mi | Fa | Fa# | Sol | Sol# | La | La# | Si |
| | C | C# | D | D# | E | F | F# | G | G# | A | A# | B |
| **0** | 0 | 1 | 2 | 3 | 4 | 5 | 6 | 7 | 8 | 9 | 10 | 11 |
| **1** | 12 | 13 | 14 | 15 | 16 | 17 | 18 | 19 | 20 | 21 | 22 | 23 |
| **2** | 24 | 25 | 26 | 27 | 28 | 29 | 30 | 31 | 32 | 33 | 34 | 35 |
| **3** | 36 | 37 | 38 | 39 | 40 | 41 | 42 | 43 | 44 | 45 | 46 | 47 |
| **4** | 48 | 49 | 50 | 51 | 52 | 53 | 54 | 55 | 56 | 57 | 58 | 59 |
| **5** | 60 | 61 | 62 | 63 | 64 | 65 | 66 | 67 | 68 | 69 | 70 | 71 |
| **6** | 72 | 73 | 74 | 75 | 76 | 77 | 78 | 79 | 80 | 81 | 82 | 83 |
| **7** | 84 | 85 | 86 | 87 | 88 | 89 | 90 | 91 | 92 | 93 | 94 | 95 |
| **8** | 96 | 97 | 98 | 99 | 100 | 101 | 102 | 103 | 104 | 105 | 106 | 107 |
| **9** | 108 | 109 | 110 | 111 | 112 | 113 | 114 | 115 | 116 | 117 | 118 | 119 |
| **10** | 120 | 121 | 122 | 123 | 124 | 125 | 126 | 127 | | | | |

A string instrument (or stringed instrument) is a musical instrument that produces sounds by means of vibrating strings.

**Definition 1 (String Instrument).** *We define a string instrument with $m$ strings $SI_m$ as a pair $(S_m, n_f)$ where:*

- *$S_m = (note_1, \cdots, note_m) \in \daleth^m$ represents the notes of the corresponding strings;*
- *$n_f \in \mathbb{N}$ represents the number of frets in this instrument.*

*Example 1.* For convention, the enumeration of the strings begins from the highest note to the most bass note. Therefore, we can define the *classical guitar [3]* as follows:

$$S_6 = (64, 59, 55, 50, 45, 40)$$
$$n_f = 19$$

instead for the *bass guitar [1]* we have:

| note/fret | I | II | III | ... | XXII |
|-----------|---|----|-----|-----|------|
| G | | | | | |
| D | | | | | |
| A | | | | | |
| E | | | | | |

$$S_4 = (31, 26, 21, 16)$$
$$n_f = 22$$

In music and music theory a *chord* is any collection of notes that appear simultaneously, or near-simultaneously over a period of time. A chord consists of three or more notes. Most often, in European influenced music, chords are tertian sonorities that can be constructed as stacks of thirds relative to some underlying scale. Two-note combinations are typically referred to as dyads or intervals. For the sake of simplicity, we use the following:

**Definition 2 (Chord).** *We say that $\xi$ is a* chord, *if it belongs to* $2^{\daleth}$.

*Example 2.* A famous chord is the *G Major* (or *Sol Major*). It is characterized by the following notes:

*D, G, B* (or *Re, Sol, Si* in Italian notation).

Using the previous definitions, $\xi = \{50, 55, 59\}$ corresponds to a *G Major* on the fourth octave. In the same way, *C Major* (formed by *C, E, G*) on the fourth octave is representable with $\xi = \{48, 52, 55\}$, whereas $\xi = \{60\}$ is a simple note *C* (or *Do*) on the fifth octave.

**Definition 3 (Chord for a String Instrument).** *Let $\mathcal{SI}_m = (S_m, n_f)$ be a string instrument with $m$ strings and $\xi = \{a_1, \ldots, a_j\}$ a chord with $j \leq m$. $\xi$ is a chord for $\mathcal{SI}_m$ iff there exists an injective function $f : \xi \rightarrow \{1, \ldots, m\}$ such that $\forall i = 1, \ldots, j$ we have $0 \leq a_i - note_{f(a_i)} \leq n_f$. We indicate with $\Gamma_{\mathcal{SI}}$ the set of these chords.*

A chord progression (also chord sequence, harmonic progression or sequence) is a series of chords played in order. In this paper, we give the following:

**Definition 4 (Chord Progression).** *Given a string instrument $\mathcal{SI}_m$, we call chord progression for the instrument $\mathcal{SI}_m$, every finite sequence $\xi_1, \ldots, \xi_n$ such that $\forall i = 1, \ldots, n$ we have $\xi_i \in \Gamma_{\mathcal{SI}}$.*

*Example 3.* If we use a classical guitar, a chord progression can be defined as follows:

*C Major, A Minor, D Minor, G 7th*
*(or Do Major, La Minor, Re Minor, Sol 7th)*

where:

| Chord name | Notes of chord | Numeric representation |
|:---:|:---|:---:|
| C Major | C, G, C, E (or Do, Sol, Do, Mi) | $\{48, 55, 60, 64\}$ |
| A Minor | A, A, C, E (or La, La, Do, Mi) | $\{45, 57, 60, 64\}$ |
| D Minor | D, A, D, F (or Re, La, Re, Fa) | $\{50, 57, 62, 65\}$ |
| G 7th | G, G, B, F (or Sol, Sol, Si, Fa) | $\{43, 55, 59, 65\}$ |

While standard musical notation represents the rhythm and duration of each note and its pitch relative to the scale based on a twelve tone division of the octave, tablature (or tabulatura) is instead operationally based, indicating where and when a finger should be depressed to generate a note, so pitch is denoted implicitly rather than explicitly. Tablature for plucked strings is based upon a diagrammatic representation of the strings and frets of the instrument. In a formal way, we give the following:

**Definition 5 (Position).** *Given an intrument $SI_m$ with m strings, we call* position *the following function TAB:*

$$TAB : N_s \rightarrow N_f \cup \{\square\}$$

*where*

- $N_s = \{1, \ldots, m\}$ *is the set of strings in the instrument $SI_m$;*
- $N_f = \{1, \ldots, n_f\}$ *is the set of frets in the instrument $SI_m$;*
- $\square$ *is the null position.*

*Example 4.* We can also describe a position using a graphic representation. In this way, we represents only the notes that the musician must pluck. For example, given a classical guitar we can represent the chord progression described in the Example 3 as follows:

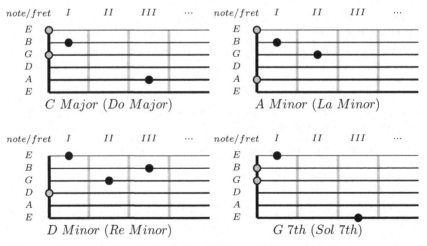

We use the gray dots to indicate that the string must be used without fret pressure.

**Definition 6 (Tablature).** *Given a string instrument* $SI_m$, *we call* tablature *a finite sequence of position* $TAB_1, ..., TAB_n$.

A position will be *realizable* on a string instrument $SI$, iff a common hand can realize such position on $SI$. Otherwise, the position will be bad for the instrument. Given an instrument $SI$, let $T_{SI}$ be the set of the positions on $SI$. $T_{SI} = \widehat{T}_{SI} \bigcup \widetilde{T}_{SI}$ where $\widehat{T}_{SI}$ is the set of the realizable positions and $\widetilde{T}_{SI}$ is the set of the bad positions.

**Definition 7 (Expansion function).** *Given an instrument* $SI_m$ *we define the* expansion function *as follows:*

$$\delta : \Gamma_{SI} \to 2^{\widehat{T}_{SI}}$$
$$\text{where } \forall \xi \in \Gamma_{SI} \text{ we have}$$
$$\delta(\xi) = \{TAB^{[\xi]} \mid TAB^{[\xi]} \in \widehat{T}_{SI} \text{ corresponds to the chord } \xi\}$$

With the previous definition we understand that, given a string instrument, we can associate a set of positions to the same chord. Extending the concept, we can associate to a specific chord progression a set of tablatures.

*Example 5.* Given a classical guitar we can play the same *C Major* defined in the Example 3 using the following positions:

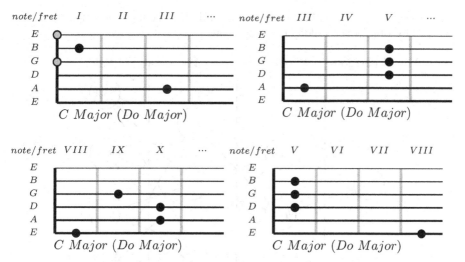

**Proposition 1 (Automaton of chord progression).** *Given a string instrument* $SI_m$ *and a chord progression* $\xi_1, ..., \xi_n$, *we can define a deterministic finite state automaton* $A = (q_{0,0}, \Omega, \Omega_f, F)$ *which represents the set of all possible tablatures for the progression in* $SI_m$, *where:*

- $q_{0,0}$ *is the initial state;*
- $\Omega$ *is the set of the states;*
- $\Omega_f \subseteq \Omega$ *is the set of the final states;*
- $F$ *is the transition function with* $F : \Omega \times \widehat{T}_{SI} \to \Omega$.

*Proof.* We associate the null position to the initial state. Moreover, $|\Omega-\{q_{0,0}\}|=$ $\sum_{i=1}^{n}|\delta(\xi_i)|$ where $\forall \xi_i$ we introduce the subset of states $\{q_{i,0},\ldots,q_{i,|\delta(\xi_i)|-1}\}$. The transition function will be $F = \{(q_{i-1,j},TAB_k^{[\xi_i]},q_{i,k})\}_{i,j,k\in N}$ with $TAB_k^{[\xi_i]} \in \delta(\xi_i)$.

*Example 6.* We take into consideration the sequence $C$, $E$ and $G$ in the fourth octave, in other words $Do = \{48\}$, $Mi = \{52\}$, $Sol = \{55\}$. Using the classical guitar and the expansion function on these notes, we have the following positions:

$$\delta(\{48\}) = \{\langle(1,\square),(2,\square),(3,\square),(4,\square),(\mathbf{5},\mathbf{3}),(6,\square)\rangle,$$
$$\langle(1,\square),(2,\square),(3,\square),(4,\square),(5,\square),(\mathbf{6},\mathbf{8})\rangle\}$$
$$\delta(\{52\}) = \{\langle(1,\square),(2,\square),(3,\square),(\mathbf{4},\mathbf{2}),(5,\square),(6,\square)\rangle,$$
$$\langle(1,\square),(2,\square),(3,\square),(4,\square),(\mathbf{5},\mathbf{7}),(6,\square)\rangle$$
$$\langle(1,\square),(2,\square),(3,\square),(4,\square),(5,\square),(\mathbf{6},\mathbf{12})\rangle\}$$
$$\delta(\{55\}) = \{\langle(1,\square),(2,\square),(\mathbf{3},\square),(4,\square),(5,\square),(6,\square)\rangle,$$
$$\langle(1,\square),(2,\square),(3,\square),(\mathbf{4},\mathbf{5}),(5,\square),(6,\square)\rangle$$
$$\langle(1,\square),(2,\square),(3,\square),(4,\square),(\mathbf{5},\mathbf{10}),(6,\square)\rangle$$
$$\langle(1,\square),(2,\square),(3,\square),(4,\square),(5,\square),(\mathbf{6},\mathbf{15})\rangle\}$$

where every pair $(c,t)$ indicate the string and the fret number respectively. In this way we obtain the following automaton:

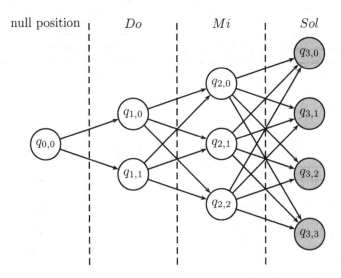

**Definition 8 (Distance between two positions).** *Given a string instrument* $SI_m = (S_m, n_f)$, *we define the* distance function *as follows:*

$$d : \widehat{T}_{SI} \times \widehat{T}_{SI} \to \{1, \ldots, n_f\}$$

*where* $\forall \, TAB_1, TAB_2 \in \widehat{T}_{SI}$ *we set*

$$\lambda_1 = \min\{fret \neq \square \mid \exists j = 1,\dots,m \ \ such \ that \ TAB_1(j) = fret\},$$
$$\lambda_2 = \min\{fret \neq \square \mid \exists j = 1,\dots,m \ \ such \ that \ TAB_2(j) = fret\},$$

then $d(TAB_1, TAB_2) = |\lambda_1 - \lambda_2|$. If $\forall j \ TAB_1(j) = \square$ or $TAB_2(j) = \square$ then $d(TAB_1, TAB_2) = 0$.

*Example 7.* We take into consideration the following positions:

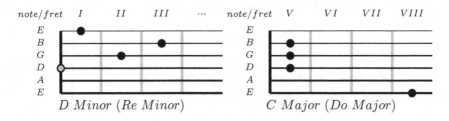

In this case the distance between the two positions is equal to four.

The concept of distance is important to determinate the complexity of a tablature. In fact, a greater distance requires a great deal of experience on the part of the performer. Therefore, we give the following:

**Definition 9 (Tablature complexity).** *Given a string instrument $SI_m$, let $TAB_1,\dots,TAB_n$ be a tablature for this instrument. We call complexity of the tablature the following quantity:*

$$\sum_{j=1}^{n-1} d(TAB_j, TAB_{j+1}).$$

**Proposition 2.** *Given a string instrument $SI_m$ and a chord progression $\xi_1,\dots,\xi_j$ for the instrument, let $A$ be its associated automaton. We can obtain the following generating function:*

$$\Xi(t) = \sum_n \Xi_n t^n$$

*which counts the number $\Xi_n$ of tablatures having complexity equal to $n$.*

*Proof.* We use the Schützenberger's methodology (or the symbolic method) to associate the indeterminate $t$ to the distance between two sequential positions. Therefore, when we change the position from $TAB_k^{[\xi_i]}$ to $TAB_h^{[\xi_{i+1}]}$, the transition $q_{i,k} \to q_{i+1,h}$ becomes a term of the generating function $\Xi_{i,k}(t)$ in the form $t^{d(TAB_k^{[\xi_i]}, TAB_h^{[\xi_{i+1}]})}\Xi_{i+1,h}(t)$ where $d(TAB_{i,k}, TAB_{i+1,h})$ is the distance between the two positions. By solving the obtained equations system in the unknown $\Xi_{0,0}(t) = \Xi(t)$ we have the desired generating function. $\quad\blacksquare$

*Example 8.* A very famous song is *Knocking on Heaven's Door* by Bob Dylan (see [2]). This song is a good example, because every strophe is characterized by the following chord progression:

*G Major, D Major, A Minor, G Major, D Major, C Major*

**Table 2.** Knocking on Heaven's Door by Bob Dylan

| |
|---|
| Introduction<br>*G Major, D Major, A Minor, G Major, D Major, C Major*<br><br>*G Major   D Major      A Minor*<br>Mama, take this badge off of me<br><br>*G Major      D Major    C Major*<br>I can't use it anymore.<br><br>*G Major      D Major      A Minor*<br>It's gettin' dark, too dark for me to see<br><br>*G Major            D Major          C Major*<br>I feel like I'm knockin' on heaven's door.<br>. . . |

or, in Italian notation:

*Sol Major, Re Major, La Minor, Sol Major, Re Major, Do Major.*

We want to study the complexity of a single strophe of this song. Using the Definition 5 we can give the following realizable positions (you can find C Major positions in Example 4):

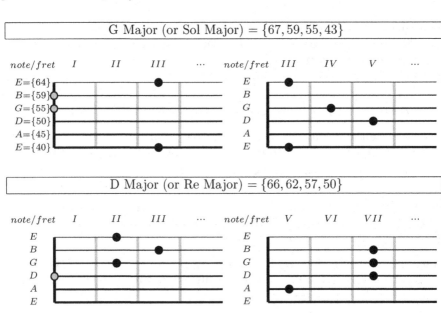

$$\text{G Major (or Sol Major)} = \{67, 59, 55, 43\}$$

$$\text{D Major (or Re Major)} = \{66, 62, 57, 50\}$$

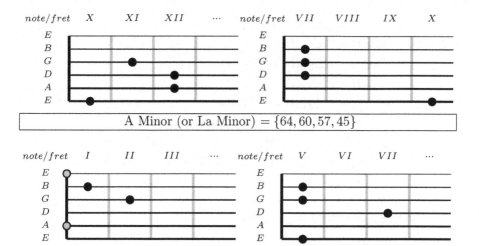

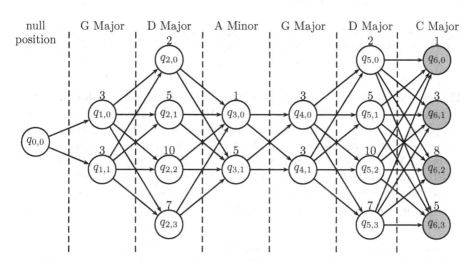

Using Proposition 1 we can generate the following automaton:

In the previous figure, for all $q_{i,k}$ we have written also the

$$min\{fret \neq \square \mid \exists j = 1, \ldots, m \text{ such that } TAB_k^{[\xi_i]}(j) = fret\}.$$

These values are necessary to compute the distance between two positions. Moreover, we can observe as, in this simple case, there are 512 different tablatures for the same chord progression. Using Proposition 2 we obtain the following system of equations:

$$\Xi_{0,0}(t) = \Xi_{1,0}(t) + \Xi_{1,1}(t)$$
$$\Xi_{1,0}(t) = t\Xi_{2,0}(t) + t^2\Xi_{2,1}(t) + t^7\Xi_{2,2}(t) + t^4\Xi_{2,3}(t)$$
$$\Xi_{1,1}(t) = t\Xi_{2,0}(t) + t^2\Xi_{2,1}(t) + t^7\Xi_{2,2}(t) + t^4\Xi_{2,3}(t)$$

$$\Xi_{2,0}(t) = t\Xi_{3,0}(t) + t^3\Xi_{3,1}(t)$$
$$\Xi_{2,1}(t) = t^4\Xi_{3,0}(t) + \Xi_{3,1}(t)$$
$$\Xi_{2,2}(t) = t^9\Xi_{3,0}(t) + t^5\Xi_{3,1}(t)$$
$$\Xi_{2,3}(t) = t^6\Xi_{3,0}(t) + t^2\Xi_{3,1}(t)$$
$$\Xi_{3,0}(t) = t^2\Xi_{4,0}(t) + t^2\Xi_{4,1}(t)$$
$$\Xi_{3,1}(t) = t^2\Xi_{4,0}(t) + t^2\Xi_{4,1}(t)$$
$$\Xi_{4,0}(t) = t\Xi_{5,0}(t) + t^2\Xi_{5,1}(t) + t^7\Xi_{5,2}(t) + t^4\Xi_{5,3}(t)$$
$$\Xi_{4,1}(t) = t\Xi_{5,0}(t) + t^2\Xi_{5,1}(t) + t^7\Xi_{5,2}(t) + t^4\Xi_{5,3}(t)$$
$$\Xi_{5,0}(t) = t + t + t^6 + t^3$$
$$\Xi_{5,1}(t) = t^4 + t^2 + t^3 + t^0$$
$$\Xi_{5,2}(t) = t^9 + t^7 + t^2 + t^5$$
$$\Xi_{5,3}(t) = t^6 + t^4 + t + t^2$$

By solving in $\Xi_{0,0}(t)$ we obtain:

$$\Xi_{0,0}(t) = 24t^6 + 28t^8 + 16t^9 + 48t^{10} + 16t^{11} + 32t^{12} + 28t^{13} + 40t^{14} + 12t^{15} +$$
$$40t^{16} + 16t^{17} + 36t^{18} + 12t^{19} + 44t^{20} + 8t^{21} + 28t^{22} + 12t^{23} + 24t^{24} + 4t^{25} +$$
$$12t^{26} + 4t^{27} + 12t^{28} + 8t^{30} + 4t^{32} + 4t^{34}$$

This generating function counts the number of tablatures for each complexity. For example, the term $24t^6$ indicates the existence of 24 different tablatures with complexity equal to 6 to execute the chord progression. These 24 solutions correspond to tablatures requiring the shortest total movement of the hand on the fretboard. From the generating function we also find that the average complexity and the variance are: $\overline{\Xi} = 16.25$ and $\sigma = 141,195$. This means that if we play the song with a *random* tablature we make a total hand jump corresponding to 16 frets, on the average.

*Example 9.* We can use the previous proposition also with *infinite sequence of chords*. In fact, in this example we define a very simple string instrument $\mathcal{SI} = ((48, 49), n_f)$ as follows:

| note/fret | I | II | III | ... |
|-----------|---|----|-----|-----|
| C         |   |    |     |     |
| C♯        |   |    |     |     |

We study the tablatures of the following infinite notes progression:

$$C = \{48\}, D = \{50\}, C, D, C, D, \ldots$$

In this case note $C$ has only one position, but note $D$ has two positions.

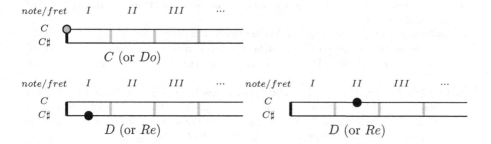

We have the following system of equations:

$$\Xi_{0,0}(t) = 1 + \Xi_{1,1}$$
$$\Xi_{1,1}(t) = t\Xi_{2,0}(t) + t^2\Xi_{2,1}$$
$$\Xi_{2,0}(t) = \Xi_{0,0}(t)$$
$$\Xi_{2,1}(t) = \Xi_{0,0}(t)$$

We set the term 1 in $\Xi_{0,0}(t)$ because we can consider the initial state as a final state. By solving the system, we obtain:

$$\Xi_{0,0}(t) = 1 + t\Xi_{0,0}(t) + t^2\Xi_{0,0}(t)$$

and this equation corresponds to the generating function of Fibonacci's numbers $F_n$, in fact:

$$\Xi_{0,0}(t) = \frac{1}{1 - t - t^2} = 1 + t + 2t^2 + 3t^3 + 5t^4 + O(t^5).$$

Therefore, in this example, there are $F_n$ different tablatures with complexity $n$.

## 3   Conclusion

In this paper we introduce the problem of tablatures for stringed instruments and explain some combinatorial properties. We don't present an exhaustive study, in fact there are some questions which require a further study. In the Example 8 we have 24 tablatures with lowest complexity. Which one is the better? A tablature can be considered better also in terms of the mechanical difficulty of any single position. This concept is linked with the problem of fingering. In a next paper we will study this kind of problems.

## References

1. Bass guitar, http://en.wikipedia.org/wiki/Bass_guitar
2. Dylan, B.: Knockin' on heaven's door.http://www.bobdylan.com/songs/knockin.html
3. Classical guitar. http://en.wikipedia.org/wiki/Classical_guitar

4. Midi committee of the association of musical electronic industry.`http://www.amei.or.jp`
5. Midi manufacturers association. `http://www.midi.org`
6. Baccherini, D.: Behavioural equivalences and generating functions. preprint (2006)
7. Baccherini, D., Merlini, D.: Combinatorial analysis of tetris-like games. preprint (2005)
8. Flajolet, Ph., Sedgewick, R.: The average case analysis of algorithms: complex asymptotics and generating functions. Technical Report 2026, INRIA (1993)
9. Flajolet, Ph., Sedgewick, R.: The average case analysis of algorithms: counting and generating functions. Technical Report 1888, INRIA (1993)
10. Goldman, J.R.: Formal languages and enumeration. Journal of Combinatorial Theory, Series A. 24, 318–338 (1978)
11. Merlini, D., Sprugnoli, R., Verri, M.C.: Strip tiling and regular grammar. Theoretical Computer Science 242(1-2), 109–124 (2000)
12. Miura, M., Hirota, I., Hama, N., Yanigida, M.: Constructiong a System for Finger-Position Determination and Tablature Generation for Playing Melodies on Guitars. System and Computer in Japan 35(6), 755–763 (2004)
13. Moore, R.F.: Elements of computer music, vol. XIV, p. 560. Prentice-Hall, Englewood Cliffs (1990)
14. Sayegh, S.: Fingering for String Instruments with the Optimum Path Paradigm. Computer Music Journal 13(6), 76–84 (1989)
15. Schützenberger, M.P.: Context-free language and pushdown automata. Information and Control 6, 246–264 (1963)
16. Sedgewick, R., Flajolet, P.: An introduction to the analysis of algorithms. Addison-Wesley, London (1996)
17. Sudkamp, T.A.: Languages and machines. Addison-Wesley, London (1997)
18. Tuohy, D.R., Potter, W.D.: A genetic algorithm for the automatic generation of playable guitar tablature. In: Proceedings of the International Computer Music Conference (2004)
19. Wilf, H.S.: Generatingfunctionology. Academic Press, San Diego (1990)

# Knitting for Fun: A Recursive Sweater

Anna Bernasconi[1], Chiara Bodei[1], and Linda Pagli[1]

Dipartimento di Informatica, Università di Pisa, Largo B. Pontecorvo, 3, I-56127,
Pisa, Italy
{annab,chiara,pagli}@di.unipi.it

**Abstract.** In this paper we investigate the relations between knitting
and computer science. We show that the two disciplines share many concepts. Computer science, in particular algorithm theory, can suggest a
lot of powerful tools that can be used both in descriptive and prescriptive
ways and that apparently have not yet been used for creative knitting.
The obtained results are short (optimal size) recursive descriptions for
complex patterns; creation of new complex recursive patterns; and the
application of three-valued algebra operations to combine and create a
wide variety of new patterns.

**Keywords:** Modeling Knitting; Pattern Knitting Diagrams; Checkerboard, Sierpinski, and Butterfly patterns; Knitting Complexity.

## 1   Knitting, Mathematics and Computer Science

Knitting is usually considered a female activity and females are usually not considered to be inclined to mathematics, or to science in general. Nevertheless
mathematical skills are necessary for knitting, because they help to realize sym-
metries, inversions, scalings and proportions; good abstraction capabilities are
indeed needed to figure the final result out and to map the idea of a pattern into
a knitted form. Therefore, even illiterate women use mathematics while knitting,
without knowing it.

Furthermore, if you think about knitting carefully, you can find a lot of formal
and abstract structures. And here, computer science may come in, by providing
tools and ways of interpreting and re-interpreting these structures, thus giving
a form to knitting.

Knitting offers us a nice chance to revisit some of the main concepts of computer science from a new perspective and, at the same time, knitting can be
better understood in the light of this theoretical tour. Computer science, especially algorithm theory, can suggest a lot of powerful tools that can be used both
in descriptive and prescriptive ways and that apparently have not yet been used
for creative knitting.

A pattern can be seen as a matrix of *stitches* (columns) and *needles* (rows),
and it is usually repeated many times horizontally or vertically, or inserted into
another pattern or interleaved with one or more other patterns. The stitches
can be chosen from a set of possible stitches, but not all the combinations are

P. Crescenzi, G. Prencipe, and G. Pucci (Eds.): FUN 2007, LNCS 4475, pp. 53–65, 2007.

allowed: we have to select them according to a set of predefined rules, which guarantee a consistent result.

Once a new pattern has been created, its description is represented through the so called *pattern knitting diagram*; in this way the pattern can be reproduced many times and communicated to others. The diagram must be read from the bottom to the top, the odd rows from right to left and the even ones in the opposite direction, i.e., a "bustrophedic" reading (from ancient greek $\beta o v \varsigma$, "ox" + $\sigma \tau \rho \epsilon \phi \epsilon \iota \nu$, "to turn", imitating the ox ploughing the field, back and forth). Rows and, in general, patterns exhibit structural regularity, which allows us to use the notion of grammars to describe them (Section 2). On the other hand, exactly the bustrophedic reading of the diagram gives the specification, row by row, of the elementary stitches to be performed and essentially represents the *algorithm* to be executed to realize the piece of work.

The relations between mathematics and knitting have been studied from many points of view. A wide review on this subject can be found in the web site *The Home of Mathematical Knitting* [8].

As computer scientists we will mostly consider other aspects of the creative knitting process, which, as we will see, are interesting and, to our knowledge, have not yet been investigated. First of all, recall that one of the first examples of an elementary computer was a mechanical loom, invented by *Joseph Jacquard* in 1801. This machine was able to execute patterns composed of several interleaved threads of different colors following the scheme of punched holes in board punch cards. In this way the Jacquard machine automatically selected the color of each stitch, allowing complex combinations. Nowadays we still use the name jacquard to indicate this kind of pattern and the idea of using punched cards as knitting diagrams to reproduce particular patterns has been used also by more modern knitting machines. Now they include very sophisticated control devices that behave as real dedicated computers and are able to reproduce any complex pattern.

We started our study by asking ourselves what applying recursion to knitting could lead to. First of all we wanted to understand if recursive motives could be employed to obtain beautiful patterns, and then to see how to exploit the power of recursion to create very short descriptions. We found some surprising results as shown by the examples of recursive patterns proposed here, which, in our opinion, show some beauty (Section 4). Moreover, the recursive patterns can be seen as schemes of patterns, from which it is possible, changing the initial conditions and the basic stitches, to obtain families of new patterns, thereby opening a new style of knitting.

In addition, recursive patterns can be defined in a very succinct way, i.e., their pattern knitting diagrams can be automatically generated with very short recursive algorithms. Applying, by analogy, the well known concept of *Kolmogorov Complexity* in this framework, we might say that recursive knitting patterns have low *"knitting complexity"* (Section 5). This result shows how recursion allows us to get an optimal compression of patterns, whose standard description would have a much higher complexity. The usual knitting instructions, described in natural language, can then be derived from the recursive algorithms, or directly from the generated pattern diagrams, in an automatic way by some sort of *knitting compiler*.

A second part of this study is still devoted to creating new patterns, but based on the combination of given patterns. Different combinations of stitches give rise to different textures, among which the most famous are *flat stockinette*, *reverse stockinette*, and *seed stitches* (Section 3). Patterns and motifs are obtained by using different textures for different areas of the knitting piece. Consequently, the elementary unit to be considered seems to be the texture unit, i.e., the stitch processed according to a particular texture. We call this unit *knitting element* or *knittel*, in analogy with pixel (picture element).

We show how considering a set of possible textures and their combinations as a *three-valued algebra*, it is possible to combine shapes and patterns in a very simple and elegant way, using the algebra operations, and to easily obtain the specifications of many nice patterns (Section 6). This is only an example of how a formal approach offers a way to enhance the design possibilities.

## 2   The Grammar of Knitting

In specialistic journals, patterns are specified both with pattern knitting diagrams and verbal descriptions. A pattern knitting diagram is a matrix, where each element corresponds to a single stitch and every row corresponds to a needle. The pattern can be repeated as many times as needed. Every kind of stitch is represented by a special knitting symbol. In other words, there is a finite alphabet $\mathcal{S}$ of knitting symbols $S$ and a precise syntax of these symbols.

$$
\begin{aligned}
S ::= \ & stitches \\
| \quad & \text{knit} \\
- \quad & \text{purl} \\
o \quad & \text{cast on} \\
... \ &
\end{aligned}
$$

The verbal description is compressed horizontally by inserting the repetitions between two stars, and vertically indicating the rows to be repeated, as in Fig. 1 (see also Fig. 5). The explanation of each row resembles the production rules in a regular grammar [3], where terminals are knitting symbols in $\mathcal{S}$ and one special non terminal symbol suffices to generate each row. Consider, e.g., the first row in Fig. 1. It is easy to rephrase it as the following production (remember that the row should be read from right to left):

$$
R ::= \quad ||| \ | \ R \ --||
$$

that, in turn, generates the following language of words $L(R) = |||\{--||\}^* = \{|||, |||--||, |||--||--||, ...\}$ (see the diagram in Fig. 1, which presents just a single repetition). It is interesting to observe that the pattern descriptions use the star with the same meaning of the Kleene star.

Cast on over a number of stitches multiple of 4 plus 3.

**Row 1:** * knit 2, purl 2; repeat from *: to last 3 sts, knit 3.

Repeat row 1.

```
| | | - - | |
| | - - | | |
| | | - - | |
| | - - | | |
| | | - - | |
| | - - | | |
| | | - - | |
```

**Fig. 1.** Standard description (top) and pattern knitting diagram (bottom) of Mistake Rib

Consequently, grammars appear a good tool for pattern modeling, as they provide simple and elegant representations of patterns. Actually, a pattern corresponds to a two-dimensional word and this calls for a generalization of formal word language theory. There are many possible models for two-dimensional languages that can be used for the knitting framework. This subject is left for future work. The nice thing is that we can define a sweater as a piece of a particular language.

## 3  Knit Textures

Different combinations of stitches give rise to knit fabrics that result in different textures (see Fig. 2). The basic knit fabric is called *stockinette* pattern and it is obtained by alternating rows of knits with rows of purls on the right side. The visual effect is a grid of V shapes. On the wrong side the pattern has a different texture, the effect is a grid of $\sim$ shapes and it is used as a pattern in itself (obtained by alternating rows of purls with rows of knits) with the name of *reverse stockinette*. Another common fabric is called *seed stitch* and it is obtained by alternating knits and purls. Generally, the visible patterns are not completely congruent to their diagrams. For instance, the visual effect of the seed stitch is that of a checkerboard, while its diagram (the right one in Fig. 2) has a different aspect. This is due to the fact that the odd rows describe the right side of the fabric, whereas even ones refer to the wrong side. When following the diagram, this corresponds to the perspective on the fabric of the person who knits it.

Patterns are usually obtained by combining different textures as in the checkerboard pattern in Fig. 3, where stockinette and reverse stockinette are combined

```
- - - -        | | | |        - | - |
| | | |        - - - -        - | - |
- - - -        | | | |        - | - |
| | | |        - - - -        - | - |
```

**Fig. 2.** Pattern Knitting Diagrams for Stockinette (left), Reverse Stockinette (center), Seed Stitch (right)

```
- - - - | | | |       a a a a b b b b
| | | | - - - -       a a a a b b b b
- - - - | | | |       a a a a b b b b
| | | | - - - -       a a a a b b b b
| | | | - - - -       b b b b a a a a
- - - - | | | |       b b b b a a a a
| | | | - - - -       b b b b a a a a
- - - - | | | |       b b b b a a a a
```

**Fig. 3.** Checkerboard: Pattern Knitting Diagram (left), Visible Pattern (right)

in a checkerboard style in four tiles (see also Section 4). Again, the diagram does not give the immediate intuition of the pattern. To better visualize the pattern, we can use a symbol for the generic point of the stockinette texture, e.g., "a" and another, e.g., the "b" for the generic point of the reverse stockinette. Sometimes a similar convention is used in pattern knitting diagrams. In other words, we are using an abstraction of the texture, namely the *knitting element* or *knittel*, the elementary texture unit. In our example, the knittel symbol "a" stands for the generic point of the stockinette, that can be either a knit or a purl, depending on the position in the stockinette area. Of course, we can also associate "a" and "b" to other pairs of textures and obtain new combinations. For ease of presentation, in Section 6 we will use a color code for textures, associating them to different scales of gray.

## 4  Recursive Knitting: An Algorithmic Description

How can we apply the powerful concept of "recursion" to the knitting world? Probably, the most natural way is that of exploiting the recursion in the description of the knitting patterns, by using recursive algorithms to automatically generate them. One additional advantage of our algorithmic description over standard ones based on instructions in natural languages as well as pattern knitting diagrams, is that with just one algorithm it is possible to obtain whole families of new patterns by simply changing the initial conditions and the basic stitches.

We propose here three examples of families of patterns of increasing *difficulty*, as well as *beauty*, that can be generated recursively. The first example concerns a pattern that could be easily defined in an iterative way; in the second example we consider a pattern based on a well known plane fractal; finally, the pattern shown in the third example is nothing other than a knitted *butterfly network*.

Without loss of generality, we consider the generation of square knitting diagrams, each represented as a matrix $a$, whose entries describe single stitches. The generation of the knitting diagram is performed in two steps: first the execution of a recursive algorithm generates the pattern, which resembles its final aspect, and then the associated knitting diagram is simply produced by inverting every other row (i.e., changing the knit stitches into purl ones, and viceversa).

**Checkerboard pattern.** As said above, a Checkerboard pattern is composed of identical squares that alternate between stockinette stitch and reverse stockinette

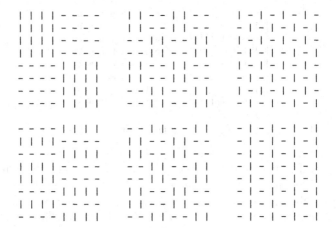

**Fig. 4.** Checkerboard visible patterns (top) and their knitting diagrams (bottom), with increasing resolution. In the visible patterns the symbol | (-) stands for a knittel of a (reverse) stockinette fabric.

stitch. The only variable is the dimension of the squares, i.e., the number of stitches across and rows long. This pattern can be seen as a generalization of the well known *seed stitch*, obtained when the dimension of the squares is one. The input of the algorithm is given by a matrix $a$, whose entries are initialized as knit stitches, its dimension $n = 2^k$, the dimension $d$ of the basic pattern, and the indexes $x$ and $y$ used to indicate the portion of matrix to fill.

**Checkerboard**$(a, n, d, x, y)$
    **if** $(n == d)$
        **for** $(i = 0; i < d/2; i{+}{+})$
            **for** $(j = 0; j < d/2; j{+}{+})$
                $a[i{+}d/2{+}x][j{+}y] = a[i{+}x][j{+}d/2{+}y] = -$ ;
    **else**
        **for** $(k = 0; k < 4; k{+}{+})$
            $i = k/2;$
            $j = k \bmod 2;$
            Checkerboard$(a, n/2, d, x + i * n/2, y + j * n/2);$

Observe that by changing the value of $d$ from its maximum value $n$ down to its minimum value 2, we obtain patterns with progressively increasing resolution. Examples of patterns of the Checkerboard family and of their corresponding knitting diagrams are shown in Fig. 4. Finally, Fig. 5 shows a standard description of a Checkerboard pattern of resolution 2, by instruction in natural languages.

**Sierpinski pattern.** The definition of this pattern is based on the plane fractal known as *Sierpinski carpet*, first described by Wacław Sierpiński in 1916. The construction of the Sierpinski carpet begins with a square. The square is cut into 9 congruent subsquares in a 3-by-3 grid, and the central subsquare is *removed*. The same procedure is then applied recursively to the remaining 8 subsquares, depending on the chosen resolution.

```
Cast on over a number n of stitches, multiple of 4.
      Row 1: * knit 2, purl 2; repeat from *.
      Row 2: repeat row 1.
      Row 3: * purl 2, knit 2; repeat from *.
      Row 4: repeat row 2.
      Repeat rows 1, 2, 3, and 4 for n/4 times.
```

**Fig. 5.** Standard description of a Checkerboard pattern for a fixed value d = 2

To realize such a pattern with our tools, i.e., needles and handwork, we only have to decide how to "remove" a square. This could be done, e.g., by using stockinette stitch as background, and reverse stockinette stitch for the removed squares, or viceversa. As before, the input of the algorithm is given by a matrix $a$, whose entries are initialized as knit stitches, its dimension $n = 3^k$, the dimension $d$ of the basic pattern, and the indexes $x$ and $y$ indicating the portion of matrix to fill.

**Sierpinski**$(a, n, d, x, y)$
    **if** $(n == d)$
        **for** $(i = d/3; i < 2*d/3; i{+}{+})$
            **for** $(j = d/3; j < 2*d/3; j{+}{+})$
                $a[i+x][j+y] = -$ ;
    **else**
        **for** $(i = n/3; i < 2*n/3; i{+}{+})$
            **for** $(j = n/3; j < 2*n/3; j{+}{+})$
                $a[i+x][j+y] = -$ ;
        **for** $(k = 0; k < 9; k{+}{+})$
            **if** $(k \neq 4)$
                $i = k/3$;
                $j = k \bmod 3$;
                Sierpinski$(a, n/3, d, x + i*n/3, y + j*n/3)$;

Again, decreasing $d$ from $n$ down to 3, we obtain patterns of progressively increasing resolution. Examples of patterns of the Sierpinski family, together with their knitting diagrams, are shown in Fig. 6. Whereas in the previous example we were able to give a concise standard description for a given resolution by verbal knitting instructions, for the present pattern a similar concision could not be attained.

**Butterfly pattern.** Our last example of recursive knitting pattern is based on the well known notion of *butterfly network* (see [5]):

**Definition 1.** *A $d$-dimensional butterfly has $(d + 1)2^d$ nodes and $d\,2^{d+1}$ edges. The nodes correspond to pairs $(w, i)$ where $i$ is the level of the node $(0 \leq i \leq d)$ and $w$ is a $d$-bit binary number that denotes the row of the node. Two nodes $(w, i)$ and $(w', i')$ are linked by an edge if and only if $i' = i + 1$ and either $w$ and $w'$ are identical (straight edge) or they differ in precisely the $i'$th bit (cross edge).*

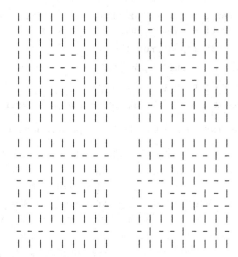

**Fig. 6.** Sierpinski visible patterns (top) and their knitting diagrams (bottom), with increasing resolution

This time, a recursive algorithm, in a classical à la *divide et impera* style, is used to generate a scheme of a $d$-dimensional butterfly (see Fig. 7). For aesthetic reasons, we only consider cross edges. Starting from such a scheme, one can easily derive a matrix describing the visible pattern, and then build the associated knitting diagram. Observe that as in the previous example, yet even more so, a request for brief standard knitting instructions cannot be satisfied for the present pattern.

The input of the algorithm is given by a matrix $a$, its dimension $n = 2^{d+1} - 2$, and the index $x$ used to indicate the portion of matrix to fill.

**Butterfly**$(a, n, x)$
 $m = (n + 2)/2$;
 **if** $(m == 2)$
  $a[0][x] = a[1][x+1] = \backslash$;
  $a[0][x+1] = a[1][x] = /$;
 **else**
  Butterfly$(a, m - 2, x)$;
  Butterfly$(a, m - 2, x + 1 + n/2)$;
  Combine$(a, m, x)$;

**Combine**$(a, m, x)$
 $r = m - 2$;
 **for** $(d = 0; d < m; d = d + 2)$
  **for** $(j = r; j < 2 * m - 2; j++)$
   $a[j][d+j-r+x] = \backslash$;
 **for** $(d = m - 1; d < 2 * m - 2; d = d + 2)$
  **for** $(j = r; j < 2 * m - 2; j++)$
   $a[j][d-j+r+x] = /$;

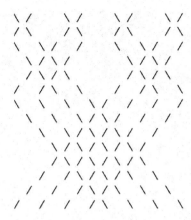

**Fig. 7.** A three-dimensional butterfly scheme recursively generated

The Butterfly pattern can be realized with needles in various way. For instance, one could play with the two basic stitches, knit and purl, and use them to realize the diagonal lines representing its edges, or, even better, the cable stitch, as shown in Fig. 8.

## 5   Knitting Complexity

In computer science, the *Kolmogorov complexity* (also known as algorithmic entropy, or program-size complexity) of an object is a measure of the computational resources needed to specify the object [6]. For the world of knitting, by analogy we introduce the following:

**Definition 2.** *The* knitting complexity *of a knitting pattern is the length in bits, expressed in order of magnitude, of the shortest description of its knitting diagram.*

We pose a basic proposition providing an immediate lower bound.

**Proposition 1.** *A knitting pattern of dimension $n \times m$ has a knitting complexity $\Omega(\log n + \log m)$.*

*Proof.* The bound easily follows by noting that $\log n + \log m$ bits are required to specify the diagram size.

Observe that the more a pattern is elementary, the more its diagram is repetitive and easy to describe with "concise" instructions. On the other hand, if a pattern does not present any structural regularity, the shortest description of its knitting diagram will consist of the diagram itself, thus requiring $\Theta(nm)$ bits.

We will now analyse the knitting complexity of the three examples described in Section 4. We have

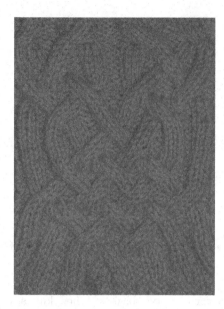

**Fig. 8.** Cable stitch realization of the Butterfly pattern

**Proposition 2.** *The Checkerboard, Sierpinski and Butterfly patterns have knit-ting complexity $\Theta(\log n)$, therefore their recursive descriptions are optimal.*

*Proof.* First of all observe that in this case $m = n$. The proof is constructively obtained from the algorithmic description of these three patterns, specified in Section 4. In fact, each algorithm consists of a constant number of instructions, and the values of the variables and input parameters are all upper bounded by $n$. Therefore $\Theta(\log n)$ bits are sufficient to describe them. The optimality immediately follows from Proposition 1.

Observe that the optimality has been obtained thanks to the power of the re-cursive description of the patterns. It can be easily seen that using the standard knitting description techniques (natural language or pattern knitting diagrams) only the complexity of the Checkerboard pattern would be of order $\Theta(\log n)$, since it takes a constant number of instructions to be described (see Fig. 5). For the Sierpinski pattern, an optimal compression can be easily obtained when the resolution is very low, e.g., $d = n$. For higher resolution ($d < n$), we can observe that the whole visible pattern can be obtained by the composition of only two basic $d \times d$ subsquares, as those shown in Fig. 9. For any $d$, these two squares can be described with $\Theta(\log d)$ bits, because of their regularity. The overall diagram can then be described as an $n/d \times n/d$ array of such subsquares. In this way we obtain a description of size $\Theta(n^2/d^2 + \log d)$.

Using the standard knitting description techniques, for the Butterfly pattern we are not able to find a better description than the whole array of stitches, requiring $\Theta(n^2)$ bits.

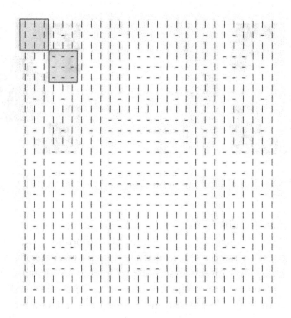

**Fig. 9.** A Sierpinsky visible pattern of size $n = 27$ and $d = 3$

# 6   Designing Patterns Using Algebra

Patterns and motifs are obtained by using different textures for different areas of the knitting piece. Previous studies in this direction can be found in [2,7]. For ease of presentation, suppose we only have three different kinds of texture in the same pattern. To obtain new patterns, we play with a three-valued algebra, i.e., an algebra over the finite field $GF(3)$. $GF(3)$ contains three elements, usually labelled with 0, 1 and 2, and arithmetic is performed modulo 3. As operations, we mainly focus on the sum and multiplication modulo 3 (typical of this ring). Note that the sum and multiplication modulo 2, i.e., the logical exclusive or (XOR) and AND operations, and the other boolean operations can be obtained as combinations of the typical $GF(3)$ operations.

We apply element-wise these operations to matrices of knittels, assuming values in $\{0, 1, 2\}$, depending on the corresponding texture. We will take two patterns, suitably coded in our algebra, and apply one or more operations. We can also combine more than two patterns. We use a color code for each value in $\{0, 1, 2\}$, by associating them to different scales of gray, to help in visualizing the patterns (0.01 for 0, 0.5 for 1, and 0.95 for 2).

Boolean operations are usually used to combine patterns and shapes in computer vision. Resorting to a third value increases the number of possible combinations. The role of values is twofold: on the one hand we can associate each value to a different texture and obtain three textures; on the other hand, their combination with different operations can play a role in the manipulation of patterns. We can think about a generalization of the masking notion, to manipulate

**Fig. 10.** Multiplication modulo 3 (right) of the two patterns A (left) and B (center)

knittel areas in bulk. It suffices to use the second pattern as a mask for the first one. If the operation applied is the multiplication modulo 3, we can obtain the following effects on the first pattern: (i) if the mask area consists of 0 the effect on the corresponding area in the first pattern is that of clearing, i.e., the area it is cleared to zero regardless of the initial value; (ii) if the mask area consists of 1, the original values in the corresponding area are not changed, while (iii) if the mask area consists of 2, the values 1 and 2 are inverted. Similar considerations can be made for the sum modulo 3, and for all the operations possible in $GF(3)$.

We present two experiments, in which we consider $n \times n$ square matrices. In the first example, in Fig. 10, we can observe the effect of the multiplication modulo 3, depending on the values.

An interesting variation could be to apply different operations to different sub-matrices, i.e., having also matrices of operations, as in Fig. 11, where each operation is applied to each $n/2 \times n/2$ sub-matrix. The matrix of operations *comb* is the following: the effect is nicely kaleidoscopic.

$$\begin{bmatrix} +_{mod\,3} & \times_{mod\,3} \\ \times_{mod\,3} & +_{mod\,3} \end{bmatrix}$$

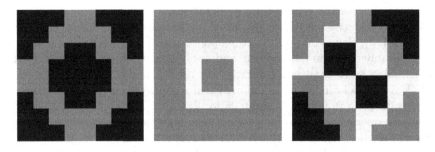

**Fig. 11.** Combination (right) of the two patterns C (left) and D (center) with the matrix of operations *comb*

## 7   Concluding Remarks

We conclude here our tour of the computational aspects of the knitting world. Summarizing, the results we have obtained are:

- Short (optimal size) recursive description for complex patterns.
- Creation of new complex patterns.
- Application of three-valued algebra operations to combine and create a wide variety of new patterns.

Beside the theoretical interest, the above results have also practical impact. In fact, using a deep level of recursion and high resolution, we can obtain automatically and in a very simple way, arbitrarily complex patterns, never designed nor produced before, to our knowledge. Their pattern knitting diagrams can be also obtained in an automatic way. Note that, even if a very complex pattern will probably require a greater skill or concentration in the executor, if human, knitting is a sequential process where one stitch is processed after the other, and therefore the overall processing time remains linear in the size of the array. A complex pattern can be obtained in approximately the same time as a simple one. This is true in particular for knitting machines whose execution time, following a program, is independent of the difficulty of the diagram.

The exploration of the knitting world with the eye of the computer scientist opens a variety of interesting topics beside those considered: this paper is only the starting point for further investigation.

It could be also nice to organize, for educational purposes, an introduction to basic concepts of computer science completely based on knitting, obviously for women only!

*Acknowledgments.* We would like to thank the numerous men who, with a mixture of diffidence and curiosity, helped us with this paper.

## References

1. Allen, P., Malcolm, T., Tennant, R., Fall, C.: Knitting for Dummies (2002)
2. Griswold, R.E.: Designing Weave Structures Using Boolean Operations, Part 1,2,3 http://www.cs.arizona.edu/patterns/weaving/webdocs.html
3. Hopcroft, J.E., Motwani, R., Ullman, J.D.: Introduction to Automata Theory, Languages, and Computation. Addison-Wesley, London (2001)
4. Knuth, D.E.: The Art of Computer Algorithms: Sorting and Searching, vol. 3. Addison-Wesley, London (1973)
5. Leighton, F.T.: Parallel Algorithms and Architectures: Array, Trees, Hypercubes. Morgan Kaufmann Publishers, San Mateo, CA (1992)
6. Li, M., Vitanyi, P.: An Introduction to Kolmogorov Complexity and Its Applications. Springer, Heidelberg (2005)
7. Suzuki, D., Miyazaki, T., Yamada, K., Nakamura, T., Itoh, H.: A Supporting System for Colored Knitting Design. In: Proc. of the 13th Int. Conf. on Industrial and engineering applications of artificial intelligence and expert systems, IEA/AIE, Springer, Heidelberg (2000)
8. The Home of Mathematical Knitting. http://www.toroidalsnark.net/mathknit.html

# Pictures from Mongolia –
# Partial Sorting in a Partial World

Paolo Boldi[1,*], Flavio Chierichetti[2,**], and Sebastiano Vigna[1]

[1] Dipartimento di Scienze dell'Informazione, Università degli Studi di Milano, Italia
[2] Dipartimento di Informatica, Università di Roma "La Sapienza", Italia

**Abstract.** You are back from that very long, marvelous journey. You have a thousand pictures, but your friends and relatives will stand just a few dozens. Choosing is a painful process, in particular when you cannot decide between the silent vastity of that desert and the idyllic picture of that tranquil, majestic lake. We are going to help.

## 1 Pictures, Pictures and More Pictures

In the summer of 2004, Paolo, Sebastiano and four more friends took part to a long journey in Mongolia, with a local guide, a driver and an old Uaz van. Five of the participants were fond of taking pictures, and Mongolia met their match: long rivers flowing through luxuriant meadows, high peaks covered with pine trees, wild horses running free, children laughing at our passage, camels in sandy deserts... On our trip back to Italy, when we were in Moscow for a short stop, we started to put together all the pictures we had been taking for three weeks and discovered that we had about six hundreds of them! (We were using mainly digital cameras and we had an old light laptop with us, so we had essentially no limit in the number of pictures we could take.)

The number was scary, not only for us but also, and importantly, for all of our friends. If you ever happened to be invited to one of those just-after-holidays dinners at one of your friend's, you know what we mean: at the end of the dinner, no doubt your friend will "propose" to look at the pictures they have been taking during their holidays in Guatemala, and those pictures turn out to be three hundreds, and most of them about a special tree growing in the jungle, and...

Yet, choosing the best representatives from a set of objects is not an easy task; even the very nature of preference is questionable, and has been largely debated by psychometrists and economists (see, e.g., [1]). In particular, a much argued point is whether personal preference is a transitive relation or not; even if some experiments actually prove that preference may indeed not be transitive, it is commonly agreed that intransitive preference lead to irrational behaviours. For this reason, it is by now accepted that preference can be modeled as a *preorder* (a.k.a. *quasiorder*), that is, a transitive reflexive relation that is not required to be antisymmetric; for sake of simplicity, though, in

---

\* Partially supported by the MIUR PRIN Project "Automi e linguaggi formali: aspetti matematici e applicativi".
\** Partially supported by the EU Project "DELIS".

P. Crescenzi, G. Prencipe, and G. Pucci (Eds.): FUN 2007, LNCS 4475, pp. 66–77, 2007.

this paper we shall mainly focus our attention on *partial orders*, although many of our algorithms can be easily extended to preorders.

So, in an abstract sense, we shall consider the problem of choosing the "best" $t$ elements from a set $X$ of $n$ elements subject to a given partial order; the exact notion of "best elements" will be discussed soon. In our intended application, $X$ is the set of all pictures, whereas $t$ is the number of pictures we are going to put in our album, and the order reflects the preference of some given, rational subject.

An algorithm for this problem will present a sequence of pairs of pictures to the subject and ask her to choose which of the two pictures she likes best, or if she is completely indifferent about the two pictures. The computational cost we are going to take into consideration is the number of comparisons the subject is requested to perform; notice, in particular, that we shall not be considering the number of instructions executed or the space occupied by the data.

Apart for pictures, the problem may find more interesting applications to other areas; for example, our algorithm suggest a new way to perform rank aggregation in Information Retrieval. Rank aggregation is the problem of combining ranking results from various sources; in the context of the Web, the main applications include building meta-search engines, combining ranking functions, selecting documents based on multiple criteria, and improving search precision through word associations. Here, by rank we mean an assignment of a relevance score to each candidate. Traditional approaches to rank aggregation [2,3] assume that ranks be aggregated in a single rank value that satisfies suitable properties. Instead, we propose to produce a partial order by intersection, and then identify suitable "best elements" of the partial order obtained in this way.

## 2 Basic Setup

A *partial order* on a set $X$ is any binary relation $\preceq$ on $X$ that is reflexive ($x \preceq x$ for all $x \in X$), antisymmetric ($x \preceq y$ and $y \preceq x$ imply $x = y$) and transitive (if $x \preceq y$ and $y \preceq z$ then $x \preceq z$); a *poset* is a set endowed with a partial order; in this paper, all posets will be finite. We let $x \prec y$ mean $x \preceq y$ and $x \neq y$.

Two elements $x, y \in X$ such that either $x \preceq y$ or $y \preceq x$ are called *comparable*, otherwise they are called *incomparable*. A set of pairwise comparable (incomparable, resp.) elements is called a *chain* (*antichain*, resp.). The *width* of a poset is the maximum cardinality of an antichain.

Antichains are "no-clue" sets: we have no freaking idea as to which picture we would prefer out of the set. Chains are "I-got-all-clues" set: we have a precise idea about which picture is better out of any pair.

We say that $x$ *covers* $y$ iff $y \prec x$ and, for every $z$, $y \preceq z \preceq x$ implies either $y = z$ or $z = x$. Intuitively, $x$ is "just better" than $y$.

A *maximal (minimal, resp.) element* in a subset $Y$ of a poset $(X, \preceq)$ is any element $y \in Y$ such that $y \preceq y' \in Y$ ($Y \ni y' \preceq y$, resp.) implies $y = y'$. Maximal elements are pictures for which we cannot find a better one (but there might be many incomparable maximal elements).

A *linear order* on $X$ is a partial order that is itself a chain. A *linear extension* of a partial order $\preceq$ is any linear order $\preceq'$ (on the same set) that extends $\preceq$ (i.e., such that $x \preceq y$ implies $x \preceq' y$).

# 3   Setting Our Goal

We wish to give a sound and meaningful definition of what are the "best" elements of a given poset; this notion is quite obvious in the case of a linear order, but turns out to be much more subtle when partial orders are involved.

As a first attempt, an *upper set* seems a good idea. Given a poset $(X, \preceq)$, a set $Y \subseteq X$ is an *upper set* iff $y \preceq z$ and $y \in Y$ imply $z \in Y$. Indeed, if we choose an upper set we are sure that no picture more important than the ones chosen has been missed.

There's however a significant problem with this definition when we try to give an intuitive, user-related meaning to incomparability. Our friends coming from Mongolia have pictures from Lake Hovsgol and also from the Gobi Desert[1]. It is reasonable to assume that it is very difficult to define preferences between pictures containing just sand and pictures containing just water, so every picture from Gobi could be incomparable with every picture from Lake Hovsgol. But in this case the pictures from Gobi would be an upper set, and a subset of Gobi pictures that is also an upper set could provide a valid answer.

In other words, if we interpret incomparability as "different setting" or "different topic" or "different subject", we would like, say, to have at least the best picture for every setting, or topic, or subject. Once we have the best pictures, if we still have room we would like to get the *second* best pictures, and so on: intuitively, we would like to *peel* the top of the poset by iteratively choosing the best pictures available. This idea suggests a more stringent definition:

**Definition 1.** *Let $(X, \preceq)$ be a poset, and define a sequence $X_0, X_1, X_2, \ldots$ of disjoint subsets of $X$, called* levels, *where $X_t$ is the set of maximal elements in $X \setminus (X_0 \cup X_1 \cup \cdots \cup X_{t-1})$ (the sequence is ultimately $\varnothing$). Let us define a new order $\sqsubseteq$ on $X$ by setting $X_0 \sqsupseteq X_1 \sqsupseteq X_2 \sqsupseteq \cdots$ (elements in each $X_i$ are incomparable). A* top set *is an upper set w.r.t. $\sqsupseteq$; a top set of size $t$ is also called a $t$-top set.*

More explicitly, top sets are exactly sets of the form $X_0 \cup X_1 \cup \cdots \cup X_{i-1} \cup X_i'$, where $X_i' \subseteq X_i$.

Note that $\sqsubseteq$ is an extension of $\preceq$ (if $x \preceq y$ and $y \in X_i$ necessarily $x \in X_j$ for $j > i$): as a consequence, all top sets are upper sets (for $\preceq$). The extension is in general proper: if we consider the set $X = \{x, y, x', y'\}$ ordered by $x \preceq x'$, $y \preceq y'$, we have that additionally $x \sqsubseteq y'$.

At this point, we are in a position to give a more precise definition of our problem: we are given a poset $(X, \preceq)$ of $n$ elements, and an integer $t \leq n$, and we must output a $t$-top set; the poset is available only implicitly, in the sense that we do not know $\preceq$ in advance, but our algorithm can use the poset as an oracle that, given two distinct

---

[1] Incidentally, "Gobi" in Mongolian means "desert", which makes the expression "Gobi Desert" at best bizarre.

elements $x, y \in X$, answers either "$x \prec y$", or "$x \succ y$" or "$x$ and $y$ are incomparable": such a call to the oracle is named *comparison*, and we want our algorithm to perform as few comparisons as possible.

To measure the loss w.r.t. a hypothetical optimal algorithm, we will say that an algorithm is $\theta$-*slow* if it never makes more than $\min_{\mathscr{A}} \max_{|P|=n} \theta q(\mathscr{A}, P, t)$ queries to return the $t$-top set of a poset of size $n$, where $q(\mathscr{A}, P, t)$ is the number of queries performed by algorithm $\mathscr{A}$ on the poset $P$ of size $n$ to return a $t$-top set.

Of course, there is a trivial algorithm that performs $\binom{n}{2}$ comparisons and rebuilds the whole poset, and in the general case we cannot hope to perform a smaller number of comparisons: every algorithm must perform at least $\binom{n}{2}$ comparison on a poset made of $n$ incomparable elements to output a top set of $n - 1$ elements. Indeed, if $x$ and $y$ are never compared and, say, $y$ is not in the output, the algorithm would fail for a poset in which $x \prec y$ (note that for this lower bound to work the choice of top sets vs. upper sets is essential). This consideration can be generalised to any $t < n$:

**Proposition 1.** *Every correct algorithm must, in the worst case, perform at least*

$$\frac{1}{2}t(2n - t - 1) \geq \frac{1}{2}tn$$

*comparisons $(0 < t < n)$ to determine a $t$-top set of a $n$-elements poset.*

**Proof.** Assume that the algorithm outputs a $t$-top set $T$ of a poset of $n$ elements without comparing every element of $T$ with every element of $P$; then, there is some $x \in T$ and some $y \in P$ that have not been compared, and the algorithm would fail on a poset in which the only comparable pair is $x \prec y$. Indeed, in that case $x$ should not be in the output, as $t < n$ and there are $n - 1$ maximal elements. Hence, the algorithm performed at least $\binom{t}{2} + t(n - t)$ comparisons, which is the left-hand side of the inequality in the statement (the right-hand side follows from $t < n$). ∎

We note by passing that a similar lower bound holds for upper sets. The previous proof can be easily amended to provide a $t(n - t)$ lower bound on the number of comparisons (if you output $t$ elements and you did not check against some of the remaining $n - t$, your output might not be an upper set). On the other hand, finding an upper set in $O(t(n - t))$ comparisons is trivial: if $t \leq n/2$ just look in a brute-force manner for a maximal element ($n - 1$ comparisons), then find another maximal element in the remaining $n - 1$ elements ($n - 2$ comparisons) and so on. This requires $(n - 1) + (n - 2) + \cdots + (n - t) = tn - t(t+1)/2 \leq \frac{3}{2}t(n - t)$ comparisons. If $t > n/2$ we can work backwards, eliminating iteratively minimal elements, and returning the remaining ones.

Maybe surprisingly, an absolutely analogous approach yields a 2-slow algorithm for top sets. The only difference is that once we find a maximal element $m$, we start searching for elements incomparable with $m$ in the poset, and we output them. Some care, however, must be exercised, as once an incomparable element has been output, all smaller elements are not good candidates for the output, even if they are incomparable with $m$. So we must keep track of which elements of the poset should be checked for incomparability with $m$, and update them each time we output a new incomparable element. The details are given in Algorithm 1.

**Algorithm 1.** A 2-slow algorithm. The notation $x \downarrow$ denotes the set of all elements smaller than or equal to $x$.

top($P$: a poset, $t$: an integer)
1:  $C \leftarrow \varnothing$ {Candidate set: elements of the level to be peeled are to be found in here.}
2:  $L \leftarrow P$ {Elements left for the next round.}
3:  **for** $t$ times **do**
4:      **if** $C = \varnothing$ **then**
5:          {We prepare to peel a new level}
6:          $C \leftarrow L$
7:          $L \leftarrow \varnothing$
8:      **end if**
9:      set $m$ to a maximal element of $C$
10:     output $m$
11:     add $m\downarrow$ minus $m$ to $L$ {Keep all elements smaller than $m$ for the next round...}
12:     remove $m\downarrow$ from $C$ {... but remove $m$ and all smaller elements from the candidates.}
13: **end for**

**Theorem 1.** *Algorithm 1 outputs a t-top set.*

**Proof.** Let $O$ be the set of already output elements. We show that at the start of the loop, $C \cup L = P \setminus O$, and the maximal elements of $C$ are the maximal elements of $P \setminus (O \cup L)$. Indeed, the execution of the if statement (i.e., when $C$ is empty, hence $L = P \setminus O$) does not change these facts. When we output $m$ we make it disappear from $C$ and appear in $O$, preserving the equation above (the elements strictly smaller than $m$ are transferred from $C$ to $L$). Finally, since we remove $m\downarrow$ from $C$, $m$ also disappears from the set of maximal elements of $C$, and no new maximal elements appear. This implies that the second property is preserved, too.

To conclude the proof, we note that just after the $i$-th execution of the if statement, $C$ will contain the entire poset $P$ minus the first $i$ levels. This is certainly true at the first reassignment, and thus the maximal elements of $C$ are exactly the first level of $P$. Since we remove them one by one, by the time we execute the second reassignment we will have output exactly the first level, so we will now assign to $C$ the entire poset minus the first level, and so on.                                                                          ∎

**Theorem 2.** *Algorithm 1 requires no more than $t(2n - t - 1)$ comparisons to output a t-top set of a poset with n elements. Thus, it is 2-slow.*

**Proof.** The algorithm performs some comparisons only when finding maximal elements, and when computing $x\downarrow$. In both cases at most $|C| - 1$ comparisons are needed, using a brute-force approach. Note that at the start of the $i$-th iteration of the loop $|C| + |L| = n - i$, as several elements are moved between $C$ and $L$, but exactly one element is output (and removed from $C$) at each iteration, so in particular $|C| \le n - i$. The total number of comparison is then bounded by $2((n - 1) + \cdots + (n - t)) = 2tn - t(t + 1) = t(2n - t - 1)$. The last claim then follows immediately from Proposition 1.                                                                          ∎

## 4    Small–Width Posets

The reader might have noticed that our lower bound uses very wide posets (i.e., with width $\Theta(n)$). It is thus possible that limiting the width we can work around the lower bound, getting a better algorithm for small-width posets. Indeed, we expect that settings, topics or subjects should be significantly fewer than the number of pictures, or it would be very difficult to choose a small best subset.

Looking into Algorithm 1, it is clear that we are somehow wasting our time by scanning for one maximal element at a time. A more efficient strategy could be building in one shot the set of maximal elements, peeling them, build the new set of maximal elements, and so on.

There are two difficulties with this approach: first of all, if we need a very small top set we could make much more queries than necessary, as the set of maximal elements could be significantly larger than the required top set. Second, rebuilding after each peeling the set of maximal elements could be expensive.

There is not much we can do about the first problem, but our next algorithm (Algorithm 2) tries to address the second one. If we divide $P$ into two subsets and compute the respective maximal elements, the latter will certainly contain all maximal elements of $P$ (plus some spurious elements). We can apply this reasoning recursively, building the set of maximal elements incrementally. To avoid an expensive recomputation after each peeling, we will arrange arbitrarily the elements of the poset $P$ on the leaves of a complete binary tree $T$. The binary tree then induces naturally a hierarchical partition of the elements of $P$—just look at the leaves of each node of given depth. We will keep track of the maximal elements of each subset of the partition, by suitably labelling each node of the tree. Initially, we will find on the root the set $M$ of maximal elements of $P$, which we can output. Then, we will remove the elements of $M$ from all labels in the tree, and recompute the new labels. The process will continue until we have output enough elements.

Note that the width of the poset does not appear explicitly in the description of the algorithm. Indeed, it will surface only during the analysis, when we shall put to good use the fact that the labels on each node cannot be of cardinality larger than the poset width.

In what follows we shall tacitly assume that the algorithm uses some data structure to keep track of the comparisons that have already been performed; in other words, for every pair $x, y \in X$, we are able to tell whether $x, y$ have ever been compared and, if so, the result of the comparison. This assumption makes it possible to bound the number of comparisons using just the *number of pairs ever compared*.

**Theorem 3.** *Algorithm 2 is correct.*

**Proof.** To prove the statement, it is sufficient to show that after the $i$-th call to **completeLabelling** the label of the root is the $i$-th level of the input poset $P$. Given a node $v$ of $T$, we define the $v$-dominated poset $v\Downarrow$ as the subset of elements of $P$ contained in leaves dominated by $v$ with the order inherited from $P$. It is immediate to show by induction that if each node $v$ of $T$ is labelled by a (possibly empty) set of maximal elements of $v\Downarrow$, then after a call to **completeLabelling** each node will be labelled by the set of *all* maximal elements of $v\Downarrow$. Indeed, during the recomputation of the labels we

---

**Algorithm 2.** An algorithm for small-width posets

---

**top**(t: an integer)

  1: let $T$ be a complete binary tree labelled on subsets of $P$ and with $n$ leaves
  2: label each leave of $T$ with a singleton containing a distinct element of $X$
  3: label the remaining nodes with $\varnothing$
  4: **while** $t > 0$ **do**
  5:    **completeLabelling**($T$)
  6:    let $A$ be the label of the root of $T$
  7:    output $u \leftarrow \min\{t, |A|\}$ elements from $A$
  8:    $t \leftarrow t - u$
  9:    remove the elements of $A$ from all the labels
10: **end while**

**completeLabelling**($T$: a binary tree)

  1: recursively consider all non-leaf nodes $v$ of $T$, bottom-up
  2: let $z_0, z_1$ be the two children of $v$
  3: **for** $k \leftarrow 0, 1$ **do**
  4:    **for** $x$ a label of $z_k$ **do**
  5:      if no label of $z_{1-k}$ is greater than $x$ add $x$ to the label of $v$
  6:    **end for**
  7: **end for**

---

compute the maximal elements of the union of the labels of the children. But the labels of the children contain (by induction) all maximal elements of the respective dominated posets, and all maximal elements of $v\!\Downarrow$ must appear in the labels of the children of $v$ (if we partition a partial order in two disjoint subsets, all maximal elements of the partial order are still maximal elements of the subset they belong).

Thus, the label of the root after the first iteration is the first level of $P$. Then, either the algorithm stops, or we remove the first level and call again **completeLabelling**. Since now the tree contains just $P$ minus its first level, and all labels are still maximal, the label of the root will be the second level, and so on. ∎

**Theorem 4.** *Algorithm 2 finds a t-top set of a poset with n elements and width w using*

$$2wn + wt(\log n - \lfloor \log w \rfloor)$$

*comparisons.*

**Proof.** We split the estimate of the number of comparisons in two parts. Assume without loss of generality that $n$ is a power of two. Define the *depth* of a node in the standard way (the root has depth 0, and the children of nodes at depth $d$ have depth $d + 1$). We shall call the nodes of depth smaller than $\log n - \lfloor \log w \rfloor$ *interesting*, and the remaining nodes *uninteresting*. Note that uninteresting nodes are those living in the last $\lfloor \log w \rfloor + 1$ levels of the tree.

Since we never repeat a comparison, all comparisons that could be ever performed on uninteresting nodes are very easily bounded:

$$\sum_{i=\log n-\lfloor \log w \rfloor}^{\log n-1} 2^i \left(2^{\log n-i-1}\right)^2 = \sum_{i=-\lfloor \log w \rfloor}^{-1} 2^{i+\log n}\left(2^{-i-1}\right)^2 = \sum_{i=1}^{\lfloor \log w \rfloor} 2^{-i}n2^{2i-2} \le wn.$$

We now focus the rest of the proof on the interesting nodes. The number of elements contained in each child of an interesting node can be just bounded by $w$, so in principle each time we need to update the labels of an interesting node we might need $w^2$ comparisons. This very rough estimate can be significantly improved noting that we cannot increase indefinitely the number of elements in a child (as it is bounded by $w$). Indeed, each time we add new elements in a node, this generates some new comparisons in its parent. Nonetheless, as long as we *add* elements, the number of overall comparisons performed by the parent approaches $w^2$, but can never get past it. So if we estimate an initial cost

$$\sum_{i=0}^{\log n-\lfloor \log w \rfloor-1} 2^i w^2 \le \frac{n}{w}w^2 = nw,$$

this estimate will cover all *additions*, as long as elements are never deleted.

What happens when elements are deleted? In this case, the deleted element can be replaced by a new one, and its cost is not included in our previous estimate. So we must fix our bound by adding $w$ comparisons for each node in which a deletion happens.

Let us make the above considerations formal. Each interesting node $v$ has a *bonus* of $w^2$ comparisons included in the estimate above. Let $v$ ambiguously denote the number of labels of a node $v$ *before* a call to **completeLabelling** (so in particular $v = 0$ for all interesting $v$ at the first iteration), and $v^*$ the number of labels of $v$ after the call. If $\ell$ and $r$ are the left and right child of $v$ we show that, provided $w^2 - \ell r$ bonus comparisons are available before the call, $w^2 - \ell^*r^*$ are still available afterwards.

When we call **completeLabelling**, the labels of each node $v$ must be updated. The update involves comparing new elements that appeared in the children of $v$; at most $(\ell^* - \ell)r + (r^* - r)\ell + (\ell^* - \ell)(r^* - r)$ comparison are needed to update $v$. But since $w^2 - \ell r - (\ell^* - \ell)r - (r^* - r)\ell - (\ell^* - \ell)(r^* - r) = w^2 - \ell^*r^*$ the invariant is preserved. The invariant is trivially true before the first call, so we conclude that the costs of **completeLabelling** are entirely covered by the bonus.

Finally, when we remove elements from the tree, each removed element potentially alters $\log n - \lfloor \log w \rfloor$ interesting nodes, reducing by 1 the number of its labels. Thus, to keep the invariant true we might need to cover some costs: with the same notation as above, if, for instance, the set of labels of $\ell$ becomes smaller we need to compensate $r + 1 \le w$ comparisons to keep the invariant true for node $v$ (symmetrically for $r$). Thus, removing an element requires patching the bonus by at most $w(\log n - \lfloor \log w \rfloor)$ additional comparisons.

All in all, emitting $t$ elements will require a fixed cost of $2nw$ comparisons, plus at most $w(\log n - \lfloor \log w \rfloor)$ comparisons for each deleted element; since the deleted elements are bounded by $t$ the result follows easily.                ∎

Note that, if $w = O(1)$ and $t = O(n/\log n)$, Algorithm 2 is asymptotically optimal, as it requires just

$$O\left(wn + wt\log\frac{n}{w}\right) = O\left(n + \frac{n}{\log n}\log n\right) = O(n)$$

comparisons and, for any $w < n$, $\Omega(n)$ is a trivial lower bound for the problem (that holds even if $t = 1$).

More generally if $w = o(t)$ Algorithm 2 is advantageous over Algorithm 1, as

$$wt \log \frac{n}{w} = nt \left(\log \frac{n}{w}\right) / \left(\frac{n}{w}\right) = o(nt).$$

The opposite happens if $t = o(w)$. To see that this is not an artifact of the analysis of the algorithm, just note that on a poset formed by $\lceil n/w \rceil$ chains if we distribute each level (which has width $w$) on a set of adjacent leaves *all* possible comparisons on uninteresting nodes will be performed during the first call to **completeLabelling**, so the algorithm actually requires $\Omega(wn) = \omega(tn)$ comparisons if $t > 0$.

## 5   A Probabilistic Algorithm

We now attack the problem from a completely different viewpoint. Since our lower bounds are based on very peculiar posets, we resort to a wonderful asymptotic structure theorem proved by Kleitman and Rotschild [4]: almost all posets of $n$ elements are *good*, that is, they are made of three levels of approximate size $n/4$, $n/2$, and $n/4$, respectively[2]. The idea is that of writing the algorithm as if *all* posets were good. In this way we will get an asymptotically very fast (albeit a bit improbable) algorithm that returns top sets on almost all posets.

Stated in a slightly more precise form, Kleitman and Rotschild prove the following property. If you draw at random a poset $(X, \preceq)$ of $n$ element uniformly among all posets of $n$ elements, then with probability $1 - O(\frac{1}{n})$ the poset is *good*, that is,[3]

- $X$ can be partitioned into three antichains $L_0$, $L_1$ and $L_2$;
- $\left| |L_i| - \frac{n}{4} \right| \le \sqrt{n} \log n$ for $i = 0, 2$;
- every element in $L_i$ only covers elements in $L_{i+1}$ (for $i = 0, 1$); hence, in particular, every element in $L_2$ is minimal and every element in $L_0$ is maximal;
- every non-maximal element is covered by at least $n/8 - n^{7/8}$ elements.

In other words, almost every poset is made by just three antichains, and contains about $n/4$ minimal and maximal elements, whereas the remaining elements are just "sandwiched" between a maximal and a minimal element. The main idea of the algorithm is that if we need no more than $n/5$ top elements[4], they are easy to find if the poset is good—and almost all posets *are* good.

To prove that probabilistic bound on Algorithm 3, we shall be using the following Chernoff-type bound on hypergeometric distributions, proved in [5], and stated in this form in [6]:

---

[2] The original paper creates levels by stripping *minimal* elements, but by duality all results are valid also in our case.

[3] The result proved in [4] is actually stronger, but we are only quoting the properties of good posets that we will be needing in our proof.

[4] In principle, it works for $\alpha n$ top elements, with $\alpha$ being any constant less than $1/4$—for simplicity, it is taken to be $1/5$. In the end, we will incidentally provide a generalization that works for any number of top elements.

---

**Algorithm 3.** An algorithm for good posets and $t \leq n/5$

---

**top**($t$: an integer, with $t \leq n/5$)

1: $s \leftarrow \min(\max(10t, \lceil 100 \log n \rceil), n)$
2: choose u.a.r. without replacement $s$ elements $x_1, x_2, \ldots, x_s$ from the poset
3: $Y \leftarrow \varnothing$
4: **for** $i = 1, 2, \ldots, s$ **do**
5:    **if** isProbablyMaximal($x_i$) **then**
6:       $Y \leftarrow Y \cup \{x_i\}$
7:    **end if**
8: **end for**
9: **return** any $\min(|Y|, t)$-elements subset of $Y$

**isProbablyMaximal**($x$)

1: $q \leftarrow \lceil 15 \log n \rceil$
2: choose u.a.r. with replacement $q$ elements $y_1, y_2, \ldots, y_q$ from the poset
3: **return** true iff no $y_i$ is larger than $x$

---

**Theorem 5.** *A bin contains $n$ balls, $r$ of which are red; $s$ balls are drawn at random without replacement, and $X$ denotes the number of red balls extracted. Then, $X$ has hypergeometric distribution with mean $E[X] = \mu = \frac{rs}{n}$. Moreover, for every $0 < \varepsilon < 1$,*

$$P[|X - \mu| > \varepsilon\mu] \leq \exp\left(-\varepsilon^2 \frac{r^2 s}{n(n-s)}\right).$$

Using this result, we obtain that:

**Theorem 6.** *If a poset $(X, \preceq)$ is extracted uniformly at random among all posets with $n$ elements, and $t \leq \frac{n}{5}$, then Algorithm 3 returns a $t$-top set with probability $1 - O(1/n)$, performing $O(\max(t, \log n) \log n)$ comparisons in the worst case.*

**Proof.** The bound on the number of comparisons is obvious. To perform the probabilistic analysis, consider the following events:

$\xi_{\text{fail}} = $ "the algorithm does not return a $t$-top set"
$\xi_{\text{good}} = $ "the poset $(X, \preceq)$ is good"
  $\xi_1 = $ "there are less than $t$ maximal elements among $x_1, \ldots, x_s$ (instr. 2)"
  $\xi_2 = $ "for some non-maximal $x_i$, **isProbablyMaximal** returned true (instr. 3)".

First observe that $\overline{\xi_1} \cap \overline{\xi_2} \implies \overline{\xi_{\text{fail}}}$, so $\xi_{\text{fail}} \implies \xi_1 \cup \xi_2$, that is, $P[\xi_{\text{fail}}] \leq P[\xi_1 \cup \xi_2] \leq P[\xi_1] + P[\xi_2]$. We are going to prove that both $P[\xi_1]$ and $P[\xi_2]$ are $O(1/n)$, whence the result follows. For $\xi_1$, first observe that

$$P[\xi_1] = P[\xi_1 \mid \xi_{\text{good}}]P[\xi_{\text{good}}] + P[\xi_1 \mid \overline{\xi_{\text{good}}}]P[\overline{\xi_{\text{good}}}] \leq$$

$$\leq P[\xi_1 \mid \xi_{\text{good}}] + P[\overline{\xi_{\text{good}}}] \leq P[\xi_1 \mid \xi_{\text{good}}] + O\left(\frac{1}{n}\right).$$

To bound $P[\xi_1 \mid \xi_{good}]$, let $r$ be the number of maximal elements of the poset, and $X$ be the number of maximal elements among $x_1, \ldots, x_s$ (those extracted at instruction 2). Obviously, if $s = n$ all the maximals are extracted from the poset (so assuming that the poset is good gives us $X \geq n/4 - \sqrt{n} \log n \geq n/5 \geq t$, for sufficiently large $n$). Otherwise, by Theorem 5, $E[X] = \mu = rs/n$ and for every $\varepsilon \in (0, 1)$

$$P[X < (1 - \varepsilon)\mu] \leq P[|X - \mu| > \varepsilon\mu] \leq \exp\left(-\varepsilon^2 \frac{r^2 s}{n(n - s)}\right).$$

If we assume $\xi_{good}$, $r \geq n/4 - \sqrt{n} \log n$ (by the very definition of good posets), and (for sufficiently large $n$) $r \geq n/5$. We can then obtain

$$\mu \geq \frac{(n/5)s}{n} = \frac{s}{5} \geq 2t$$

using the fact that $s \geq 10t$.

Taking $\varepsilon = 1/2$, we have $(1 - \varepsilon)\mu = \mu/2 \geq t$, so

$$P[\xi_1 \mid \xi_{good}] = P[X < t \mid \xi_{good}] \leq P[X < (1 - \varepsilon)\mu \mid \xi_{good}] \leq$$

$$\leq \exp\left(-\left(\frac{1}{2}\right)^2 \frac{r^2 s}{n(n - s)}\right) \leq \exp\left(-\frac{1}{4} \frac{r^2 s}{n^2}\right).$$

Since $r \geq n/5$ and $s \geq 100 \log n$,

$$P[\xi_1 \mid \xi_{good}] \leq \exp\left(-\frac{1}{4}\left(\frac{n}{5}\right)^2 \frac{100 \log n}{n^2}\right) = \exp(-\log n) = O\left(\frac{1}{n}\right).$$

As for $\xi_2$, reasoning in the same way we have that $P[\xi_2] \leq P[\xi_2 \mid \xi_{good}] + O(\frac{1}{n})$, so we can limit ourselves to proving that $P[\xi_2 \mid \xi_{good}] = O(\frac{1}{n})$. Recall that any non-maximal element $x_i$ is dominated by at least $n/8 - n^{7/8}$ elements in a good poset. Alternatively, for any positive constant $\alpha > 0$, every non-maximal element is dominated by at least $n(1/8 - \alpha)$ elements, for sufficiently large $n$; hence, the probability that among $q$ elements chosen uniformly at random there is none larger than $x_i$ is at most $(1 - (1/8 - \alpha))^q = (7/8 + \alpha)^q$. Since there are at most $s \leq n$ non-maximal elements for which the function **isProbablyMaximal** is called, we have that

$$P[\xi_2 \mid \xi_{good}] \leq n\left(\frac{7}{8} + \alpha\right)^q.$$

We have to choose $q$ so that the latter is no more than $1/n$, that is

$$n(7/8 + \alpha)^q \leq 1/n$$
$$q \log(7/8 + \alpha) \leq -2 \log n$$
$$q \geq \frac{2}{-\log(7/8 + \alpha)} \log n.$$

Since $2/\log(8/7) \approx 14.9777$, choosing $q$ to be at least $15 \log n$ is enough to guarantee the result.  ∎

Note that, in particular, if $t = O(\log n)$ then the number of comparisons performed by the algorithm is as low as $O(\log^2 n)$. Yet, this algorithm is of little use in practice, because the asymptotic bound is obtained for good posets: even if almost all posets are good, this is just useful if the preference poset is drawn uniformly at random among all possible posets, something that is unlikely at best. Moreover, differently from the deterministic algorithms presented so far, extending the probability bound to the case of preorders is not immediate. These issues will not be discussed further in this paper, but they constitute directions for future work.

We can remove the limit on $t$ with the help of another probabilistic algorithm that, for any $t$, returns a $t$-top set with $O(n \log n)$ comparisons. The algorithm partitions the (hopefully good) poset $X$ into the classes $L_0, L_1, L_2$ by calling, for each $x \in X$, the functions **isProbablyMaximal**$(x)$ and **isProbablyMinimal**$(x)$ (whose only difference from the former is returning true iff no $y_i$ is smaller—rather than larger—than $x$ at instruction 3). Having this partition, the algorithm can trivially return a $t$-top set. For this algorithm to fail it is necessary that the given poset is not good or that at least a call to **isProbablyMaximal** or **isProbablyMinimal** returns a wrong answer—along the same lines of the previous proofs, it can be shown that the probability of this event is $O(1/n)$.

Indeed, by combining the two probabilistic algorithms (and using the second only when $t$ is too large for the first), we remove the restriction on $t$ while leaving the same computational bound of $O(\max(t, \log n) \log n)$.

## Acknowledgmenents

We started chatting about this problem when we were still in Mongolia: we wish to thank all our travel companions, namely, Roberto Cuni, Corrado Manara, Roberto Radicioni and Lorenzo Virtuoso, our guide and translator (and cook!) Miss "it's-possible" Nora, and, last but not least, our tireless driver, mechanic, human-GPS, heroic Puujee.

We also want to thank Mongolia itself: among other things, for that August starry night in the desert that we shall never forget...

## References

1. Cowan, T., Fishburn, P.: Foundations of preference. In: Essays in Honor of Werner Leinfellner. Reidel, D., Dordrecht, pp. 261–271 (1988)
2. Dwork, C., Kumar, R., Naor, M., Sivakumar, D.: Rank aggregation methods for the web. In: WWW '01: Proceedings of the 10th international conference on World Wide Web, pp. 613–622. ACM Press, New York (2001)
3. Fagin, R., Kumar, R., Sivakumar, D.: Efficient similarity search and classification via rank aggregation. In: SIGMOD '03: Proceedings of the 2003 ACM SIGMOD international conference on Management of data, pp. 301–312. ACM Press, New York (2003)
4. Kleitman, D., Rothschild, B.L.: Asymptotic enumeration of partial orders on a finite set. Trans. of the American Mathematical Society 205, 205–220 (1975)
5. Chvátal, V.: The tail of the hypergeometric distribution. Discrete Mathematics 25(3), 285–287 (1979)
6. Dubhashi, D.D., Panconesi, A.: Concentration of measure for the analysis of randomised algorithms (1998) Draft available at http://www.dsi.uniroma1.it/~ale/papers.html, Oct. 1998.

# Efficient Algorithms for the
# Spoonerism Problem[*]

Hans-Joachim Böckenhauer[1], Juraj Hromkovič[1], Richard Královič[1,3],
Tobias Mömke[1], and Kathleen Steinhöfel[2]

[1] Department of Computer Science, ETH Zurich, Switzerland
{hjb,juraj.hromkovic,richard.kralovic,tobias.moemke}@inf.ethz.ch
[2] Department of Computer Science, King's College London, United Kingdom
kathleen.steinhofel@kcl.ac.uk
[3] Department of Computer Science, Comenius University, Slovakia

**Abstract.** A spoonerism is a sentence in some natural language where
the swapping of two letters results in a new sentence with a different
meaning. In this paper, we give some efficient algorithms for deciding
whether a given sentence, made up from words of a given dictionary, is
a spoonerism or not.

## 1 Introduction

It probably happened to most people that when speaking quickly one accidentally swapped two words of a sentence. If the resulting sentence still has a meaning, it might reveal a new meaning and may turn out funny. A *Spoonerism* is
such an accidentally transposition of words or parts of words in a sentence. It is
named after Reverend William Archibald Spooner (1844-1930). He was an English scholar who attended New College, Oxford, as an undergraduate in 1862,
and remained there for over 60 years in various capacities. Before he ultimately
became warden or president of the College, he was lecturing subjects such as
history, philosophy, and divinity. Spooner was famous for his talks and lectures
that are said to be full of these verbal slips in speech. The reason for these substitutions of phonetically similar parts is not silliness or nervousness but rather
that the mind is so swift the tongue cannot keep up.

For a detailed biography of Reverend Spooner see [3] and for a brief history and
some examples of the spoonerism see the February 1995 edition of the Reader's
Digest Magazine. Here it says: 'Reverend Spooner's tendency to get words and
sounds crossed up could happen at any time, but especially when he was agitated.
He reprimanded one student for "fighting a liar in the quadrangle" and another
who "hissed my mystery lecture." To the latter he added in disgust, "You have
deliberately tasted two worms and you can leave Oxford by the town drain."
(lighting a fire; missed my history lecture; wasted two terms; down train)'.

Many such examples have been attributed to Spooner. But a new biography
of Spooner [3] suggest that most of them were actually invented by Spooner's
students. So, the Oxfords Dictionary gives only one example of spoonerism

---

[*] Partially supported by VEGA 1/3106/06 and EPSRC EP/D062012/1.

P. Crescenzi, G. Prencipe, and G. Pucci (Eds.): FUN 2007, LNCS 4475, pp. 78–92, 2007.

("weight of rages") that can be tracked back to Spooner and says: 'Many other Spoonerisms, such as those given in the previous editions of O.D.Q., are now known to be apocryphal.'

In French (contrepéterie) this play with word for amusement is also very popular. However traditionally the swap results in an often indecent meaning such that only the original part should be said. The sometimes hard task to find the swap revealing the funny meaning is left to readers or listeners.

As said before, the substitutions often rhyme. In German there are short rhymes that are based on the exchange of the last two stressed syllables (or parts). These are known as "Schüttelreim" (shaken rhyme). Examples are: "Ein Schornsteinfeger gegen Ruß / am besten steht im Regenguß." (A chimney sweeper avoids soot best when standing in the rain) and "Beim Zahnarzt in den Wartezimmern / hört man oft auch Zarte wimmern." (In dentist waiting rooms one often hears tender ones whimper). The latter example (and many more) can be found at de.wikiquote.org.

In Slovak and Czech, the spoonerism is known as "výmenka" (exchange riddle). An example is "úľ bez nálady – ľúbezná lady" (A hive in bad mood – lovable lady).

In psychological tests, spoonerisms have been used to analyze phonological awareness which is related to spelling abilities [1].

In string matching and analysis, transposition of letters is used as metric for the similarity of strings. For instance, the Jaro distance metric [4, 5] and the Jaro-Winkler distance [8] are string comparators that account for transpositions of single letters.

We consider here the problem of deciding whether a given word or sentence is a spoonerism, i.e., whether there exists a transposition of two letters such that the resulting string is a valid word according to a given dictionary or can be decomposed into valid words. The problem was introduced in an unpublished presentation [7]. Some sketch of the ideas of the presentation can be found in an unpublished manuscript [2]. All the algorithms and results achieved and presented there (see the comparison in Section 2) are disjoint from our results.

The problem can be formalized in the following way. A sentence $s$ is given as a string, i.e., as a concatenation of its words. Its length $n$ is the total number of letters in $s$. The second part of the input is a dictionary $\mathcal{D}$ of valid words. The task is to decide whether there exist two positions in the string such that a swap of the letters at these position leaves a string that can be decomposed into words of the dictionary. For example, by swapping the letters $l$ and $p$ in the phrase "a lack of pies" we get another meaningful phrase "a pack of lies" which is correctly spelt in English.

In this paper, we will give some efficient algorithms for deciding if a given sentence is a spoonerism and analyze their worst-case running times. In Section 2, we will fix our notion and present a formal definition of the problem. In Section 3, we will present a first dynamic-programming approach for solving the spoonerism problem; two technically more involved algorithms with improved running times will be presented in Sections 4 and 5.

## 2    Preliminaries

Before we formally define the spoonerism problem, we fix some notation: For any alphabet $\Sigma$, we denote by a *dictionary* a finite subset $\mathcal{D}$ of $\Sigma^+$. By $\varepsilon$ we denote the empty string, by $w^R = a_l \ldots a_1$ we denote the reverse of a string $w = a_1 \ldots a_l$.

A *word* is a string $w \in \mathcal{D}$, and a *sentence* over an alphabet $\Sigma$ with respect to a dictionary $\mathcal{D}$ is a string $s \in \mathcal{D}^*$, i.e., a string that can be decomposed into dictionary words.

Using this notation, we can define the spoonerism problem as follows.

**Definition 1.** *The* spoonerism problem *is the following decision problem:*

**Input:** *A dictionary* $\mathcal{D} = \{w_1, w_2, \ldots, w_l\}$ *and a sentence* $s = s_1 s_2 \ldots s_n$ *of length* $n$ *over an alphabet* $\Sigma$.
**Output:** YES *if there exist* $i, j \in \{1, \ldots, n\}$ *such that* $s_i \neq s_j$ *and the string*

$$s' = s_1 \ldots s_{i-1} s_j s_{i+1} \ldots s_{j-1} s_i s_{j+1} \ldots s_n$$

*is a sentence over* $\Sigma$, NO *otherwise.*

Informally speaking, we are asked to find out whether we can swap exactly two different symbols in a sentence $s$ and get a new sentence $s'$.

Throughout the paper, we will use $a, b, c, \ldots$ to denote single letters from $\Sigma$ and we will use $u, v, w, \ldots$ to denote (possibly empty) strings over $\Sigma$. Furthermore, let $n = |s|$ be the length of the input sentence $s$, let $k = \max_{u \in \mathcal{D}} |u|$ be the length of the longest word in the given dictionary and let $m = \sum_{u \in \mathcal{D}} |u|$ be the total size of the dictionary.

The running time and space complexities of all of our algorithms will depend on the four parameters $|\Sigma|$, $m$, $n$ and $k$. It is obvious that $m \geq k$, furthermore, we will assume in the following that $n \geq k$ (otherwise we can just ignore longer dictionary words).

We present two algorithms for the spoonerism problem based on the idea of dynamic programming. The complexity of these algorithms (under the assumption of fixed $|\Sigma|$) is summarized in Table 1. These results improve the results claimed[1] by [7, 2], which use a graph-theoretic approach combined with a divide and conquer technique and achieve time complexity $O(n^2 k + n k^2 \log n)$ for processing the input sentence.

**Table 1.** Time complexity of presented algorithms

|  | Preprocessing the dictionary | Processing the sentence |
|---|---|---|
| Basic algorithm | $O(m)$ | $O(nk^3)$ |
| Improved algorithm | $O(mk)$ | $O(nk^2)$ |

---

[1] The complete proofs of these results are not given in [7,2] and therefore we have not been able to check their correctness.

# 3    The Basic Dynamic-Programming Algorithm

In this section, we will present an algorithm for solving the spoonerism problem that works in time $O(|\Sigma|m + nk(|\Sigma| + k)^2)$.

The main idea of the algorithm is to preprocess the input dictionary (in $O(|\Sigma|m)$ time) and to use dynamic programming to process the input sentence, processing each letter in $O(k(|\Sigma| + k)^2)$ time.

**Definition 2.** *We denote the set of all prefixes of all words from the dictionary $\mathcal{D}$ as* Pref $(\mathcal{D})$. *Formally,* Pref $(\mathcal{D}) = \{u \mid \exists v : uv \in \mathcal{D}\}$.

It is easy to see that $\varepsilon \in$ Pref $(\mathcal{D})$ and $\mathcal{D} \subseteq$ Pref $(\mathcal{D})$.

**Definition 3.** *Let $u, v \in \Sigma^*$. We say that $v$ is a* live suffix *of $u$ w.r.t. the dictionary $\mathcal{D}$ if and only if there exists a partition $u = u'v$ such that*

- *$u'$ is a sentence w.r.t. $\mathcal{D}$, i.e., $u'$ can be represented as a sequence of words from $\mathcal{D}$, and*
- *$v$ is a prefix of some word from the dictionary $\mathcal{D}$, i.e., $v \in$ Pref $(\mathcal{D})$.*

*We denote the set of all live suffixes of word $u$ w.r.t. the dictionary $\mathcal{D}$ as $\mathcal{L}_{\mathcal{D}}^S(u)$. If the dictionary is clear from the context, we also write $\mathcal{L}^S(u)$.*

Intuitively, the idea behind our algorithm can be described as follows: We want to process the input sentence $s$ sequentially from left to right, and, for any prefix $u$ of $s$, we want to keep track of all possible partitions of $u$ into dictionary words (plus one prefix of a dictionary word at the end). Actually, we will not need to remember the complete partition, two partitions ending with the same live suffix can be treated as equivalent; thus, we only need to store information about the live suffixes. Here, we have to distinguish between three possible situations: The desired swap of two letters can occur completely inside $u$, completely outside $u$, or it can exchange a letter from $u$ with a letter from the remainder of $s$. In the latter case, we also need to remember the letters exchanged. Formally, we can define these sets of live suffixes as follows.

**Definition 4.** *Let $u$ be a string over some alphabet $\Sigma$, let $\mathcal{D}$ be some dictionary over $\Sigma$.*

- *$\mathcal{S}_0(u)$ is the set of all live suffixes of $u$, i.e. $\mathcal{S}_0(u) = \mathcal{L}^S(u)$.*
- *For all $a, b \in \Sigma$ such that $a \neq b$, $\mathcal{S}_1^{a \to b}(u)$ is the set of all live suffixes of all strings $u'$ obtained by replacing a single letter $a$ by letter $b$ in the string $u$. More formally, $\mathcal{S}_1^{a \to b}(u) = \bigcup_{vav'=u} \mathcal{L}^S(vbv')$.*
- *$\mathcal{S}_2(u)$ is the set of all live suffixes of all strings $u'$ obtained by swapping two letters $a \neq b$ in the string $u$. Formally, $\mathcal{S}_2(u) = \bigcup_{vav'bv''=u \wedge a \neq b} \mathcal{L}^S(vbv'av'')$.*

*We will call the sets $\mathcal{S}_0(u)$, $\mathcal{S}_1^{a \to b}(u)$ for all $a \neq b$, and $\mathcal{S}_2(u)$ the $\mathcal{S}$-sets for $u$ or the $\mathcal{S}(u)$-sets for short.*

It is easy to see that there is a solution for the spoonerism problem on the input sentence $s$ if and only if $\varepsilon \in S_2(s)$.

The idea of our algorithm is to compute the sets $S_0$, $S_1^{a \to b}$, and $S_2$ incrementally by using the following equations:

- To compute $S_0(uc)$, it is sufficient to augment all possible elements of $S_0(u)$ with letter $c$. Furthermore, if the obtained set contains a word from the dictionary, then we can add $\varepsilon$ into $S_0(uc)$. Formally,

$$S_0(uc) = \begin{cases} \{vc \mid v \in S_0(u), vc \in \mathrm{Pref}\,(\mathcal{D})\} \cup \{\varepsilon\} & \text{iff } \exists v \in S_0(u) : vc \in \mathcal{D} \\ \{vc \mid v \in S_0(u), vc \in \mathrm{Pref}\,(\mathcal{D})\} & \text{otherwise} \end{cases}$$

(1)

- For computing $S_1^{a \to b}(uc)$, we distinguish two cases. Either the replacement of the letters occurs in $u$, or (in case $a = c$) the letter $c$ is replaced. Each of these cases yields a subset of $\mathrm{Pref}\,(\mathcal{D})$ and the resulting set $S_1^{a \to b}(uc)$ is the union of these subsets. In the former case, the situation is analogous to the one described in previous paragraph. Let $X^{a \to b}(uc)$ be the subset of $\mathrm{Pref}\,(\mathcal{D})$ obtained from the first case:

$$X^{a \to b}(uc) = \{vc \mid v \in S_1^{a \to b}(u), vc \in \mathrm{Pref}\,(\mathcal{D})\}$$

In the latter case (occurring only if $c = a$), we use our knowledge of $S_0(u)$. Let $Y^{a \to b}(uc)$ be the subset of $\mathrm{Pref}\,(\mathcal{D})$ obtained from the second case:

$$Y^{a \to b}(uc) = \begin{cases} \{vb \mid v \in S_0(u), vb \in \mathrm{Pref}\,(\mathcal{D})\} & \text{iff } a = c \\ \emptyset & \text{otherwise} \end{cases}$$

If the obtained set $X^{a \to b}(uc) \cup Y^{a \to b}(uc)$ contains a word from the dictionary, we have to add the empty string:

$$S_1^{a \to b}(uc) = \begin{cases} X^{a \to b}(uc) \cup Y^{a \to b}(uc) \cup \{\varepsilon\} & \text{iff } \exists v \in X^{a \to b}(uc) \cup Y^{a \to b}(uc) : \\ & \qquad v \in \mathcal{D} \\ X^{a \to b}(uc) \cup Y^{a \to b}(uc) & \text{otherwise} \end{cases}$$

(2)

- Computing $S_2(uc)$ is analogous to computing $S_1^{a \to b}(uc)$. Again, we distinguish two cases. Either the swap occurs in $u$, or the last letter $c$ is one of the swapped letters. For the first case, we have

$$X(uc) = \{vc \mid v \in S_2(u), vc \in \mathrm{Pref}\,(\mathcal{D})\}.$$

For the second case, we have

$$Y(uc) = \{va \mid a \in \Sigma, v \in S_1^{a \to c}(u), va \in \mathrm{Pref}\,(\mathcal{D})\}.$$

Finally, we augment the set by $\varepsilon$ if necessary:

$$S_2(uc) = \begin{cases} X(uc) \cup Y(uc) \cup \{\varepsilon\} & \text{iff } \exists v \in X(uc) \cup Y(uc) : v \in \mathcal{D} \\ X(uc) \cup Y(uc) & \text{otherwise} \end{cases}$$

(3)

---

**Algorithm 1.** Basic dynamic programming for the spoonerism problem

---

**Input:** A dictionary $\mathcal{D}$ of total size $m$ with maximum word length $k$ over an alphabet $\Sigma$ and a sentence $s = s_1 \ldots s_n$ w.r.t. $\mathcal{D}$.

1. Construct a trie from $\mathcal{D}$.
2.     **for** $i := 1$ **to** $n$ **do**
        Compute the $\mathcal{S}$-sets for $s_1 \ldots s_i$ according to Equations (1), (2), and (3), using the trie for the look-up operations in the dictionary.
3.     **if** $\varepsilon \in \mathcal{S}_2(s_1 \ldots s_n)$ **then**
        Output YES
    **else**
        Output NO

**Output:** YES if there exists a sentence $s'$ that can be constructed from $s$ by swapping exactly two different letters, NO otherwise.

---

It is easy to see that these equations are correct. Hence, after computing $\mathcal{S}_2(s)$, the algorithm can decide if there exists a solution for the given input sentence $s$ just by checking if $\varepsilon \in \mathcal{S}_2(s)$.

The resulting algorithm is summarized as Algorithm 1.

**Theorem 1.** *Algorithm 1 can be implemented to run in $O(|\Sigma|m + nk(|\Sigma|+k)^2)$ time and $O(|\Sigma|m + k(|\Sigma|+k)^2)$ space.*

*Proof.* For implementing the algorithm efficiently, we use a trie $T$ representing all words from the dictionary $\mathcal{D}$.[2] Each vertex of this trie uniquely represents one element from Pref $(\mathcal{D})$, the root vertex represents $\varepsilon$ and the parent of the vertex representing $ua$ represents $u$. For each vertex, it is sufficient to remember pointers to its children and a flag whether it represents a word from the dictionary. The total size of $T$ is $O(|\Sigma|m)$ and it can also be built in $O(|\Sigma|m)$ time. Moreover, once built, the trie can be reused for different runs of the main part of the algorithm on different input sentences.

The main part of the algorithm processes each letter from the input sentence and computes the corresponding $\mathcal{S}$-sets. Each of these sets (there are $|\Sigma|^2 - |\Sigma| + 2$ of them) can be represented as a list of vertices of the trie $T$. Suppose the algorithm has computed the $\mathcal{S}(u)$-sets for some prefix $u$ of the input sentence. To enumerate all members of $\mathcal{S}(uc)$-sets for the prefix augmented by one letter $c$, it is sufficient to iterate through all elements of the $\mathcal{S}(u)$-sets and apply the rules described in Equations (1), (2), and (3). Using the trie representation, it is possible to process one element of the $\mathcal{S}(u)$-sets in constant time as follows: Since a string $v$ is in the $\mathcal{S}(u)$-sets represented by a pointer to a vertex in the trie, also any string $vd$, for $d \in \Sigma$, can be looked up in the trie by traversing only one of its edges. This way, the time complexity required to process the letter $c$ of the input word is linear in the total size of the $\mathcal{S}(u)$-sets.

---

[2] For a detailed description of the trie data structure see e.g. Section 6.3 in [6].

However, there may be some duplicate elements in the newly created $\mathcal{S}(uc)$-sets, which have to be removed. This can be done in the following way. For each set (possibly containing duplicates), we iterate through its elements and mark them directly in the trie. When finding an already marked element, we remove it as a duplicate. After finishing this, we iterate through the elements once more and unmark them in the trie. Such duplication removal requires only linear running time with respect to the number of elements of the $\mathcal{S}(uc)$-sets, regardless of the size of the dictionary.

Since each element can be processed in constant time, the key part of the complexity analysis of Algorithm 1 is to find an upper bound on the size of the $\mathcal{S}$-sets. We will give such a bound in what follows.

All elements in $\mathcal{S}_0(u)$ are suffixes of the word $u$ of length at most $k$ (recall that $k$ is the length of the longest word in the dictionary). Hence, $|\mathcal{S}_0(u)| \in O(k)$.

Similarly, each element of $|\mathcal{S}_2(u)|$ is a suffix of $u$ of length at most $k$, possibly with two letters swapped. Since there are $O(k^2)$ possibilities for this swap, $|\mathcal{S}_2(u)| \in O(k^3)$.

Now we analyze $\sum_{a,b\in\Sigma\wedge a\neq b}|\mathcal{S}_1^{a\rightarrow b}(u)|$ by considering each word from the set $\bigcup_{a,b\in\Sigma\wedge a\neq b}\mathcal{S}_1^{a\rightarrow b}(u)$ and estimating in how many $\mathcal{S}_1$-sets it can be: There are at most $k$ different words that are suffixes of $u$ and each of them can be in at most $O(|\Sigma|^2)$ sets: A suffix of $u$ of length $l \leq k$ can be included in at most $|\Sigma|\cdot(|\Sigma|-1)$ sets corresponding to the $|\Sigma|\cdot(|\Sigma|-1)$ possible letter replacements in one of the first $|u|-l$ positions of $u$. There are at most $|\Sigma|k^2$ other strings in $\bigcup_{a,b\in\Sigma\wedge a\neq b}\mathcal{S}_1^{a\rightarrow b}(u)$, since there are $k$ possibilities for the length of the word, at most $k$ possibilities for the location of replacement and $|\Sigma|$ possibilities for the replacement letter. Each of these strings can belong to only one $\mathcal{S}_1(u)$-set, since both the replaced and the replacing letter are uniquely determined by the string itself and the string $u$. Hence, $\sum_{a,b\in\Sigma\wedge a\neq b}|\mathcal{S}_1^{a\rightarrow b}(u)| \in O(|\Sigma|^2k+|\Sigma|k^2)$.

Putting this together, there can be at most $O(k + |\Sigma|^2k + |\Sigma|k^2 + k^3) = O\left(k(|\Sigma| + k)^2\right)$ elements in the $\mathcal{S}(u)$-sets, which proves the time complexity $O\left(nk\left(|\Sigma| + k\right)^2\right)$ of step 2 of Algorithm 1.

As for the memory complexity, observe that only the $\mathcal{S}$-sets for the latest prefix have to be stored, which requires $O(k\left(|\Sigma| + k\right)^2)$ space. Furthermore, the trie representing the dictionary has to be stored, thus the total memory complexity is $O(|\Sigma|m + k\left(|\Sigma| + k\right)^2)$.    □

It is not difficult to show that the complexity analysis presented in this theorem is tight. To show that $|\mathcal{S}_2(u)| = \Theta(k^3)$, consider a word $u = a^ib^ic^i$ such that $i = k/3$. There are exactly $i^2$ different words $w$ that can be obtained from $b^ic^i$ by switching one letter $b$ with a letter $c$. Hence there are exactly $i^3 = \Theta(k^3)$ different words $a^jw$ such that $j \leq i$. All of these words (if contained in the dictionary) also belong to the set $\mathcal{S}_2(u)$, so, for an appropriate dictionary, the size of $\mathcal{S}_2(u)$ is indeed $\Theta(k^3)$. Similar reasoning can be used for the $\mathcal{S}_1(u)$-sets, too.

# 4   An Improved Algorithm

In this section, we will present an improved algorithm with a faster running time for processing the input sentence at the expense of a slower preprocessing phase. This algorithm will consider two separate cases, depending on whether the swapped letters occur inside the same dictionary word or not.

It will turn out that the case of letters that occur in two different dictionary words after swapping can be handled with asymptotically the same preprocessing time as in the previous algorithm and $O(n(|\Sigma|^2 k + |\Sigma|k^2))$ time for processing the input sentence.

But handling the case of swapping inside a dictionary word with an improved time complexity will require a more expensive preprocessing in $O(mk^2|\Sigma|)$ time.

## 4.1   Swapping Across Dictionary Word Boundaries

First we present an algorithm which can detect in $O(n(|\Sigma|^2 k + |\Sigma|k^2))$ time, after a preprocessing in $O(|\Sigma|m)$ time, if there is a possibility to swap two letters in the input sentence that are in different dictionary words in some partition of the so constructed sentence. We can formally define this special case of the spoonerism problem as follows.

**Definition 5.** *The* separated spoonerism problem *is the following decision problem:*

**Input:** *A dictionary* $\mathcal{D} = \{w_1, w_2, \ldots, w_l\}$ *and a sentence* $s = s_1 s_2 \ldots s_n$ *over an alphabet* $\Sigma$.
**Output:** YES *if there exist* $i, j, l \in \{1, \ldots, n\}$ *such that* $i < l < j$, $s_i \neq s_j$, *and the strings* $x' = s_1 \ldots s_{i-1} s_j s_{i+1} \ldots s_l$ *and* $y' = s_{l+1} \ldots s_{j-1} s_i s_{j+1} \ldots s_n$ *are sentences over* $\Sigma$, NO *otherwise.*

To solve the separated spoonerism problem, we will use a similar idea as in the previous section, processing the input sentence not only in forward direction but also backwards.

The algorithm tries to find a solution for all possible pairs of swapped letters $a, b \in \Sigma$ separately. Such a solution exists if and only if it is possible to decompose the input sentence $s = xy$ into parts $x$ and $y$ such that the following holds:

**C1.** It is possible to replace some letter $a$ with $b$ in the part $x$ such that the result can be decomposed into dictionary words.
**C2.** It is possible to replace some letter $b$ with $a$ in the part $y$ such that the result can be decomposed into dictionary words.

It is obvious that condition *C1* can be easily checked using the $\mathcal{S}$-sets described in Algorithm 1: *C1* holds if and only if $\varepsilon \in \mathcal{S}_1^{a \to b}(x)$. To check condition *C2*, we need to process the word $s$ backwards and compute sets analogous to $\mathcal{S}$, but with reversed roles of prefixes and suffixes.

**Definition 6.** *We denote the dictionary containing reverses of all words from the dictionary $\mathcal{D}$ as $\mathcal{D}^R$. Formally, $\mathcal{D}^R = \{u^R \mid u \in \mathcal{D}\}$.*

*We denote the set of all suffixes of all words from the dictionary $\mathcal{D}$ as $\mathrm{Suff}\,(\mathcal{D})$. Formally, $\mathrm{Suff}\,(\mathcal{D}) = \{u \mid \exists v : vu \in \mathcal{D}\}$. It is easy to see that $\mathrm{Suff}\,(\mathcal{D}) = \left(\mathrm{Pref}\,(\mathcal{D}^R)\right)^R$ and $\varepsilon \in \mathrm{Suff}\,(\mathcal{D}) \supseteq \mathcal{D}$.*

*Let $u, v \in \Sigma^*$. We call $v$ a live prefix of $u$ w.r.t. the dictionary $\mathcal{D}$, if and only if $v^R$ is a live suffix of $u^R$ w.r.t. the dictionary $\mathcal{D}^R$. We denote the set of all live prefixes of word $u$ as $\mathcal{L}^P(u)$.*

*By $\mathcal{P}_0(u)$ we denote the set of all live prefixes of $u$, i. e., $\mathcal{P}_0(u) = \mathcal{L}^P(u)$.*

*For all $a \neq b \in \Sigma$, $\mathcal{P}_1^{a \to b}(u)$ is the set of all live prefixes of all strings $u'$ obtained by replacing a single letter $a$ by letter $b$ in the string $u$. Formally, $\mathcal{P}_1^{a \to b}(u) = \bigcup_{vav'=u \wedge a \neq b} \mathcal{L}^P(vbv')$.*

Hence, for checking condition $C2$, it is sufficient to check if $\varepsilon \in \mathcal{P}_1^{b \to a}(y)$. To do so, the elements of the $\mathcal{P}$-sets can be computed similarly as the elements of the $\mathcal{S}$-sets, processing the input word backwards.

- To compute $\mathcal{P}_0(cu)$, it is sufficient to prepend all possible elements of $\mathcal{P}_0(u)$ with letter $c$. Furthermore, if the obtained set contains a word from the dictionary, then we can add $\varepsilon$ into $\mathcal{P}_0(cu)$. Formally,

$$\mathcal{P}_0(cu) = \begin{cases} \{cv \mid v \in \mathcal{P}_0(u), cv \in \mathrm{Suff}\,(\mathcal{D})\} \cup \{\varepsilon\} & \text{iff } \exists v \in \mathcal{P}_0(u) : cv \in \mathcal{D} \\ \{cv \mid v \in \mathcal{P}_0(u), cv \in \mathrm{Suff}\,(\mathcal{D})\} & \text{otherwise} \end{cases}$$
$$(4)$$

- For computing $\mathcal{P}_1^{a \to b}(cu)$, we distinguish two cases. Either the replacement of the letters occurs in $u$, or (in case $a = c$) the letter $c$ is replaced. Each of these cases yields a subset from $\mathrm{Suff}\,(\mathcal{D})$ and the resulting set $\mathcal{P}_1^{a \to b}(cu)$ is the union of these subsets. In the former case, the situation is analogous to the one described in the previous paragraph. Let $X^{a \to b}(cu)$ be the subset of $\mathrm{Suff}\,(\mathcal{D})$ obtained from the first case:

$$X^{a \to b}(cu) = \{cv \mid v \in \mathcal{P}_1^{a \to b}(u), cv \in \mathrm{Suff}\,(\mathcal{D})\}$$

In the latter case (occurring only if $c = a$), we use our knowledge of $\mathcal{P}_0(u)$. Let $Y^{a \to b}(cu)$ be the subset of $\mathrm{Suff}\,(\mathcal{D})$ obtained from the second case:

$$Y^{a \to b}(cu) = \begin{cases} \{bv \mid v \in \mathcal{P}_0(u), bv \in \mathrm{Suff}\,(\mathcal{D})\} & \text{iff } c = a \\ \emptyset & \text{otherwise} \end{cases}$$

If the obtained set $X^{a \to b}(cu) \cup Y^{a \to b}(cu)$ contains a word from the dictionary, we have to add the empty string:

$$\mathcal{P}_1^{a \to b}(cu) = \begin{cases} X^{a \to b}(cu) \cup Y^{a \to b}(cu) \cup \{\varepsilon\} & \text{iff } \exists v \in X^{a \to b}(cu) \cup Y^{a \to b}(cu) : \\ & \qquad v \in \mathcal{D} \\ X^{a \to b}(cu) \cup Y^{a \to b}(cu) & \text{otherwise} \end{cases}$$
$$(5)$$

The resulting strategy is summarized in Algorithm 2.

---

**Algorithm 2.** Solving the separated spoonerism problem

---

**Input:** A dictionary $\mathcal{D}$ of total size $m$ with maximum word length $k$ over an alphabet $\Sigma$ and a sentence $s = s_1 \ldots s_n$ w.r.t. $\mathcal{D}$.

1. Construct a trie $T$ from $\mathcal{D}$ and a trie $T^R$ from $\mathcal{D}^R$.
2. **for** $i := 1$ **to** $n$ **do**
      Compute the $\mathcal{S}_0$- and $\mathcal{S}_1$-sets for $s_1 \ldots s_i$ according to Equations (1) and (2), using $T$ for the look-up operations in the dictionary.
      Compute the $\mathcal{P}_0$- and $\mathcal{P}_1$-sets for $s_{n-i} \ldots s_n$ according analogous equations, using $T^R$ for the look-up operations in the dictionary.
3. **for** $a, b \in \Sigma$, $a \neq b$ **do**
      **for** $l := 1$ **to** $n - 1$ **do**
         **if** $\varepsilon \in \mathcal{S}_1^{a \to b}(s_1 \ldots s_l)$ **and** $\varepsilon \in \mathcal{P}_1^{b \to a}(s_{l+1} \ldots s_n)$ **then**
            Output YES and stop.
      Output NO

**Output:** YES if there exists a solution to the separated spoonerism problem, NO otherwise.

---

**Lemma 1.** *Algorithm 2 solves the separated spoonerism problem in* $O(m|\Sigma| + n(|\Sigma|^2 k + |\Sigma|k^2))$ *time and* $O(m|\Sigma| + |\Sigma|k^2 + n|\Sigma|^2)$ *space.*

*Proof.* Obviously, Algorithm 2 correctly solves the separated spoonerism problem.

The preprocessing of the dictionary in step 1 can be done in $O(m|\Sigma|)$ time as already explained in the analysis of Algorithm 1.

Calculating the $\mathcal{S}_0$- and $\mathcal{S}_1$-sets in step 2 is possible in $O(n(|\Sigma|^2 k + |\Sigma|k^2))$ time, as already shown in the proof of Theorem 1. Calculating the $\mathcal{P}_0$- and $\mathcal{P}_1$-sets obviously takes the same time.

The test in each single iteration of step 3 can be implemented to take constant time; thus, step 3 needs $O(n|\Sigma|^2)$ time overall.

Summarizing, the total time required by Algorithm 2 is in $O(m|\Sigma| + n(|\Sigma|^2 k + |\Sigma|k^2))$. To prove the claimed space complexity, we note that the tries need $O(m|\Sigma|)$ space, and the $\mathcal{S}(u)$- and $\mathcal{P}(u)$-sets need $O(|\Sigma|^2 k + |\Sigma|k^2)$ space for one prefix $u$. There is no reason in storing all these sets for every prefix; only the information whether $\varepsilon \in \mathcal{S}_1^{a \to b}(u)$ and $\varepsilon \in \mathcal{P}_1^{a \to b}(u)$ needs to be stored for each prefix $u$ and for each $a, b \in \Sigma$, hence requiring $O(n|\Sigma|^2)$ space. $\square$

## 4.2 Swapping Inside a Dictionary Word

If the separated spoonerism problem has no solution, we proceed to the case where the swapped letters are located in the same dictionary word. The main idea of the algorithm is to try all possible decompositions of the input sentence $s = w_1 w_2 w_3$ into three parts $w_1$, $w_2$, and $w_3$ such that $w_1$ and $w_3$ can be decomposed into dictionary words and $w_2$ is a string such that a swap of two different letters makes it a dictionary word.

It is easy to see that $w_1$ and $w_3$ can be decomposed if and only if $\varepsilon \in S_0(w_1)$ and $\varepsilon \in P_0(w_3)$. Since the sets $S_0$ and $P_0$ have already been precomputed in Algorithm 2 for the separated spoonerism problem, each of these checks can be made in constant time.

Now we describe how to check whether two different letters in $w_2$ can be swapped as to yield a dictionary word.

Our algorithm constructs another trie $T'$ which contains all strings that can be reached from a dictionary word by swapping two different letters. For a dictionary word $w$ of length $l$, there are obviously at most $l^2 \leq k^2$ different reachable strings. Thus, for each dictionary word, the resulting trie $T'$ contains at most $k^2$ strings of the same length. The total size of $T'$ is thus in $O(k^2 m |\Sigma|)$.

After we have constructed this additional trie, we can process an input sentence as follows: There are $O(nk)$ possible partitions $s = w_1 w_2 w_3$ as described above, the consistency check for $w_1$ and $w_3$ can be done in constant time for each of these partitions. By enumerating the possible partitions in a suitable way, also the look-up of $w_2$ in the additional trie $T'$ can be done in amortized constant time.

This strategy is summarized in Algorithm 3.

---

**Algorithm 3.** Solving the spoonerism problem with extensive preprocessing

---

**Input:** A dictionary $\mathcal{D}$ of total size $m$ with maximum word length $k$ over an alphabet $\Sigma$ and a sentence $s = s_1 \ldots s_n$ w.r.t. $\mathcal{D}$.

1. Use Algorithm 2 to check whether the separated spoonerism problem for $\mathcal{D}$ and $s$ has a solution. If so, output YES and stop.
2. Construct a trie $T'$ for the set $\mathcal{D}'$ of all strings that can be reached from a dictionary word by swapping two different letters.
3.     **for** $i := 0$ **to** $n - 1$ **do**
            **for** $l := 1$ **to** $\min(k, n - i)$ **do**
                **if** $\varepsilon \in S_0(s_1 \ldots s_i)$ **and** $\varepsilon \in P_0(s_{i+l+1} \ldots s_n)$ **and** $s_{i+1} \ldots s_{i+l} \in \mathcal{D}'$ **then**
                    Output YES and stop.
        Output NO

**Output:** YES if there exists a solution to the spoonerism problem, NO otherwise.

---

**Theorem 2.** *Algorithm 3 solves the spoonerism problem in $O(mk^2 |\Sigma| + n(|\Sigma|^2 k + |\Sigma| k^2))$ time and $O(mk^2 |\Sigma| + |\Sigma| k^2 + n|\Sigma|^2)$ space.*

*Proof.* From the discussion above it is clear that Algorithm 3 solves the spoonerism problem.

We will now analyze its time complexity. Step 1 is possible in $O(m|\Sigma| + n(|\Sigma|^2 k + |\Sigma|k^2))$ time according to Lemma 1. The trie $T'$ has a size of $O(k^2 m|\Sigma|)$ as discussed above, and it can obviously also be constructed in $O(k^2 m|\Sigma|)$ time. The tests in each iteration of step 3 can be performed in constant time: We have already discussed this for the tests on the sets $\mathcal{S}_0$ and $\mathcal{P}_0$ above; for testing the membership of $s_{i+1} \ldots s_{i+l}$ in $\mathcal{D}'$ after having tested the membership of $s_{i+1} \ldots s_{i+l-1}$ in the previous iteration, only one step along one edge of the trie is needed, hence this can also been done in constant time. This leads to an overall running time in $O(nk)$ for step 3. The total running time of the algorithm is thus in $O(m|\Sigma| + n(|\Sigma|^2 k + |\Sigma|k^2) + k^2 m|\Sigma| + nk) = O(k^2 m|\Sigma| + n(|\Sigma|^2 k + |\Sigma|k^2))$.

The space complexity of Algorithm 3 is obviously determined by the size of the additional trie $T'$ and the space complexity of Algorithm 2. $\qquad \square$

## 5   A Further Improvement for Small Alphabets

In this section, we will describe another algorithm which can, at least in the case where $k \gg |\Sigma|$, save some preprocessing time at the expense of a slightly higher time complexity for processing an input sentence.

This algorithm in a first phase also uses Algorithm 2 to solve the separated spoonerism problem. In a second phase, it again considers all possible partitions $s = w_1 w_2 w_3$ of the input sentence $s$ into two sentences $w_1$ and $w_3$ and a string $w_2$ which becomes a dictionary word by swapping two different letters.

For testing whether the swapping of letters may occur inside $w_2$, this algorithm will combine a dynamic programming approach similar to the one of Algorithm 1 with the idea of an expanded preprocessed trie from Algorithm 3. More precisely, in a preprocessing step, the algorithm will construct $O(|\Sigma|^2)$ new dictionaries, where the dictionary $\mathcal{D}^{a \to b}$ contains all strings that can be reached from a dictionary word from $\mathcal{D}$ by replacing a letter $a$ by a letter $b$, i.e., $\mathcal{D}^{a \to b} = \{xby \mid xay \in \mathcal{D}\}$. Using dynamic programming, the algorithm further constructs the set $\mathcal{T}(w_2)$ of all strings from $\mathcal{D}^{a \to b}$ that can be obtained from $w_2$ by replacing an $a$ by a $b$. In other words, the set $\mathcal{T}(w_2)$ contains all strings which can be obtained both from $w_2$ and from some dictionary word $w \in \mathcal{D}$ by replacing a letter $a$ by a letter $b$.

If $\mathcal{T}^{a \to b}(w_2)$ is non-empty for some $a, b \in \Sigma$, it contains some string $u = xby = x'by'$, such that $w_2 = xay$ and $v = x'ay' \in \mathcal{D}$. As long as we can guarantee that $x \neq x'$, this string from $\mathcal{T}^{a \to b}(w_2)$ gives us a positive solution to the spoonerism problem. To ensure this, we have to store some information about the positions where the replacements may occur. This will be done both while constructing the tries representing the dictionaries $\mathcal{D}^{a \to b}$ and while computing the set $\mathcal{T}^{a \to b}(w_2)$. For every string in $\mathcal{T}^{a \to b}(w_2)$, the replacement position has to be unique, since we are starting with the unique string $w_2$.

For the strings in $\mathcal{D}^{a \to b}$, the situation is slightly more complicated. If there are two dictionary words $v = x'ay'$ and $z = x''ay''$, both mapped to the same string $u = x'by' = x''by'' \in \mathcal{D}^{a \to b}$ where $x' \neq x''$, i.e, via replacements in different positions, the presence of $u$ in $\mathcal{T}^{a \to b}(w_2)$ already ensures a positive solution to the spoonerism problem. This means that we have to store, for each string in $\mathcal{D}^{a \rightarrow b}$, either the unique replacement position or just the information that the position is not unique.

The $T^{a \rightarrow b}$-sets can be computed in a similar way as the $\mathcal{S}^{a \rightarrow b}$-sets in Algorithms 1 and 2. Creating the modified dictionaries $\mathcal{D}^{a \rightarrow b}$ in the preprocessing phase can be implemented in $O(mk|\Sigma|^2)$ time and space. The processing of the input sentence itself takes $O(nk|\Sigma|^2)$ time – considering $O(|\Sigma|^2)$ different letter pairs, $O(nk)$ partitions $s = w_1 w_2 w_3$ and $O(k)$ elements of the $T^{a \rightarrow b}$-sets and processing each element in constant time. The space complexity (not counting the preprocessed dictionaries and information reused from Algorithm 2) is $O(k)$, required for storing the $T^{a \rightarrow b}$-sets. Adding the complexity of Algorithm 2, which is used for deciding whether the swapping of letters may occur across dictionary word boundaries, a time complexity in $O(mk|\Sigma|^2 + nk^2|\Sigma|^2)$ and a space complexity in $O(mk|\Sigma|^2 + |\Sigma|k^2 + n|\Sigma|^2)$ is obtained.

Now we present the description and analysis of the algorithm in more detail. At first we provide the formal definition of the set $T^{a \rightarrow b}(u)$ and describe how to compute it for any string $u \in \Sigma^*$. The idea is analogous to that of computing the set $\mathcal{S}_1^{a \rightarrow b}(u)$ used in Algorithms 1 and 2.

**Definition 7.** *For all $a, b \in \Sigma$, $a \neq b$, the set $T^{a \rightarrow b}(u)$ is the set of all pairs $(w, l)$, where $w$ is a string obtained by replacing a single letter $a$ by a letter $b$ in the string $u$ that also is a prefix of some word from the dictionary $\mathcal{D}^{a \rightarrow b}$, and where $l \in \{1, \ldots, |u|\}$ denotes the position of the letter replacement. Formally,*

$$T^{a \rightarrow b}(u) = \{(xby, |x| + 1) \mid u = xay \land xby \in \mathrm{Pref}\left(\mathcal{D}^{a \rightarrow b}\right)\}.$$

Slightly abusing notation, in the following we will also say that $w \in T^{a \rightarrow b}(u)$, if there exists an $l$ such that $(w, l) \in T^{a \rightarrow b}(u)$.

The sets $T^{a \rightarrow b}$ can be computed similarly as the sets $\mathcal{S}_1^{a \rightarrow b}$, except that the empty string $\varepsilon$ is not added to the sets and the dictionary $\mathcal{D}^{a \rightarrow b}$ is used instead of $\mathcal{D}$. For computing $T^{a \rightarrow b}(uc)$, we distinguish two cases. Either the replacement of the letters occurs in $u$, or (in case $a = c$) the letter $c$ is replaced. Let $X^{a \rightarrow b}(uc)$ be the set of strings corresponding to the former case:

$$X^{a \rightarrow b}(uc) = \{(vc, l) \mid (v, l) \in T^{a \rightarrow b}(u), vc \in \mathrm{Pref}\left(\mathcal{D}^{a \rightarrow b}\right)\}$$

In the latter case (that occurs only if $c = a$), we obtain at most one string:

$$Y^{a \rightarrow b}(uc) = \begin{cases} \{(ub, |ub|)\} & \text{iff } c = a \\ \emptyset & \text{otherwise} \end{cases}$$

Putting this together yields

$$T^{a \rightarrow b}(uc) = X^{a \rightarrow b}(uc) \cup Y^{a \rightarrow b}(uc). \tag{6}$$

The complete strategy of this approach is summarized in Algorithm 4.

**Algorithm 4.** Refined dynamic programming for the spoonerism problem

**Input:** A dictionary $\mathcal{D}$ of total size $m$ with maximum word length $k$ over an alphabet $\Sigma$ and a sentence $s = s_1 \ldots s_n$ w.r.t. $\mathcal{D}$.

1. Use Algorithm 2 to check whether the separated spoonerism problem for $\mathcal{D}$ and $s$ has a solution. If so, output YES and stop.
2. For all pairs $(a, b)$ of different letters from $\Sigma$, construct a trie for the dictionary $\mathcal{D}^{a \rightarrow b}$ where the vertices of the trie are labeled with either the unique position of letter replacement that led to the corresponding string or with MULT if this position is not unique.
3.    **for** $a, b \in \Sigma$, $a \neq b$ **do**
        **for** $i := 0$ **to** $n - 1$ **do**
            **for** $l := 1$ **to** $\min(k, n - i)$ **do**
                **if** $\varepsilon \in \mathcal{S}_0(s_1 \ldots s_i)$ **and** $\varepsilon \in \mathcal{P}_0(s_{i+l+1} \ldots s_n)$ **then**
                    Construct the set $\mathcal{T}^{a \rightarrow b}(s_{i+1} \ldots s_{i+l})$ from the set
                    $\mathcal{T}^{a \rightarrow b}(s_{i+1} \ldots s_{i+l-1})$ using Equation (6)
                    **for all** $(w, l) \in \mathcal{T}^{a \rightarrow b}(s_{i+1} \ldots s_{i+l})$ **do**
                        Check if the vertex in $\mathcal{D}^{a \rightarrow b}$ corresponding to $w$ is labeled with some
                        $l' \neq l$ or with MULT. If so, output YES and stop.
    Output No

**Output:** YES if there exists a solution to the spoonerism problem, NO otherwise.

---

**Theorem 3.** *Algorithm 4 solves the spoonerism problem in $O(mk|\Sigma|^2 + nk^2|\Sigma|^2)$ time and $O(|\Sigma|^2 km + |\Sigma|k^2 + n|\Sigma|^2)$ space.*

*Proof.* It is clear from our discussion above that the algorithm correctly solves the spoonerism problem.

Step 1 again takes $O(m|\Sigma| + n(|\Sigma|^2 k + |\Sigma|k^2))$ time according to Lemma 1. We now analyze the time complexity needed to create and preprocess the modified dictionaries in step 2. Each word $u \in \mathcal{D}$ yields $|\Sigma||u|$ different strings belonging to various of the modified dictionaries. Each of these strings can be inserted into the appropriate dictionary represented by a trie in $O(|\Sigma||u|)$ time. Labeling any vertex of any of the tries requires only constant additional time and space, hence also the overall time and space complexity of step 2 is

$$O\left(\sum_{u \in \mathcal{D}} |\Sigma|^2 |u|^2\right) = O\left(|\Sigma|^2 k \sum_{u \in \mathcal{D}} |u|\right) = O\left(|\Sigma|^2 km\right).$$

The inner loop in step 3 is performed $O(nk|\Sigma|^2)$ times. Since obviously $|\mathcal{T}^{a \rightarrow b}(u)| \leq k$ for each $u$, the construction of the $\mathcal{T}$-sets according to Equation (6) can be performed in $O(k)$ time using the trie for the modified dictionary $\mathcal{D}^{a \rightarrow b}$. Looking up the middle part $s_{i+1} \ldots s_{i+l}$ in the corresponding trie can be done in amortized constant time, with a proof analogous to the one of Theorem 2. Thus, step 3 has a total time complexity in $O(nk^2|\Sigma|^2)$.

Overall, Algorithm 4 has a time complexity in $O(m|\Sigma| + n(|\Sigma|^2 k + |\Sigma|k^2) + mk|\Sigma|^2 + nk^2|\Sigma|^2) = O(mk|\Sigma|^2 + nk^2|\Sigma|^2)$.

Considering the space complexity, the algorithm needs to store the tries for the modified dictionaries, requiring $O(mk|\Sigma|^2)$ space. Moreover, the space requirements of step 1 exceed those for step 3 (not counting the tries). Thus, the overall space complexity is in $O(mk|\Sigma|^2 + |\Sigma|^2 k + |\Sigma|k^2 + n|\Sigma|^2) = O(mk|\Sigma|^2 + |\Sigma|k^2 + n|\Sigma|^2)$. □

## 6   Conclusion

We have presented some efficient algorithms for the spoonerism problem. The worst-case running time of the basic dynamic-programming algorithm (Algorithm 1) is $O(m|\Sigma|)$ for preprocessing and $O(nk(|\Sigma| + k)^2)$ for processing the input. The improved algorithm (Algorithm 3) reduces the input processing time to $O(nk(|\Sigma|^2 + |\Sigma|k))$, which is asymptotically better even for the case $k = \Theta(n)$. Finally, we have presented a possible improvement of the preprocessing time of Algorithm 3 to $O(mk|\Sigma|^2)$ at the expense of a slightly worse input processing time $O(nk^2|\Sigma|^2)$.

For a variant of the spoonerism problem, where substrings of length greater than 1 may be swapped, these algorithms obviously yield a running time that is exponential in the length of the interchanged substrings. We leave it as an open problem to find more efficient algorithms for this problem.

## Acknowledgment

We would like to thank Michael Bender who pointed us to this interesting problem and gave us comments and suggestions helpful for improving the presentation of our paper.

## References

1. Allyn, F.A., Burt, J.S.: Pinch my wig or winch my pig: Spelling, spoonerisms and other language skills. Reading and Writing 10, 51–74 (1998)
2. Bender, M., Clifford, R., Steinföfel, K., Tsichlas, K.: The Spoonerism Problem, Unpublished manuscript
3. Hayter, W.: Spooner: A Biography, W. H. Allen, ISBN 0-491-01658-1 (1976)
4. Jaro, M.A.: Advances in record linking methodology as applied to the 1985 census of Tampa Florida. Journal of the American Statistical Society 64, 1183–1210 (1989)
5. Jaro, M.A.: Probabilistic linkage of large public health data file. Statistics in Medicine 14, 491–498 (1995)
6. Knuth, D.E.: The Art of Computer Programming – Sorting and Searching, vol. 3. Addison-Wesley, London (1973)
7. Tsichlas, K., Bender, M.: ADG in King's College: The Spoonerism Problem, Unpublished presentation at London Stringology Day (2005)
8. Winkler, W.E.: The state of record linkage and current research problems, Statistics of Income Division, Internal Revenue Service Publication R99/04 (1999)

# High Spies
# (or How to Win a Programming Contest)

André Deutz, Rudy van Vliet, and Hendrik Jan Hoogeboom

Leiden Institute of Advanced Computer Science (LIACS), Leiden University,
Niels Bohrweg 1, 2333 CA Leiden, The Netherlands
rvvliet@liacs.nl

**Abstract.** We analyse transports between leaves in an edge-weighted tree. We prove under which conditions there exists a transport matching the weights of a given tree. We use this to compute minimum and maximum values for the transport between a given pair of leaves.

## 1 Introduction

*You have been approached by a spy agency to determine the amount of contraband goods that are being traded among several nefarious countries. After being shipped from its country of origin, each container of goods is routed through at least one neutral port. At the port, the containers are stored in a warehouse before being sent on their way, so you cannot trace individual containers from their country of origin to their final destination. Satellite cameras can tell you the number of containers travelling in each direction on each leg of the journey. They cannot distinguish individual containers, nor do you have information on the times individual containers have been observed. You know, however, that every container takes the shortest possible route to its destination.*

*The task is to determine both the maximum and minimum number of containers that could have been travelling from one country to another. The transport network that is observed is an unrooted tree, with countries as leaves and ports as internal nodes. For each edge the number of containers is given in two directions. You know from the description above that no container leaves a port in the direction it came from.*

This is the description of one of the problems at the Benelux Algorithm Programming Contest 2006, which was held in Leiden on 21 October 2006, see www.bapc2006.nl. The name of the problem was *High Spies*. The problem and some phrases in its description were taken from [Shasha, 2003].

We want to emphasize that High Spies is not just a maximum network flow problem, to which we can apply, e.g., the well-known Ford-Fulkerson algorithm (see [Ford and Fulkerson, 1957]). We do not so much consider networks with (maximum) capacities on the edges, but networks with numbers (of containers) on the edges that are *actually observed*, meaning that the numbers must really be met by the transport. Moreover, no container is allowed to travel from one node to another and back again.

P. Crescenzi, G. Prencipe, and G. Pucci (Eds.): FUN 2007, LNCS 4475, pp. 93–107, 2007.

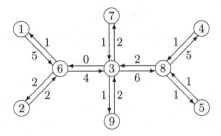

**Fig. 1.** A transport network with six countries and three ports

Consider, for example the network depicted in Fig. 1, and suppose we are interested in the transport from country 1 to country 4. The (standard) maximum flow from 1 to 4 consists of four containers, as this is the minimum number observed on the route from 1 to 4. It is, however, not possible to send four containers along this route. In that case, in port 6, one of the containers arriving from country 2 would be forced to return, which is forbidden.

Similarly, one might think that the minimum flow from 1 to 4 has value 0. Indeed, one may send the five containers leaving country 1 to countries 2, 5, 7 and 9. [1] However, in that case, at least one of the containers arriving at port 3 from port 8 would be forced to return.

In this paper, we present and analyse an algorithm for a generalized version of High Spies. In this version, the transport observed between two neighbouring nodes of the tree may be any non-negative real number.

The intuition behind the algorithm is simple. For each port on the (unique) route from a certain country (the source) to another country (the target), it determines the minimum and maximum number of containers that can be passed on from the direction of the source in the direction of the target. The minimum number results from passing on as many containers as possible in directions different from that of the target. In all this, we make sure that no container entering the port has to go back in the direction it came from. We use the (local) minima and maxima from the individual ports to compute the (global) minimum and maximum number of containers transported from source to target.

The paper is organized as follows. In Sect. 2, we model the problem in terms of weighted trees. In Sect. 3 and 4, we analyse which weighted trees actually correspond to valid transports. We will see that global conditions for this can be translated into easily checkable, local conditions. We need these conditions to justify our algorithm, which we describe in Sect. 5. There, we use the local conditions to determine the local maxima and minima just mentioned, and we combine these into a global solution. Finally, we make some concluding remarks, including some more remarks about the use of standard max-flow algorithms to solve the problem.

---

[1] Also this flow can be found with a standard max-flow algorithm, by introducing a special target node, which is only reachable from countries 2, 5, 7 and 9.

## 2   Problem Model

An unrooted tree can be denoted by an ordered pair $T = (V, E)$, where $V$ is the (non-empty) set of nodes, and $E$ is the set of edges of the tree. In High Spies, the direction of edges is important. Therefore, we consider an edge as an ordered pair of nodes $(i, j)$. To reflect the undirected nature of the tree as a whole, we have $(i, j) \in E$, if and only if $(j, i) \in E$.

**Definition 1.** *A weighted tree is an ordered pair $(T, c)$, where $T = (V, E)$ is an unrooted tree and $c$ is a non-negative function (a weight function) on $E$.*

The weight $c(i, j)$ can be considered as the observed number of containers travelling from node $i$ to node $j$. Note, however, that it does not have to be an integer number.

From now on, we assume that a tree $T = (V, E)$ contains at least two nodes. This allows for the identification of leaves and internal nodes in the tree. For ease of notation, we also assume that $V = \{1, \ldots, n\}$ for some $n \geq 2$. We let Leaves($T$) denote the set of leaves of $T$.

**Definition 2.** *A transport Tr on an unrooted tree $T$ is a non-negative function on the ordered pairs $(l_1, l_2)$ with $l_1, l_2 \in Leaves(T)$ and $l_1 \neq l_2$.*

The number $\mathrm{Tr}(l_1, l_2)$ can be considered as the number of containers shipped from leaf (country) $l_1$ to leaf (country) $l_2$ via the edges of the tree. Despite this interpretation, $\mathrm{Tr}(l_1, l_2)$ does not have to be an integer number. Note that $\mathrm{Tr}(l_1, l_2)$ is not necessarily equal to $\mathrm{Tr}(l_2, l_1)$.

We are interested in the total transport over a certain edge $(i, j)$ of the tree. This total transport comes from the leaves on one side of $(i, j)$ and goes to the leaves on the other side of $(i, j)$. We now define this formally.

Each edge $(i, j)$ of the tree is a 'cut'. It partitions the set $V$ of nodes into two subsets: the nodes on $i$'s side of the tree, and the nodes on $j$'s side of the tree. Let us call these subsets of nodes Left$(i, j)$ and Right$(i, j)$, respectively. This partitioning induces a partitioning of Leaves($T$) into a subset of leaves on $i$'s side of the tree, and a subset of leaves on $j$'s side of the tree. Let us call these subsets LLeaves$(i, j)$ and RLeaves$(i, j)$, respectively. Clearly, LLeaves$(i, j) =$ RLeaves$(j, i)$ and RLeaves$(i, j) =$ LLeaves$(j, i)$.

For example, in the tree in Fig. 1, LLeaves$(6, 3) = \{1, 2\}$ and RLeaves$(6, 3) = \{4, 5, 7, 9\}$.

**Definition 3.** *A matching transport Tr on a weighted tree $(T, c)$ with $T = (V, E)$ is a transport on $T$, such that for each edge $(i, j) \in E$,*

$$\sum_{l_1 \in LLeaves(i, j)} \sum_{l_2 \in RLeaves(i, j)} Tr(l_1, l_2) = c(i, j) \ . \tag{1}$$

Indeed, if we assume that the containers travelling from one leaf to another take the shortest route in the tree (i.e., they do not travel in two directions over the same undirected edge), and that each container observed is on its way from one

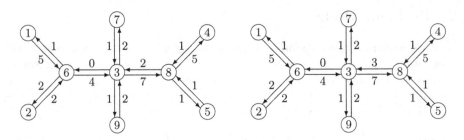

**Fig. 2.** Two weighted trees that do not have a matching transport

leaf to another (i.e., its origin or destination is not an internal node), then (1) must hold. Each container travelling from a leaf in LLeaves$(i, j)$ to a leaf in RLeaves$(i, j)$ passes edge $(i, j)$ exactly once, and there are no other containers travelling along this (directed) edge.

For example, a matching transport Tr for the tree in Fig. 1 is given by

$$\text{Tr}(1, 2) = 2, \text{Tr}(1, 4) = 2, \text{Tr}(1, 5) = 1, \text{Tr}(2, 1) = 1, \text{Tr}(2, 9) = 1,$$
$$\text{Tr}(4, 7) = 1, \text{Tr}(5, 7) = 1, \text{Tr}(7, 4) = 1, \text{Tr}(9, 4) = 2$$

(and Tr$(l_1, l_2) = 0$ for pairs of leaves $(l_1, l_2)$ not mentioned).

Now, the problem High Spies can be rephrased as follows: given a weighted tree for which at least one matching transport exists, and given two different leaves $l_1$ and $l_2$ of this tree, determine the minimum value and the maximum value for Tr$(l_1, l_2)$ over all matching transports Tr on the tree.

## 3    Existence of a Matching Transport

Before we start solving High Spies, we consider the question under which conditions a weighted tree actually permits a matching transport. We have already seen a matching transport for the weighted tree from Fig. 1. If we slightly modify the tree, then there may not exist a matching transport. For example, there do not exist matching transports for the two weighted trees in Fig. 2, in which we have only altered the weights on edges between nodes 3 and 8.

It is intuitively clear that there cannot be a matching transport for the left tree, because nodes 3 and 8 do not satisfy the graph analogue of Kirchhoff's current law: the total weight on the edges entering node 3 is unequal to the total weight on the edges leaving that node, and similarly for node 8.

Although the right tree does satisfy Kirchhoff's law, it is also intuitively clear that there cannot be a matching transport for that tree. The weight 7 on edge $(3, 8)$ should be carried along to nodes 4 and 5. However, the total weight on the edges $(8, 4)$ and $(8, 5)$ is only 6. Also, the weight 3 on edge $(8, 3)$ must have arrived at node 8 from nodes 4 and 5. However, the total weight on the edges $(4, 8)$ and $(5, 8)$ is only 2.

In Sect. 4, we prove that these are exactly the types of arguments determining the existence of a matching transport on a weighted tree. For that purpose, we reformulate the (global) definition of a matching transport into local terms.

**Definition 4.** *Let $(T, c)$ be a weighted tree, and let $j$ be an internal node of $T$. Let $h_1, \ldots, h_m$ for some $m \geq 2$ be the neighbours of $j$ in $T$. Then $j$ is a transport node, if there exists an $m \times m$ matrix $A = (a_{i,k})$ satisfying*

$$a_{i,k} \geq 0 \qquad\qquad (i, k = 1, \ldots, m) \qquad\qquad (2)$$
$$a_{i,i} = 0 \qquad\qquad (i = 1, \ldots, m) \qquad\qquad (3)$$
$$\sum_{k=1}^{m} a_{i,k} = c(h_i, j) \qquad (i = 1, \ldots, m) \qquad\qquad (4)$$
$$\sum_{i=1}^{m} a_{i,k} = c(j, h_k) \qquad (k = 1, \ldots, m) . \qquad\qquad (5)$$

*In this case, the matrix $A$ is called a witness matrix for node $j$.*

Intuitively, a witness matrix determines the transport on a local scale. The entry $a_{i,k}$ can be considered as the number of containers that is shipped from node $h_i$ to node $h_k$ (via node $j$).

For example, let us consider node $j = 6$ in the tree from Fig. 1. This node, whose neighbours are nodes 1, 2 and 3, is a transport node. If we let $h_1 = 1$, $h_2 = 2$ and $h_3 = 3$, then a witness matrix is

$$A = \begin{pmatrix} 0 & 2 & 3 \\ 1 & 0 & 1 \\ 0 & 0 & 0 \end{pmatrix} \begin{matrix} c(1, 6) = 5 \\ c(2, 6) = 2 \\ c(3, 6) = 0 \end{matrix}$$
$$c(6, 1) = 1 \quad c(6, 2) = 2 \quad c(6, 3) = 4$$

Indeed, when we calculate the sums of the individual rows and columns of $A$, we obtain the weights of the corresponding edges in the tree, as indicated to the right of the matrix and below the matrix.

We now establish that local properties guarantee the existence of a (global) matching transport:

**Theorem 5.** *Let $(T, c)$ be a weighted tree. There exists a matching transport on $(T, c)$, if and only if each internal node of $(T, c)$ is a transport node.*

**Proof.** $\Longrightarrow$ Assume that there exists a matching transport Tr on $(T, c)$. Then let $j$ be an arbitrary internal node of $T$ and let $h_1, \ldots, h_m$ for some $m \geq 2$ be the neighbours of $j$ in $T$. Then the $m \times m$ matrix $A = (a_{i,k})$ defined by

$$a_{i,i} = 0 \qquad\qquad (i = 1, \ldots, m)$$
$$a_{i,k} = \sum_{l_1 \in \mathrm{LLeaves}(h_i, j)} \sum_{l_2 \in \mathrm{RLeaves}(j, h_k)} \mathrm{Tr}(l_1, l_2) \qquad (i, k = 1, \ldots, m; i \neq k)$$

is a witness for $j$ being a transport node.

$\Longleftarrow$ Assume that each internal node of $(T, c)$ is a transport node. We use induction on the number $p$ of internal nodes to prove that there exists a matching transport Tr on $(T, c)$.

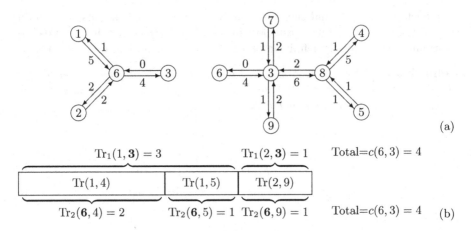

Tr₁(1, 3) = 3        Tr₁(2, 3) = 1        Total=c(6, 3) = 4

| Tr(1, 4) | Tr(1, 5) | Tr(2, 9) |
|---|---|---|

Tr₂(**6**, 4) = 2     Tr₂(**6**, 5) = 1  Tr₂(**6**, 9) = 1     Total=c(6, 3) = 4     (b)

**Fig. 3.** Constructions used in the proof of Theorem 5. (a) Subtrees $T_1$ (left) and $T_2$ (right) resulting when we break up the weighted tree from Fig. 1. (b) Greedy distribution of values $\mathrm{Tr}_1(l_1, 3)$ and $\mathrm{Tr}_2(6, l_2)$ over values $\mathrm{Tr}(l_1, l_2)$.

If $p = 1$, then $T$ is a 'star graph'. Let $j$ be the only internal node, and let $h_1, \ldots, h_m$ for some $m \geq 2$ be the neighbours of $j$. These neighbours are exactly all leaves of the tree. By assumption, $j$ is a transport node. The $m \times m$ matrix $A = (a_{i,k})$ which is a witness for this, directly defines a matching transport:

$$\mathrm{Tr}(h_i, h_k) = a_{i,k} \quad (i, k = 1, \ldots, m; i \neq k) \ .$$

Induction step. Let $p \geq 1$, and suppose that for every weighted tree $(T, c)$ with at most $p$ internal nodes, each of which is a transport node, there exists a matching transport on $(T, c)$ (induction hypothesis). Now consider a weighted tree $(T, c)$ with $p+1$ internal nodes, each of which is a transport node. Let $i$ and $j$ be two arbitrary adjacent, internal nodes.

We use the edge $(i, j)$ to break up $(T, c)$ into two smaller trees with some overlap. Let $T_1$ be the subtree of $T$ consisting of node $j$ and all nodes in $\mathrm{Left}(i, j)$, together with the edges connecting these nodes. Let $T_2$ be the subtree of $T$ consisting of node $i$ and all nodes in $\mathrm{Right}(i, j)$, together with the edges connecting these nodes. The weight functions $c_1$ and $c_2$ of $T_1$ and $T_2$ are equal to $c$, restricted to $T_1$ and $T_2$, respectively. In Fig. 3(a), we have illustrated this construction for edge $(6, 3)$ of the tree from Fig. 1.

Then node $j$ is a leaf in $T_1$ and $T_1$ contains at most $p$ internal nodes. Moreover, each internal node in $T_1$ is also an internal node in $T$, with the same neighbours and with the same weigths on the edges to and from these neighbours. Consequently, each internal node of $T_1$ is a transport node, simply because it is one in $T$. By the induction hypothesis, there exists a matching transport $\mathrm{Tr}_1$ on $(T_1, c_1)$. Analogously, there exists a matching transport $\mathrm{Tr}_2$ on $(T_2, c_2)$.

These two matching transports can be combined into one matching transport $\mathrm{Tr}$ on $(T, c)$. For pairs of leaves in $T$ at the same side of edge $(i, j)$, $\mathrm{Tr}$ simply inherits the value of $\text{Tr}_1$ or $\text{Tr}_2$. For pairs of leaves in $T$ at different sides of edge $(i, j)$, we observe that (for a transport from left to right)

$$\sum_{l_1 \in \text{LLeaves}(i,j)} \text{Tr}_1(l_1, j) = c(i, j) = \sum_{l_2 \in \text{RLeaves}(i,j)} \text{Tr}_2(i, l_2), \qquad (6)$$

as $\text{Tr}_1$ and $\text{Tr}_2$ are matching transports on $T_1$ and $T_2$, respectively. We now distribute every value $\text{Tr}_1(l_1, j)$ and every value $\text{Tr}_2(i, l_2)$ over values $\text{Tr}(l_1, l_2)$ (and analogously from right to left). This can be done in a greedy way, as follows.

We take an arbitrary ordering of the values $\text{Tr}_1(l_1, j)$ for $l_1 \in \text{LLeaves}(i, j)$, and partition the total quantity $c(i, j)$ according to these values. We label each resulting fragment with the corresponding leaf $l_1$. We also take an arbitrary ordering of the values $\text{Tr}_2(i, l_2)$ for $l_2 \in \text{RLeaves}(i, j)$, and partition the same total quantity $c(i, j)$ according to these values, again labelling the fragments with the corresponding leaves. The resulting, double labelling determines $\text{Tr}$.

For example, let us consider matching transports $\text{Tr}_1$ and $\text{Tr}_2$ for the trees $T_1$ and $T_2$ in Fig. 3(a), given by

$$\text{Tr}_1(1, 2) = 2, \text{Tr}_1(2, 1) = 1, \text{Tr}_1(1, 3) = 3, \text{Tr}_1(2, 3) = 1 \text{ and}$$
$$\text{Tr}_2(6, 4) = 2, \text{Tr}_2(6, 5) = 1, \text{Tr}_2(6, 9) = 1, \text{Tr}_2(4, 7) = 1,$$
$$\text{Tr}_2(5, 7) = 1, \text{Tr}_2(7, 4) = 1, \text{Tr}_2(9, 4) = 2,$$

respectively. The values $\text{Tr}_1(l_1, l_2)$ with $l_1, l_2 \in \{1, 2\}$ and $l_1 \neq l_2$ are inherited by $\text{Tr}$, and similarly for $\text{Tr}_2(l_1, l_2)$ with $l_1, l_2 \in \{4, 5, 7, 9\}$ and $l_1 \neq l_2$.

Figure 3(b) illustrates how the values $\text{Tr}_1(1, 3)$ and $\text{Tr}_1(2, 3)$ (in this order) and the values $\text{Tr}_2(6, 4)$, $\text{Tr}_2(6, 5)$ and $\text{Tr}_2(6, 9)$ (in this order) are distributed over $\text{Tr}(1, 4) = 2$, $\text{Tr}(1, 5) = 1$ and $\text{Tr}(2, 9) = 1$. It should be obvious from the picture and (6) that this algorithm always yields a valid matching transport.

Note that in this example, edge $(3, 6)$ has weight 0. so that $\text{Tr}_1(3, l_1)$, $\text{Tr}_2(l_2, 6)$ and $\text{Tr}(l_2, l_1)$ must be 0 for all $l_1 \in \text{LLeaves}(6, 3)$ and $l_2 \in \text{RLeaves}(6, 3)$.

Note also that there usually exist different distributions of $\text{Tr}_1(l_1, j)$ and $\text{Tr}_2(i, l_2)$ over values $\text{Tr}(l_1, l_2)$. For example, if we apply the same greedy algorithm to a different ordering of the values $\text{Tr}_1(l_1, j)$ and/or to a different ordering of the values $\text{Tr}_2(i, l_2)$, then we may obtain a different distribution. $\qquad \square$

## 4 Conditions for Being a Transport Node

By Theorem 5, in order to decide whether or not a matching transport for a weighted tree exists, it suffices to check locally if each internal node is a transport node. By definition, an internal node $j$ is a transport node, if and only if there exists a witness matrix $A$, which matches the weights $c(i, j)$ and $c(j, i)$ for all neighbours $i$ of $j$.

Instead of actually constructing such a witness matrix, we now prove that such a witness matrix exists, if and only if the weights $c(i, j)$ and $c(j, i)$ satisfy certain conditions. Then to check if node $j$ is a transport node, we only have to check these conditions.

In this section and in Theorem 11, we consider individual internal nodes $j$, together with their neighbours and a weight function $c$ on the edges between $j$ and its neighbours. We also use the terms 'transport node' and 'witness matrix', as if they have been defined for this local context. The results that we obtain, however, can be applied in the context of complete weighted trees. We will do that at the end of Sect. 5.

In Definition 4, we denoted the neighbours of an internal node $j$ by $h_1, \ldots, h_m$ for some $m \geq 2$. From now on, for notational convenience, we assume that these neighbours are nodes $1, \ldots, m$, respectively.

**Lemma 6.** *If node $j$ is a transport node, then*

$$\sum_{i=1}^{m} c(i,j) = \sum_{k=1}^{m} c(j,k), \tag{7}$$
$$c(i,j) \leq \sum_{k \neq i} c(j,k) \qquad (i = 1, \ldots, m) \quad and \tag{8}$$
$$c(j,k) \leq \sum_{i \neq k} c(i,j) \qquad (k = 1, \ldots, m) \ . \tag{9}$$

Equation (7) is Kirchhoff's law, meaning that the total weight entering node $j$ equals the total weight leaving node $j$. Equation (8) expresses the fact that the weight coming in from a neighbour $i$ can go out to the other neighbours of node $j$. Finally, (9) expresses the fact that the weight going out to a neighbour $k$ can have come from the other neighbours of node $j$. Exactly these equations were violated by nodes 3 and 8 in the weighted trees in Fig. 2.

Before we prove Lemma 6, we show that there is some redundancy in (7)–(9). For $i_0 = 1, \ldots, m$, let

$$\mathrm{Margin}_c(i_0, j) = \sum_{k \neq i_0} c(j,k) - c(i_0, j) \quad \text{and}$$
$$\mathrm{Margin}_c(j, i_0) = \sum_{i \neq i_0} c(i,j) - c(j, i_0) \ .$$

Hence, $\mathrm{Margin}_c(i_0, j)$ and $\mathrm{Margin}_c(j, i_0)$ denote the differences between the right hand side and the left side of (8) and (9), respectively. If the weight function $c$ is clear from the context, we will simply write $\mathrm{Margin}(i_0, j)$ and $\mathrm{Margin}(j, i_0)$. We then have:

**Lemma 7.** *If (7) holds for node $j$, then for $i_0 = 1, \ldots, m$, $\mathrm{Margin}(i_0, j) = \mathrm{Margin}(j, i_0)$.*

This result follows directly from the definitions and (7). It implies in particular that if (7) holds, then (8) and (9) are equivalent. Therefore, in the rest of the paper, when we have to prove that (7)–(9) are valid for an internal node $j$ and a weight function $c$, we will not mention (9).

**Proof of Lemma 6.** Assume that node $j$ is a transport node. Then by definition, there exists an $m \times m$ matrix $A = (a_{i,k})$ satisfying (2)–(5).

When we add up all entries of $A$, row by row or column by column, we find

$$\sum_{i=1}^{m} c(i,j) = \sum_{i=1}^{m} \sum_{k=1}^{m} a_{i,k} = \sum_{k=1}^{m} \sum_{i=1}^{m} a_{i,k} = \sum_{k=1}^{m} c(j,k) \ .$$

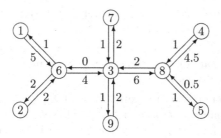

**Fig. 4.** Tree resulting from the tree in Fig. 1, if we assign a quantity $a = 0.5$ from edge $(5, 8)$ to edge $(8, 4)$

Indeed, (7) holds. Let $i_0$ be an arbitrary neighbour of $j$. Then

$$c(i_0, j) = \sum_{k=1}^{m} a_{i_0, k} = \sum_{k \neq i_0} a_{i_0, k} \leq \sum_{k \neq i_0} \left( \sum_{i=1}^{m} a_{i,k} \right) = \sum_{k \neq i_0} c(j, k) \ .$$

Hence, also (8) holds.                                                      □

At first glance, it is not obvious that Lemma 6 can be reversed. That is, that if (7)–(9) hold for an internal node $j$, then $j$ is a transport node. In particular, it is imagineable that for each individual neighbour $i$ of $j$, the weight on edge $(i, j)$ can be carried along to the other neighbours of $j$, but that this is not possible for all neighbours of $j$ simultaneously. Simply, because the weights on incoming edges $(i, j)$ for different neighbours $i$ are competing for the same outgoing edges $(j, k)$.

For example, let $j$ be node 8 in the tree from Fig. 1. In Fig. 4 we have depicted the tree resulting from assigning a quantity $a = 0.5$ from the incoming edge $(5, 8)$ to the outgoing edge $(8, 4)$. Now, the weight on edge $(3, 8)$ can no longer be passed on to the other neighbours of node 8.

We will see that this problem can be avoided, and that we can indeed reverse Lemma 6. First, however, we state a result on the values of Margin$(i, j)$ for competing edges $(i_0, j)$ and $(i_1, j)$, which follows directly from the definitions and (7).

**Lemma 8.** *Assume that (7)–(9) hold for node $j$. Let $i_0$ and $i_1$ be two different neighbours of $j$. Then*

$$Margin(i_0, j) + Margin(i_1, j) = \sum_{i \neq i_0, i_1} c(i, j) + \sum_{k \neq i_0, i_1} c(j, k) \ . \tag{10}$$

We now state the converse of Lemma 6 and provide its proof.

**Lemma 9.** *If (7)–(9) hold for node $j$, then $j$ is a transport node.*

**Proof.** Assume that $c$ satisfies (7)–(9). We prove that we can find an $m \times m$ matrix $A = (a_{i,k})$ satisfying conditions (2)–(5). If this is true, then node $j$ is a transport node.

We use induction on the number $p_c$ of weights on the incoming and outgoing edges of $j$, which are strictly positive.

If $p_c = 0$, then for $i = 1, \ldots, m$, $c(i, j) = 0$, and for $k = 1, \ldots, m$, $c(j, k) = 0$. It is easily verified that for this case, the $m \times m$ matrix $A = (a_{i,k})$ with all 0-entries satisfies conditions (2)–(5).

Induction step. Let $p \geq 0$, and suppose that for every non-negative weight function $c$ on the incoming and outgoing edges of $j$, which satisfies (7)–(9) and for which $p_c \leq p$, we can find an $m \times m$ matrix $A$ as specified (induction hypothesis). Now, consider a non-negative weight function $c$ on the same edges and satisfying the same conditions, with $p_c = p + 1$.[2]

Without loss of generality, assume that $c(i_0, j) > 0$ for some neighbour $i_0$ of $j$. Let $k_0$ be an arbitrary neighbour of $j$ which satisfies $k_0 \neq i_0$ and $c(j, k_0) > 0$. By (8), such a neighbour exists.

We now proceed to assign a maximum quantity $a$ of the weight $c(i_0, j)$ of the incoming edge $(i_0, j)$ to the outgoing edge $(j, k_0)$. In order to specify $a$, we introduce the following quantity:

$$\text{MinMargin} = \min_{i \neq i_0, k_0} \text{Margin}_c(i, j) \ .$$

Let $i_1 \neq i_0, k_0$ be a neighbour of $j$ for which this value is achieved. If $m = 2$, then MinMargin is set to infinity and $i_1$ is not defined.

Obviously, the value $a$ is bounded by $c(i_0, j)$ and $c(j, k_0)$. It is, however, also bounded by MinMargin. If we let $a$ be larger than MinMargin, then the total weight on outgoing edges, that is available to the incoming edge $(i_1, j)$ would become smaller than $c(i_1, j)$. We therefore set

$$a = \min \big( c(i_0, j), c(j, k_0), \text{MinMargin} \big) \ .$$

We now distinguish two cases:

- If $a = c(i_0, j)$ or $a = c(j, k_0)$, then we define a new weight function $c'$ by subtracting $a$ from $c(i_0, j)$ and $c(j, k_0)$ and leaving all other weights unchanged.

It is easily verified that $c'$ is non-negative and satisfies (7)–(9). Moreover, as either $c'(i_0, j) = 0$, or $c'(j, k_0) = 0$ (or both), $p_{c'} \leq p$. Hence, by the induction hypothesis, there exists an $m \times m$ matrix $A' = (a'_{i,k})$ satisfying conditions (2)–(5) for the weight function $c'$. It follows immediately that the $m \times m$ matrix $A = (a_{i,k})$ defined by

$$a_{i_0,k_0} = a'_{i_0,k_0} + a$$
$$a_{i,k} = a'_{i,k} \qquad\qquad ( \ (i, k) \neq (i_0, k_0) \ )$$

satisfies the same conditions for the original weight function $c$.

- If $a = \text{MinMargin}$ and $\text{MinMargin} < \min(c(i_0, j), c(j, k_0))$, then in particular $m \geq 3$ and node $i_1$ is defined. Now, we define a new weight function $c'$ by

$$c'(i_0, j) = 0$$

---

[2] Note that, because of (7), it is impossible that $p_c = 1$. This, however, does not harm our argument.

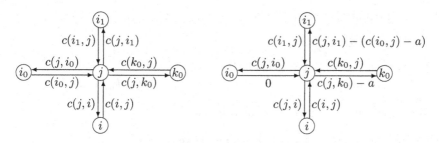

**Fig. 5.** Old (left) and new (right) weights for the second case in the proof of Lemma 9

$$c'(j, k_0) = c(j, k_0) - a$$
$$c'(j, i_1) = c(j, i_1) - (c(i_0, j) - a)$$
$$c'(i, j) = c(i, j) \qquad\qquad (i \neq i_0)$$
$$c'(j, k) = c(j, k) \qquad\qquad (k \neq k_0, i_1)$$

(see Fig. 5). By definition and by Lemma 7,

$$a = \mathrm{Margin}_c(i_1, j) = \mathrm{Margin}_c(j, i_1) = \sum_{i \neq i_1} c(i, j) - c(j, i_1) \geq c(i_0, j) - c(j, i_1) \ .$$

Hence, $c'(j, i_1) \geq 0$, which implies that $c'$ is non-negative for every edge.

It is easily verified that the weight function $c'$ satisfies (7). It remains to be proved that $c'$ also satisfies (8), i.e., that $\mathrm{Margin}_{c'}(i_2, j) \geq 0$ for each neighbour $i_2$ of $j$. For this, we can distinguish four subcases: $i_2 = i_0$, $i_2 = i_1$, $i_2 = k_0$ and $i_2 \neq i_0, i_1, k_0$. The proofs for the first two cases are straightforward. The proofs for the last two cases are more involved, but similar. We give the details for the case that $i_2 \neq i_0, i_1, k_0$:

$$\mathrm{Margin}_{c'}(i_2, j) = \sum_{k \neq i_2} c'(j, k) - c'(i_2, j)$$

$$= \sum_{k \neq i_2} c(j, k) - a - (c(i_0, j) - a) - c(i_2, j) = \mathrm{Margin}_c(i_2, j) - c(i_0, j) \ .$$

By Lemma 8,

$$\mathrm{Margin}_c(i_2, j) + \mathrm{Margin}_c(i_1, j) = \mathrm{Margin}_c(i_2, j) + a$$
$$= \sum_{i \neq i_2, i_1} c(i, j) + \sum_{k \neq i_2, i_1} c(j, k) \geq c(i_0, j) + c(j, k_0) > c(i_0, j) + a \ .$$

This implies that $\mathrm{Margin}_{c'}(i_2, j) \geq 0$.

Because by definition, $c'(i_0, j) = 0$, we have $p_{c'} \leq p$. Hence, by the induction hypothesis, there exists an $m \times m$ matrix $A' = (a'_{i,k})$ satisfying conditions (2)–(5) for the weight function $c'$. It follows immediately that the $m \times m$ matrix

$A = (a_{i,k})$ defined by

$$a_{i_0,k_0} = a'_{i_0,k_0} + a$$
$$a_{i_0,i_1} = a'_{i_0,i_1} + (c(i_0,j) - a)$$
$$a_{i,k} = a'_{i,k} \qquad \qquad (\,(i,k) \neq (i_0,k_0),(i_0,i_1)\,)$$

satisfies the same conditions for the original weight function $c$. □

When we combine Lemma 6 and Lemma 9, we obtain

**Theorem 10.** *Node $j$ is a transport node, if and only if (7)–(9) hold for $j$.*

## 5  Minimum and Maximum Transport

A transport node in a weighted tree may have many different witness matrices. We examine the minimum and maximum values for each of the entries in these witness matrices.

**Theorem 11.** *Assume that node $j$ is a transport node and that $A = (a_{i,k})$ is a witness matrix for this. Let $i_0$ and $k_0$ be two arbitrary, different neighbours of $j$. Then*

$$a_{i_0,k_0} \geq \max\left(\,0,\; c(i_0,j) - \sum_{k \neq i_0,k_0} c(j,k),\; c(j,k_0) - \sum_{i \neq i_0,k_0} c(i,j)\,\right) \quad and \quad (11)$$

$$a_{i_0,k_0} \leq \min\left(\,c(i_0,j),\; c(j,k_0),\; \min_{i \neq i_0,k_0} Margin_c(i,j)\,\right), \qquad (12)$$

*and each value satisfying these equations can be achieved.*

It would be a nice exercise to prove that the right-hand side of (11) is at most as large as the right-hand side of (12), without using these equations themselves.

Note that by (4) and Lemma 7, (11) is equivalent to the following inequality:

$$\sum_{k \neq k_0} a_{i_0,k} \leq \min\left(\,c(i_0,j),\; \sum_{k \neq i_0,k_0} c(j,k),\; Margin_c(k_0,j)\,\right) \,. \qquad (13)$$

**Proof.** We first prove (11). By definition, $a_{i_0,k_0} \geq 0$. By successively applying (3), (2) and (5), we find

$$\sum_{k \neq k_0} a_{i_0,k} = \sum_{k \neq i_0,k_0} a_{i_0,k} \leq \sum_{k \neq i_0,k_0} \sum_{i=1}^{m} a_{i,k} = \sum_{k \neq i_0,k_0} c(j,k) \,.$$

Hence, by (4), $a_{i_0,k_0} \geq c(i_0,j) - \sum_{k \neq i_0,k_0} c(j,k)$. Analogously, we find $a_{i_0,k_0} \geq c(j,k_0) - \sum_{i \neq i_0,k_0} c(i,j)$.

In the proof of Lemma 9, we have seen that also (12) holds.

We now prove that each value $a_{i_0,k_0} = a$ satisfying (11) and (12) can be achieved. If $m = 2$, then $\sum_{k \neq i_0,k_0} c(j,k) = \sum_{i \neq i_0,k_0} c(i,j) = 0$. Hence, by (11) and (12), the only possible value for $a_{i_0,k_0}$ is $a_{i_0,k_0} = c(i_0,j) = c(j,k_0)$. The existence of a witness matrix $A$ implies that this value can indeed be achieved.

Now assume that $m \geq 3$ and let $a_{i_0,k_0} = a$ be an arbitrary value satisfying (11) and (12).

By Lemma 6, we know that the weight function $c$ satisfies (7)–(9). Let the weight function $c'$ be defined by

$$c'(i_0, j) = c(i_0, j) - a$$
$$c'(j, k_0) = c(j, k_0) - a$$
$$c'(i, j) = c(i, j) \qquad (i \neq i_0)$$
$$c'(j, k) = c(j, k) \qquad (k \neq k_0) .$$

Because $a \leq c(i_0, j)$ and $a \leq c(j, k_0)$, $c'$ is non-negative. Further, $c'$ satisfies (7), because $c$ does. Finally, because $a \leq \min_{i \neq i_0, k_0} \text{Margin}_c(i, j)$, $\text{Margin}_{c'}(i, j) \geq 0$ for $i = 1, \ldots, m$. Hence, $c'$ also satisfies (8). By Theorem 10, $j$ is still (in the context of the new weight function $c'$) a transport node.

Now, let $k_1, \ldots, k_{m-2}$ be the neighbours of $j$ different from $i_0$ and $k_0$. We must assign the remaining weight $c'(i_0, j) = c(i_0, j) - a$ on the incoming edge $(i_0, j)$ to the outgoing edges $(j, k_1), \ldots (j, k_{m-2})$. That is, we must find a witness matrix $A' = (a'_{i,k})$ for $j$ and $c'$ for which $a'_{i_0,k_0} = 0$.

We can prove that such a matrix exists by induction on the number of neighbours $k \in \{k_1, \ldots, k_{m-2}\}$ for which $\text{Margin}_{c'}(k, j) < c'(i_0, j)$. The intuition behind this is, that if for some neighbour $k_l$ with $1 \leq l \leq m - 2$, $\text{Margin}_{c'}(k_l, j) \geq c'(i_0, j)$, then after any partitioning of $c'(i_0, j)$ over $a'_{i_0,k_1}, \ldots, a'_{i_0,k_{m-2}}$, we can still pass the weight $c'(k_l, j)$ to the edges $(j, k)$ with $k \neq k_l$. If, on the other hand, $\text{Margin}_{c'}(k_l, j) < c'(i_0, j)$, then we must make sure that $a'_{i_0,k_l}$ gets at least a share $c'(i_0, j) - \text{Margin}_{c'}(k_l, j)$ of the weight $c'(i_0, j)$.

As the details of this proof are rather technical, we do not carry out this proof here. We just make two observations. First, by (11),

$$c'(i_0, j) = c(i_0, j) - a \leq \sum_{k \neq i_0, k_0} c(j, k) = \sum_{k \neq i_0, k_0} c'(j, k) .$$

Hence, the total weight available on the edges $(j, k_1), \ldots, (j, k_{m-2})$ is enough to receive the remaining weight on edge $(i_0, j)$. Second, by (11) and Lemma 7,

$$c'(i_0, j) = c(i_0, j) - a \leq c(i_0, j) - c(j, k_0) + \sum_{i \neq i_0, k_0} c(i, j)$$

$$= \text{Margin}_c(j, k_0) = \text{Margin}_c(k_0, j) = \text{Margin}_{c'}(k_0, j) .$$

Hence, after assigning $c'(i_0, j)$ to the edges $(j, k_1), \ldots, (j, k_{m-2})$, we can still pass the weight $c'(k_0, j)$ of edge $(k_0, j)$ to neighbours of $j$ different from $k_0$.

Assuming that we can find the matrix $A'$, the matrix $A^* = (a^*_{i,k})$ defined by

$$a^*_{i_0,k_0} = a$$
$$a^*_{i,k} = a'_{i,k} \qquad ( (i, k) \neq (i_0, k_0) )$$

is a witness matrix for $j$ and $c$, with the desired value $a$ for $a^*_{i_0,k_0}$. $\qquad \square$

## Algorithm for the Global Problem

We return to the context of a complete weighted tree $(T, c)$. We can use Theorem 5 and Theorem 10 to decide whether or not there exists a matching transport Tr on $(T, c)$. Assume that this is the case. For two different leaves $l_1$ (the source) and $l_2$ (the target), let $\text{MinTr}(l_1, l_2)$ and $\text{MaxTr}(l_1, l_2)$ denote the minimum and maximum possible values for $\text{Tr}(l_1, l_2)$ respectively. We use Theorem 11 to determine these values. The algorithm for this is simple.

The transport from $l_1$ to $l_2$ follows the (unique) path in the tree from $l_1$ to $l_2$. Let $j_1, \ldots, j_p$ for some $p \geq 0$ be the internal nodes on this path, in the order of occurrence on the path. Let us define $j_0 = l_1$ and $j_{p+1} = l_2$.

If $p = 0$ (which is only possible if $l_1$ and $l_2$ are the only nodes in the tree), then obviously, $\text{MinTr}(l_1, l_2) = \text{MaxTr}(l_1, l_2) = c(l_1, l_2)$. It is impossible for containers from $l_1$ to 'escape' to other destinations, as there are no other destinations.

Now assume that $p \geq 1$. Let us denote the lower bound and upper bound for $a_{i_0, k_0}$ from Theorem 11 by $\text{LB}(i_0, j, k_0)$ and $\text{UB}(i_0, j, k_0)$, respectively.

As announced in the introduction, to obtain $\text{MinTr}(l_1, l_2)$, we proceed in a greedy way. In each internal node $j_k$ on the path from $l_1$ to $l_2$, we direct as much weight as possible from the preceding node $j_{k-1}$ to neighbours of $j_k$ other than $j_{k+1}$. As we have to pass at least $\text{LB}(j_{k-1}, j_k, j_{k+1})$ of the weight from $j_{k-1}$ to $j_{k+1}$, we can direct at most $c(j_{k-1}, j_k) - \text{LB}(j_{k-1}, j_k, j_{k+1})$ to the other neighbours (see also (13)). Then $\text{MinTr}(l_1, l_2)$ equals the weight remaining from $c(l_1, j_1)$ after passing by all internal nodes:

$$\text{MinTr}(l_1, l_2) = \max \left( 0, \ c(l_1, j_1) - \sum_{k=1}^{p} (c(j_{k-1}, j_k) - \text{LB}(j_{k-1}, j_k, j_{k+1})) \right). \quad (14)$$

To obtain $\text{MaxTr}(l_1, l_2)$, for each internal node $j_k$ on the path from $l_1$ to $l_2$, we pass as much weight as possible from $j_{k-1}$ via $j_k$ to $j_{k+1}$. The minimum value $\text{UB}(j_{k-1}, j_k, j_{k+1})$ we encounter on the path determines the maximum weight that can be transported from $l_1$ to $l_2$:

$$\text{MaxTr}(l_1, l_2) = \min_{k=1}^{p} \text{UB}(j_{k-1}, j_k, j_{k+1}) \ . \quad (15)$$

The complete algorithm consists of first determining the path from $l_1$ to $l_2$, and then (if $p \geq 1$) calculating (14) and (15). It is not hard to see that the time complexity of this algorithm is linear in the size of the input.

Let us consider the weighted tree from Fig. 1, and let $l_1 = 1$ and $l_2 = 4$. Then the path from $l_1$ to $l_2$ contains $p = 3$ internal nodes 6, 3, 8, and our algorithm yields the following sequence of values:

| $k$ | | 0 | 1 | 2 | 3 | 4 |
|---|---|---|---|---|---|---|
| $j_k$ | | 1 | 6 | 3 | 8 | 4 |
| $c(j_{k-1}, j_k) - \text{LB}(j_{k-1}, j_k, j_{k+1})$ | | | 2 | 1 | 1 | |
| $\text{UB}(j_{k-1}, j_k, j_{k+1})$ | | | 3 | 4 | 5 | |

As a result, $\text{MinTr}(1, 4) = 5 - (2 + 1 + 1) = 1$ and $\text{MaxTr}(1, 4) = \min(3, 4, 5) = 3$.

## 6   Concluding Remarks

A large fraction of this paper dealt with transport nodes and corresponding witness matrices $A = (a_{i,k})$. In the resulting algorithm, these concepts do not play any role. However, we needed them to prove that the algorithm is correct, i.e., to prove that the minimum (or maximum) value computed for the transport from one leaf (the source) to another leaf (the target) in a weighted tree, can be extended to a complete matching transport on the tree, and that this value is indeed minimal (maximal, respectively).

In the introduction, we explained why standard max-flow algorithms such as the Ford-Fulkerson algorithm could not be applied directly to High Spies. For the integer-valued case, however, there does exist an approach to the problem, based on such algorithms. In this approach, the lower bound and upper bound for the (local) transport at an internal node $j$ are found by fixing a candidate value $a_{i_0,k_0} = a$ (for proper choices of $a$), and trying to extend this to a complete $m \times m$ witness matrix $A = (a_{i,k})$. With two copies of every neighbour $i$ of $j$ (corresponding to the edges $(i,j)$ and $(j,i)$, respectively), two additional nodes (source and target) and proper connections between the nodes, $A$ can be derived from a maximum flow with value $\sum_i^m c_{i,j} - a$. However, as the number of connections is quadratic in $m$ and the complexity of max-flow algorithms is at least linear in this number, this 'max-flow approach' is far more time-consuming than applying Theorem 11.

The 'max-flow approach' can be extended in a natural way to an approach for finding the maximum (global) transport between two leaves, which does not rely on (local) transports at internal nodes. Again, however, this would be far less efficient than our algorithm. Moreover, this extension does not work for the minimum (global) transport.

Both the original definition of a matching transport (Definition 3) and the equivalent formulation in terms of transport nodes from Theorem 5 are well suited to generate instances of High Spies. Due to space limitations, we could not elaborate on that here, but we plan to do this in a forthcoming report. In that report, we will also give more detailed proofs for some of the results in this paper.

We want to emphasize, that we did not expect the teams that participated in the Benelux Algorithm Programming Contest 2006 to prove that their solutions for High Spies were correct. A correct implementation of the algorithm described in Sect. 5 was sufficient.

## References

Ford Jr., L.R., Fulkerson, D.R.: A simple algorithm for finding maximal network flows and an application to the Hitchcock problem. Canadian Journal of Mathematics 9, 210–218 (1957)

Shasha, D.E.: Puzzling adventures: High spies. Scientific American 289(1), 80 (July 2003)

# Robots and Demons
# (The Code of the Origins)*

Y. Dieudonné and F. Petit

LaRIA CNRS, Université de Picardie Jules Verne
Amiens, France

**Abstract.** In this paper, we explain how Robert Langdon, a famous Harvard Professor of Religious Symbology, brought us to decipher the Code of the Origins. We first formalize the problem to be solved to understand the Code of the Origins. We call it the Scatter Problem (SP). We then show that the SP cannot be deterministically solved. Next, we propose a randomized algorithm for this problem. The proposed solution is trivially self-stabilizing. We then show how to design a self-stabilizing version of any deterministic solution for the Pattern Formation and the Gathering problems.

**Keywords:** Fun with distributed algorithms, fun with mobile robot networks, fun with stabilization.

## 1 Introduction

*End of April 2007.*
The corpse of a distinguished scientist was found in one of the corridors at the CERN (European Organization for Nuclear Research), in Switzerland. On the body of the unfortunate scientist, an odd symbol has been branded—refer to Figure 1.

Quickly, the Swiss Judicial Police called for Captain Fache, a French policeman who is known for having solved the odd Louvre's case [4]. Fache felt taken aback when he saw the symbol. He guessed he had already seen the symbol in the Rosslyn Chapel—often called the Cathedral of Codes—which was built by the Knights Templar in 1446 near Edinburgh, the capital of Scotland. Nobody knew more than Fache who to call for. He summoned Robert Langdon, a famous Professor of Religious Symbology at Harvard University. Fache and Langdon met in Paris during the Louvre's case [4]. They became friends afterwards.

Right from his arrival at CERN, Langdon took a close look at the body of the scientist. At first glance, Langdon recognized the sign.

'*The Four Elements!*' Langdon yelled. '*The Greek classical elements are Fire, Earth, Air, and Water*' Langdon said to Fache. '*In Greek philosophy and science, they represent the realms of the cosmos wherein all things exist and whereof all things consist*'.

---

\* Title and story freely inspired from [3].

P. Crescenzi, G. Prencipe, and G. Pucci (Eds.): FUN 2007, LNCS 4475, pp. 108–119, 2007.
© Springer-Verlag Berlin Heidelberg 2007

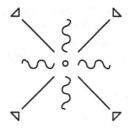

**Fig. 1.** The symbol branded on the body of the scientist

'*Good heavens!*' Langdon huffed by pointing suddenly the little circle in the center of the figure. '*How come I almost missed the circle? A circle in the middle generally means the negative form of the sign.*' '*In other words*', Langdon said glancing at Fache, '*this symbol is the original icon for the Antimatter!*'

Fache was dumbfounded. Once again, he was impressed by the huge knowledge of his friend. Fache stared at Langdon for a moment and then went to a nearby toolbag from which he took a papyrus roll. Langdon smiled. '*One clue after another, my friend!*' he said, while he was already looking into the old document. After a couple of seconds, he turned pale and looked overwhelmed. '*My friend*', he said turning to Fache, '*I am holding in my hands the most ancient manuscript of Humanity!*'

Later, Langdon's conjecture was confirmed by a carbon dating.

The manuscript was probably written upwards of 9000 years B.C. By contrast, the most ancient Egyptian papyrus known until now is dated between 3580 and 3536 B.C.

After some investigations, the papyrus roll disclosed its secrets. Langdon found that this astonishing document included tens of thousands of unbelievable clues about the making of the Universe. In particular, he found a passage containing the sign branded on the body of the unfortunate scientist. He understood that the fragment seemed to describe how the Antimatter was formed during the Big Bang. The description is followed by a kind of informal algorithm showing the relativistic dilatation of the Universe. That is, how starting from the primitive soup, the matter clusters were self-scattered and self-organized in star clusters.

Fache had never seen his friend so exited. For three whole days, Langdon did nothing else than study the ancient document to figure out the informal algorithm which he called the Code of the Origins. However, Langdon met with difficulties. They lie in the fact that the Code of the Origins is made of a string of clues or riddles. Despite his incredible knowledge of ancient myths and symbols, Langdon eventually got stuck with the following odd phrase in the Code of the Origins:

> '*Lord Grapefruit condenses the Antimatter while Lord Onion Skin scatters the Matter.*'

Langdon thought '*The onion skin is an example of the Voronoi diagram. I should call for scientists who are currently working on Voronoi diagrams in dynamic systems.'* Langdon heard from a colleague at Harvard that some computer scientists were working on that sort of problems, mainly in the area of the distributed coordination of autonomous mobile robots.

That is how we soon received an email from Robert Langdon with the following odd message:

> *This is an urgent and quite important message I would like to bring to your attention. Your approach in the design of algorithms for a team of weak robots (agents, or whatever. . . ) should be helpful in the understanding of a major paradigm of the making of the Universe. A key mechanism of this paradigm is called the Code of the Origins. I definitely need your help to figure the formal code out, and also to prove its correctness. Again, this is a very important message. The Code of the Origins is a keystone in the origins of the Universe and of course, for Humanity. Can you please help? I look forward to receiving an agreement from your side.*

We accepted the challenge. We first figured out from the explanations given by Langdon that in fact, the Code of the Origins was not strictly speaking an algorithm or a code, but rather a paradigm describing the scatter of matter clusters. Moreover, the paradigm does not assume that the matter is initially concentrated in a sort of "single point" as one can imagine.

Below are the results of our research.

In the next section, we describe the model considered in this paper and the formal definition of the problem to be solved, i.e., the *Scatter Problem*. Next, in Section 3, we consider how this problem can be solved. We first show that the Scatter Problem cannot be deterministically solved in the considered model. We then give a probabilistic algorithm for this problem along with its correctness proof. In Section 4, we put the result of Section 3 back into the context of distributed coordination of autonomous mobile robots. In this area, two classes of problems have received a particular attention[1]:

1. The *Pattern Formation Problem* (PFP) which includes the *Circle Formation Problem*, e.g. [21,13,15,7,17,8,9,10];
2. The *Gathering Problem* (GP), e.g., [1,21,14,21], and also in [6,5].

We consider these two major classes of problems into self-stabilization settings. Regardless of the initial states of the computing units, a self-stabilizing system is guaranteed to converge to the intended behavior in finite time [11,12]. To our best knowledge, all the above solutions assume that in the initial configuration, no two robots are located at the same position. As already noticed [7,16], this implies that none of them is self-stabilizing. In Section 4, we show that, being

---

[1] Note that some of the following solutions are in a model called CORDA [18] allowing more asynchrony among the robots than the semi-synchronous model (SSM) used in this paper. However, it is showed in [19] that any algorithm that correctly solves a problem $P$ in CORDA, correctly solves $P$ in $SSM$. So, any algorithm described in CORDA also works in $SSM$.

self-stabilizing, the proposed algorithm can be used to provide a self-stabilizing version of any deterministic solution for PFP and GP, i.e., assuming any arbitrary initial configuration—*including* configurations where two or more robots can be located at the same position. Finally, we conclude the story in Section 5.

## 2   Preliminaries

In this section, we define the distributed system, basic definitions and the problem considered in this paper.

*Distributed Model.* We adopt the model introduced in [20], below referred to as *SSM*—SSM stands for *Semi-Synchronous Model*. The *distributed system* considered in this paper consists of $n$ mobile *robots* (*entity, agent,* or *element*) $r_1, r_2, \cdots, r_n$—the subscripts $1, \ldots, n$ are used for notational purpose only. Each robot $r_i$, viewed as a point in the Euclidean plane, moves on this two-dimensional space unbounded and devoid of any landmark. When no ambiguity arises, $r_i$ also denotes the point in the plane occupied by that robot. It is assumed that the robots never collide and that two or more robots may simultaneously occupy the same physical location. Any robot can observe, compute and move with infinite decimal precision. The robots are equipped with sensors enabling to detect the instantaneous position of the other robots in the plane. Each robot has its own local coordinate system and unit measure. There is no kind of explicit communication medium. The robots implicitly "communicate" by observing the position of the other robots in the plane, and by executing a part of their program accordingly.

The considered robots are *uniform, oblivious,* and *anonymous*. Uniform indicates that they all follow the same program. Obliviousness means that the robots cannot remember any previous observation nor computation performed in any previous step. Anonymous means that no local parameter (such as an identity) could be used in the program code to differentiate any of them.

In this paper, we also discuss some capabilities the robots are able to have or not:

*Multiplicity Detection*: The robots are able to distinguish whether there is more than one robot at a given position;

*Localization Knowledge*: The robots share a common coordinate system, i.e., a common Cartesian coordinate system with a common origin and common $x$-$y$ axes with the same orientations.

Time is represented as an infinite sequence of time instants $t_0, t_1, \ldots, t_j, \ldots$ Let $P(t_j)$ be the set of the positions in the plane occupied by the $n$ robots at time $t_j$ ($j \geq 0$). For every $t_j$, $P(t_j)$ is called the *configuration* of the distributed system in $t_j$. $P(t_j)$ expressed in the local coordinate system of any robot $r_i$ is called a *view*, denoted $v_i(t_j)$. At each time instant $t_j$ ($j \geq 0$), each robot $r_i$ is either *active* or *inactive*. The former means that, during the computation step $(t_j, t_{j+1})$, using a given algorithm, $r_i$ computes in its local coordinate system a position $p_i(t_{j+1})$ depending only on the system configuration at $t_j$, and moves towards $p_i(t_{j+1})$—$p_i(t_{j+1})$ can be equal to $p_i(t_j)$, making the location of $r_i$ unchanged. In the latter case, $r_i$ does not perform any local computation and remains at the same position. In every single activation, the distance traveled by any robot $r$ is bounded by $\sigma_r$. So, if the destination point computed by $r$ is farther than $\sigma_r$, then $r$ moves toward a point of at most $\sigma_r$. This distance may be different between two robots.

The concurrent activation of robots is modeled by the interleaving model in which the robot activations are driven by a *fair scheduler*. At each instant $t_j$ ($j \geq 0$), the scheduler arbitrarily activates a (non empty) set of robots. Fairness means that every robot is infinitely often activated by the scheduler.

*Specification.* The *Scatter Problem* (SP) is to design a protocol for $n$ mobile autonomous robots so that the following properties are true in every execution:

*Convergence*: Regardless of the initial position of the robots on the plane, no two robots are eventually located at the same position.

*Closure*: Starting from a configuration where no two robots are located at the same position, no two robots are located at the same position thereafter.

## 3   Algorithm

The scope of this section is twofold. We first show that there exists no deterministic algorithm solving $SP$. The result holds even if the robots are not oblivious, share a common coordinate system, or are able to detect multiplicity. Next, we propose a randomized algorithm which converges toward a distribution where the robots have distinct positions.

### 3.1   Impossibility of a Deterministic Algorithm

**Lemma 1.** *There exists no deterministic algorithm that solves the Scatter Problem in SSM, even if the robots have the localization knowledge or are able to detect multiplicity.*

*Proof.* Assume, by contradiction, that a deterministic algorithm $A$ exists solving SP in SSM with robots having the localization knowledge and being able to detect multiplicity. Assume that, initially ($t_0$), all the robots are located at the same position. So, it does not matter whether the robots have the localization knowledge, are able to detect multiplicity or not, all the robots have the same view of the world. Assume that at $t_0$, all the robots are active and execute $A$. Since $A$ is a deterministic algorithm and all the robots have the same view, then all the robots choose the same behavior. So, at time $t_1$, all of them share the same position on the place. Again, they all have the same view of the world. By induction, we can deduce that there exists at least one execution of $A$ where the robots always share the same position. This contradicts the specification of SP. Hence, such Algorithm $A$ does not exist.

Note that Lemma 1 also holds whether the robots are oblivious or not. Indeed, assume non-oblivious robots, i.e., any robot moves according to the current and previous configurations. So, each robot $r_i$ is equipped with a (possibly infinite) history register $\mathcal{H}_i$. At time $t_0$, for each robot $r_i$, the value in $\mathcal{H}_i$ depends on whether the registers are assumed to be initialized or not.

First assume that, at $t_0$, $\mathcal{H}_i$ is initialized for every robot. Since the robots are assumed to be uniform and anonymous, the values stored in the history registers cannot be different. So, for every pair of robots $(r_i, r_{i'})$, $\mathcal{H}_i = \mathcal{H}_{i'}$ at $t_0$. Then, all the robots have the same view of the world. This case leads to the proof of Lemma 1.

Now, assume that, for every robot $r_i$, $\mathcal{H}_i$ is not assumed to be initialized at time $t_0$. Note that this case captures the concept of self-stabilization. In such a system, at $t_0$, one possible initialization of the history registers can be as follows: $(r_i, r_{i'})$, $\mathcal{H}_i = \mathcal{H}_{i'}$ for every every pair $(r_i, r_{i'})$. This case is similar to the previous case.

## 3.2   Randomized Algorithm

We use the following concept, *Voronoi diagram*, in the design of Algorithm 1.

**Definition 1 (Voronoi diagram).** *[2,7] The Voronoi diagram of a set of points $P = \{p_1, p_2, \cdots, p_n\}$ is a subdivision of the plane into $n$ cells, one for each point in $P$. The cells have the property that a point $q$ belongs to the Voronoi cell of point $p_i$ iff for any other point $p_j \in P$, $dist(q, p_i) < dist(q, p_j)$ where $dist(p, q)$ is the Euclidean distance between $p$ and $q$. In particular, the strict inequality means that points located on the boundary of the Voronoi diagram do not belong to any Voronoi cell.*

We now give an informal description of Procedure $SP$, shown in Algorithm 1. Each robot uses Function $Random()$, which returns a value randomly chosen over $\{0, 1\}$ : 0 with a probability $\frac{3}{4}$ and 1 with a probability $\frac{1}{4}$. When any robot $r_i$ becomes active at time $t_j$, it first computes the Voronoi Diagram of $P_i(t_j)$, i.e., the set of points occupied by the robots, $P(t_j)$, computed in its own coordinate system. Then, $r_i$ moves toward a point inside its Voronoi cell $Cell_i$ if $Random()$ returns 0.

---

**Algorithm 1.** Procedure $SP$, for any robot $r_i$

---

Compute the Voronoi Diagram;
$Cell_i :=$ the Voronoi cell where $r_i$ is located;
$Current\_Pos :=$ position where $r_i$ is located;
**if** $Random()=0$
**then**  Move toward an arbitrary position in $Cell_i$, which is different from $Current\_Pos$;
**else**  Do not move;

---

**Lemma 2 (Closure).** *For any time $t_j$ and for every pair of robots $(r_i, r_{i'})$ having distinct positions at $t_j$ $(p_i(t_j) \neq p_{i'}(t_j))$, then by executing Procedure SP, $r_i$ and $r_{i'}$ remain at distinct positions, thereafter $(\forall j' > j, \ p_i(t_{j'}) \neq p_{i'}(t_{j'}))$.*

*Proof.* Clearly, if at time $t_j$, $r_i$ and $r_{i'}$ have distinct positions, then $r_i$ and $r_{i'}$ are in two different Voronoi cells, $V_i$ and $V_j$, respectively. From Definition 1, $V_i \cap V_j = \emptyset$. Furthermore, each robot can move only in its Voronoi cell. So, we deduce that $r_i$ and $r_{i'}$ have distinct positions at time $t_{j+1}$. The lemma follows by induction on $j'$, $j' > j$.

In the following, we employ notation $Pr[A] = v$ to mean that $v$ is the probability that event $A$ occurs. Two events $A$ and $B$ are said to be *mutually exclusive* if and only if $A \cap B = \emptyset$. In this case, $Pr[A \cup B] = Pr[A] + Pr[B]$. The probability that event $A$ occurs given the known occurrence of event $B$ is the conditional probability of $A$ given $B$, denoted by $Pr[A|B]$. We have $Pr[A \cap B] = Pr[A|B]Pr[B]$.

**Lemma 3 (Convergence).** *For any time $t_j$ and for every pair of robots $(r_i, r_{i'})$ such that $p_i(t_j) = p_{i'}(t_j)$. By executing Procedure SP, we have*

$$\lim_{k \to \infty} Pr[p_i(t_{j+k}) \neq p_{i'}(t_{j+k})] = 1$$

*Proof.* Consider at time $t_j$, two robots $r_i$ and $r_{i'}$ such that $p_i(t_j) = p_{i'}(t_j)$. Let $X_{t_j}$ (respectively, $Y_{t_j}$) be the random variable denoting the number of robots among $r_i$ and $r_{i'}$ which are activated (respectively, move). $Pr[X_{t_j} = z]$ (resp. $Pr[Y_{t_j} = z']$) indicates the probability that $z \in [0..2]$ (resp. $z' \in [0..2]$) robots among $r_i$ and $r_{i'}$ are active (resp.move) at time $t_j$. Note that we make no other assumption than fairness on the activation schedule. In particular, the probability $Pr[X_{t_j} = z]$ is not assumed to be time invariant.

Robot $r_i$ (resp $r_{i'}$) can move only if $r_i$ (resp $r_{i'}$) is active. Both $r_i$ and $r_{i'}$ are in a single position at time $t_{j+1}$ only if one of the following four events arises in computation step $(t_j, t_{j+1})$:

- **Event1:** "Both $r_i$ and $r_{i'}$ are inactive." In this case:

$$Pr[Event1] = Pr[X_{t_j} = 0] \leq 1 \tag{1}$$

- **Event2:** "There is exactly one active robot which does not move and one inactive robot." Then, we get:

$$Pr[Event2] = Pr[X_{t_j} = 1 \cap Y_{t_j} = 0]$$

So,

$$Pr[Event2] = Pr[Y_{t_j} = 0|X_{t_j} = 1]Pr[X_{t_j} = 1]$$

$$Pr[Event2] = \frac{1}{4}Pr[X_{t_j} = 1]$$

Thus,

$$Pr[Event2] \leq \frac{1}{4} \tag{2}$$

- **Event3:** "There are exactly two active robots and both of them move toward the same location." The probability that both robots are activated and move (not necessarily at the same location) is given by:

$$Pr[X_{t_j} = 2 \cap Y_{t_j} = 2]$$

But,

$$Pr[X_{t_j} = 2 \cap Y_{t_j} = 2] = Pr[Y_{t_j} = 2|X_{t_j} = 2]Pr[X_{t_j} = 2]$$

That is,

$$Pr[X_{t_j} = 2 \cap Y_{t_j} = 2] = (\frac{3}{4})^2 Pr[X_{t_j} = 2]$$

Thus,

$$Pr[X_{t_j} = 2 \cap Y_{t_j} = 2] \leq \frac{9}{16}$$

Since the probability that all the robots are activated and move (not necessary at the same location) is lower than or equal to $\frac{9}{16}$, the probability of Event3 (i.e both move toward the same location) is also lower than or equal to $\frac{9}{16}$, i.e.

$$Pr[Event3] \leq \frac{9}{16} \tag{3}$$

- **Event4:** "There are exactly two active robots and both of them do not move." The probability that both robots are activated and do not move is given by:

$$Pr[X_{t_j} = 2 \cap Y_{t_j} = 0]$$

We have

$$Pr[X_{t_j} = 2 \cap Y_{t_j} = 0] = Pr[Y_{t_j} = 0|X_{t_j} = 2]Pr[X_{t_j} = 2]$$

Hence

$$Pr[Event4] \leq (\frac{1}{4})^2 = \frac{1}{16}$$

Let $\Omega$ be a sequence of time instants starting from $t_j$. Denote by $k$ the number of time instants in $\Omega$. Value $a$ (resp. $na$) indicates the number of instant in $\Omega$ where at least one robot is active (resp. both $r_i$ and $r_{i'}$ are inactive) among $r_i$ and $r_{i'}$. Obviously, $a + na = k$. From Equations (2) and (3) and the fact that Event2, Event3 and Event4 are mutually exclusive, we have:

$$Pr[Event2 \cup Event3 \cup Event4] = Pr[Event2] + Pr[Event3] + Pr[Event4]$$

So,

$$Pr[Event2 \cup Event3 \cup Event4] \leq \frac{1}{4} + \frac{9}{16} + \frac{1}{16} = \frac{7}{8} \tag{4}$$

From Equation (4), the probability that $r_i$ and $r_{i'}$ are located at the same position after $k$ time instant is

$$Pr[p_i(t_{j+k}) = p_{i'}(t_{j+k})] \leq (\frac{7}{8})^a$$

By fairness, both $r_i$ and $r_{i'}$ are infinitely often activated. Therefore, $\lim_{k \to \infty} a = \infty$, and then

$$\lim_{k \to \infty} Pr[p_i(t_{j+k}) = p_{i'}(t_{j+k})] = 0$$

The lemma follows from the fact that $Pr[p_i(t_{j+k}) \neq p_{i'}(t_{j+k})] = 1 - Pr[p_i(t_{j+k}) = p_{i'}(t_{j+k})]$.

From Lemma 2 and 3 follows:

**Theorem 1.** *Procedure $SP$ solves the Scatter Problem in SSM with a probability equal to 1.*

Note that as a result of Theorem 1 and by the specification of the Scatter Problem, Procedure $SP$ provides a self-stabilizing solution in SSM.

# 4     Related Problems and Self-stabilization

The acute reader should have noticed that by executing Procedure $SP$ infinitely often, the robots never stop moving inside their Voronoi cells, even if no two robots are located at the same position. This comes from the fact that Procedure $SP$ does not require robots to have the multiplicity detection capability. Henceforth, in this section, let us assume that the robots are equipped of such an ability. This assumption trivially allows the robots to stop if there exists no position with more than one robot. So, with the multiplicity detection, Procedure $SP$ provides a valid initial configuration for every solution for PFP and GP. In the next two subsections, we show how Procedure $SP$ can be used to provide self-stabilizing algorithms for PFP and GP.

## 4.1     Pattern Formation Problem

This problem consists in the design of protocols allowing the robots to form a specific class of patterns.

Let Procedure $A_{PF}(C)$ be a deterministic algorithm in $SSM$ allowing the robots to form a class of pattern $C$. Algorithm 2 shows Procedure $SSA_{PF}(C)$, which can form all the patterns in $C$ starting from any arbitrary configuration.

---

**Algorithm 2.** Procedure $SSA_{PF}(C)$ for any robot $r_i$

---

**if** there exists at least **one** position with a strict multiplicity
**then**   $SP$;
**else**    $A_{PF}$;

---

**Theorem 2.** *Procedure $SSA_{PF}(C)$ is a self-stabilizing protocol for the Pattern Formation Problem in SSM with a probability equal to 1.*

## 4.2   Gathering Problem

This problem consists in making $n \geq 2$ robots gathering in a point (not predetermined in advance) in a finite time. In [19], it has been proved that GP is deterministically unsolvable in $SSM$ and CORDA. In fact, one feature that the robots must have in order to solve GP is multiplicity detection [21,6,5]. Nevertheless, even with the ability to detect multiplicity, GP remains unsolvable, in a deterministic way, for $n = 2$ in $SSM$ [21]. For all the other cases ($n \geq 3$), GP is solvable. So, when $n \geq 3$, the common strategy to solve GP is to combine two subproblems which are easier to solve. In this way, GP is separated to two distinct steps:

1. Starting from an arbitrary configuration wherein all the positions are distinct, the robots must move in such a way as to create exactly one position with at least two robots on it;
2. Then, starting from there, all the robots move toward that unique position with a strict multiplicity.

As for the deterministic algorithms solving PFP, the deterministic algorithm solving GP ($n \geq 3$) requires that the robots are arbitrarily placed in the plane but with no two robots in the same position. Let Procedure $A_{GP}$ be a deterministic algorithm solving GP, for $n \geq 3$, with multiplicity detection in $SSM$. Algorithm 3 shows Procedure $SSA_{GP}$, which solves GP with multiplicity detection starting from any arbitrary configuration if $n \geq 3$. Note that it is paradoxical that to make GP self-stabilizing, the robots must scatter before gathering.

---

**Algorithm 3.** Procedure $SSA_{GP}$ for any robot $r_i$, $n \geq 3$

**if** there exist at least **two** positions with a strict multiplicity
**then**  $SP$;
**else**  $A_{GP}$;

---

**Theorem 3.** *Procedure $SSA_{GP}$ is a self-stabilizing protocol for the Gathering Problem in SSM with a probability equal to 1 whether $n \geq 3$.*

Note that, for case $n = 2$, we can provide a randomized algorithm solving GP. Informally, when any robot becomes active, it chooses to move to the position of the other robot with a probability of $\frac{1}{2}$. By using a similar idea as in the proof of Lemma 3, we can prove that both robots eventually occupy the same position with a probability of 1. By combining our basic routine for $n = 2$ with Procedure $SSA_{GP}$, we obtain a procedure which solves the self-stabilizing GP with multiplicity detection starting from any arbitrary configuration. It follows that:

**Theorem 4.** *There exists a self-stabilizing protocol for the Gathering Problem in SSM with a probability equal to 1 for any $n \geq 2$.*

## 5  Epilogue

We have shown that the Scatter Problem cannot be deterministically solved. We have proposed a randomized self-stabilizing algorithm for this problem. We have used it to design a self-stabilizing version of any deterministic solution for the Pattern Formation and the Gathering problems.

*June 5th, 2007, 23:20, Beau Rivage Hotel, Geneva.*

**Fig. 2.** The signs drawn on the piece of paper

Robert Langdon was lying on his bed, staring into space. He had just put down the manuscript revealing the above results down on the bedside table. Landgon felt confused. He had not yet made the connection between the murder of the scientist and the results about the Scatter Problem. Disappointed and tired, he was prepared to go to sleep when someone pounded on his door. *'Mr Langdon? I've got a message from Mr Fache for you.'* a female voice said behind the door. Landgon rushed to open the door. He took the piece of paper given by the women. Odd signs were drawn one one side of the paper—refer to Figure 2. On the other side, Langdon read :

*Pasquini Castel, Castiglioncello.*

*'My God! The Etruscans!'* Langdon felt his hairs raise on his arms. *'I should have thought of this before!'*

Langdon picked up the telephone. The male receptionist in the lobby answered. *'Good evening. May I help you, Sir?'*

*'Could you please book a seat on the next flight for Pisa, Italy? Thanks.'*

## Acknowledgments

We are grateful to the anonymous reviewers for their valuable comments. Also, we would like to thank Lionel Guyot who helped to improve the presentation of the paper.

## References

1. Ando, H., Oasa, Y., Suzuki, I., Yamashita, M.: A distributed memoryless point convergence algorithm for mobile robots with limited visibility. IEEE Transaction on Robotics and Automation 15(5), 818–828 (1999)
2. Aurenhammer, F.: Voronoi diagrams- a survey of a fundamental geometric data structure. ACM Comput. Surv. 23(3), 345–405 (1991)

3. Brown, D.: Angels and Demons. Pocket Star (2001)
4. Brown, D.: Da Vinci Code. Doubleday (2003)
5. Cieliebak, M., Flocchini, P., Prencipe, G., Santoro, N.: Solving the robots gathering problem. In: Baeten, J.C.M., Lenstra, J.K., Parrow, J., Woeginger, G.J. (eds.) ICALP 2003. LNCS, vol. 2719, pp. 1181–1196. Springer, Heidelberg (2003)
6. Cieliebak, M., Prencipe, G.: Gathering autonomous mobile robots. In: 9th International Colloquium on Structural Information and Communication Complexity (SIROCCO 9), pp. 57–72 (2002)
7. Defago, X., Konagaya, A.: Circle formation for oblivious anonymous mobile robots with no common sense of orientation. In: 2nd ACM International Annual Workshop on Principles of Mobile Computing (POMC 2002), pp. 97–104 (2002)
8. Dieudonné, Y., Labbani, O., Petit, F.: Circle formation of weak mobile robots. In: Datta, A.K., Gradinariu, M. (eds.) SSS 2006. LNCS, vol. 4280, pp. 262–275. Springer, Heidelberg (2006)
9. Dieudonné, Y., Petit, F.: Circle formation of weak robots and Lyndon words. Information Processing Letters 104(4), 156–162 (2007)
10. Dieudonné, Y., Petit, F.: Swing words to make circle formation quiescent. In: Fourteenth Colloquium on Structural Information and Communication Complexity (SIROCCO '07), Springer, Heidelberg (2007)
11. Dijkstra, E.W.: Self stabilizing systems in spite of distributed control. Communications of the Association of the Computing Machinery 17, 643–644 (1974)
12. Dolev, S.: Self-Stabilization. The MIT Press, Cambridge (2000)
13. Flocchini, P., Prencipe, G., Santoro, N., Widmayer, P.: Hard tasks for weak robots: The role of common knowledge in pattern formation by autonomous mobile robots. In: Aggarwal, A., Pandu Rangan, C. (eds.) ISAAC 99. LNCS, vol. 1741, pp. 93–102. Springer, Heidelberg (1999)
14. Flocchini, P., Prencipe, G., Santoro, N., Widmayer, P.: Gathering of autonomous mobile robots with limited visibility. In: Ferreira, A., Reichel, H. (eds.) STACS 2001. LNCS, vol. 2010, pp. 247–258. Springer, Heidelberg (2001)
15. Flocchini, P., Prencipe, G., Santoro, N., Widmayer, P.: Pattern formation by autonomous robots without chirality. In: VIII International Colloquium on Structural Information and Communication Complexity (SIROCCO 2001), pp. 147–162 (2001)
16. Herman, T.: Private communication (2006)
17. Katreniak, B.: Biangular circle formation by asynchronous mobile robots. In: Pelc, A., Raynal, M. (eds.) SIROCCO 2005. LNCS, vol. 3499, pp. 185–199. Springer, Heidelberg (2005)
18. Prencipe, G.: Corda: Distributed coordination of a set of autonomous mobile robots. In: ERSADS 2001, pp. 185–190 (2001)
19. Prencipe, G.: Instantaneous actions vs. full asynchronicity: Controlling and coordinating a set of autonomous mobile robots. In: Restivo, A., Ronchi Della Rocca, S., Roversi, L. (eds.) ICTCS 2001. LNCS, vol. 2202, pp. 185–190. Springer, Heidelberg (2001)
20. Suzuki, I., Yamashita, M.: Agreement on a common x-y coordinate system by a group of mobile robots. Intelligent Robots: Sensing, Modeling and Planning, pp. 305–321 (1996)
21. Suzuki, I., Yamashita, M.: Distributed anonymous mobile robots - formation of geometric patterns. SIAM Journal of Computing 28(4), 1347–1363 (1999)

# The Traveling Beams
# Optical Solutions for Bounded NP-Complete Problems[*]
## (Extended Abstract)

Shlomi Dolev and Hen Fitoussi

Department of Computer Science, Ben-Gurion University of the Negev, Israel
{dolev,henf}@cs.bgu.ac.il

**Abstract.** Architectures for optical processors designed to solve bounded instances of NP-Complete problems are suggested. One approach mimics the traveling salesman by traveling beams that simultaneously examine the different possible paths. The other approach uses a pre-processing stage in which $O(n^2)$ masks are constructed, each representing a different edge in the graph. The choice and combination of the appropriate (small) subset of these masks yields the solution. The solution is rejected in cases where the combination of these masks totally blocks the light and accepted otherwise. We present detailed designs for basic primitives of the optical processor. We propose designs for solving Hamiltonian path, Traveling Salesman, Clique, Independent Set, Vertex Cover, Partition, 3-SAT, and 3D-matching.

## 1   Introduction

The basic element used for computing is a switching element [7]. One such basic element in the scope of electronic circuitry is the transistor that is used to implement basic logic gates, such as logical *and* and logical *or*. The technology today seeks multi-core solutions in order to cope with the clock frequency limitations of VLSI technology. Namely, to implement on a single chip parallel/distributed system where a bus is used for communication among the processing units. Therefore, the communication overhead associated with distributed/parallel processing would be reduced dramatically. One may take the multi-core technology to the extreme — having a very large number of cores that are incorporated into the processing by sending signals over high-speed buses, maybe using optical/laser communication instead of traditional buses [10].

Optical communication may be chosen due to the free space transmission capabilities of laser beams or the need to transmit signals from/to the processing unit through fiber optic channels. In such cases one may try to avoid the optical to digital conversion (and the digital to optical conversion) and use optical

---

[*] Partially supported by the Lynne and William Frankel Center for Computer Science and the Rita Altura Trust Chair in Computer Science.

P. Crescenzi, G. Prencipe, and G. Pucci (Eds.): FUN 2007, LNCS 4475, pp. 120–134, 2007.

switches for computing. The straightforward solution is to implement optical logic gates, such as logical *and* gate and logical *or* gate, a design that directly maps the current VLSI design to an all optical processor [4,9].

On the other hand, electronic computers are not structured as mechanical computers, such as the Babbage machine [6], and it is possible that optical computers should be designed differently as well. In fact, some success in using many beams in free space for computing has been recently reported [8]. The design of [8] is based on parallel optical multiplication. The use of similar multiplication devices to solve bounded NP-Complete problems is suggested in [12], where a device which is designed to solve NP-Complete problems bounded by $n$ is able to solve instances of size smaller or equal to $n$. Still, use of the fact that beams propagate in three dimensions is limited in the architectures of [8,12] as the propagation of beams is in approximately the same direction. The beam traversal time is not used in the architectures that use Multiplication; we propose to use the space in multi directional fashion and use the time dimension as well.

The seminal work of [11] demonstrates the mapping between beam propagation and the computation of the deterministic Turing machine. In [2] use of a mapping similar to the non-deterministic Turing machine by (amplifying and) splitting beams is suggested. The mapping can be viewed as a theoretical existence proof for a solution, rather than an efficient solution. Knowing that a solution for a (bounded) NP-Complete problem instance exists, we seek for the most efficient solution in terms of the number of beams used, the number of optical elements (or location in space used to represent a computation state), the energy needed in terms of the maximum number of beams that should be split from a single source beam (fan-out), and the number of locations a beam needs to visit (and possibly split) from its creation until its final detection of arrival. Note that we are not solving every instance of an NP-Complete problem in a polynomial time, but we suggest that an optical approach may be promising in solving larger instances of hard combinatorial problems. We extend the design suggested in [2] and present a design for all (six) basic NP-Complete problems listed in [5]: Hamiltonian Path, Clique, Vertex Cover, Partition, 3-SAT, and 3D-matching. Note that polynomial reduction between NP-Complete problems is not used here, since we are concerned with the constant blowups in the instances size; solving the largest possible instance of each of these problems.

There are two main approaches used for the traveling beam Architecture; the first is a mapping of the graph nodes to physical locations in space and propagation of beams according to the edges of the input graph instance. The second approach propagates beams along a computation tree such that the leaves represent all possible solutions and the delay in propagation from the root to each leaf corresponds to the "value" of the specific combination. We also present a totally different architecture called the *coordinated holes in mask-made-blackbox*. In this architecture a set of masks with "holes" are chosen from $n^2 - n$ precomputed masks, according to the input instance. A solution exists only if the combined masks do not block all beams.

**Paper organization.** The next section describes the settings used for our designs. Section 3 details the design for the six basic NP-Complete problems. In particular, we extend the discussion on solving the Hamiltonian-path to gain some intuitive insight that is later used in describing the solutions for the next five problems. In Section 4 we present a totally different architecture for solving the Hamiltonian-path. Section 5 concludes the paper. Proofs and details are omitted from this extended abstract and can be found in [3].

## 2    Settings of the Optical Computing Device

The optical micro-processor simulates a non-deterministic Turing machine. In a deterministic Turing machine, the next configuration is uniquely defined by applying the transition function $\delta$ to the current configuration (the transition function $\delta$ defines for each configuration the next configuration it yields in the computation). In a non-deterministic Turing machine the transition relation $\delta$ is a set of pairs of configurations such that $(c_1, c_2) \in \delta$ if and only if a configuration $c_1$ yields configuration $c_2$, meaning it is possible to go from the first configuration to a number of different configurations in a single move.

We will think of the non-deterministic computation as a directed graph. In this graph, each configuration is represented by a node, a directed edge connects two nodes $v_1$ and $v_2$ if it is possible to go from the configuration $v_1$ represents to the configuration $v_2$ represents in a single move. The graph constructed is a tree, in which the initial configuration is the root, and all the final configurations are the leaves.

In the optical micro-processor, each configuration corresponds to a 3D *location*. Laser or other beam creator (i.e., electronic beam) is used as a source at the initial configuration. The transition from one configuration to all the following configurations simultaneously is simulated by splitting the light at each *location* and propagating it in parallel to all the following configurations, as determined by the transition relation $\delta$. According to the input, we may use barriers to block light between two configurations so that the transition between them is not allowed, or we may use an arrangement similar to an almost parallel arrangement of mirrors that will delay the transition of light between two configurations.

The constructed optical micro-processor enables us to use the parallel qualities of light to explore all possible paths of computations simultaneously. Light at a certain *location* indicates the feasibility of the correlated computation, while the absence of light indicates that the computation is not feasible. We will (possibly use a prism to direct all outputs to a single *location*, and) use light detectors at the leaves *locations*, in order to check whether and when light arrived at one of the final accepting configurations. According to this information, we will decide if the output is *accepted*, or the output is *rejected*.

In the sequel, we will use the term *column* to describe a group of *locations* that share the same $(x,y)$ coordinates in space. These *locations* usually represent configurations that share a common attribute in the computation.

In the following section we describe the architectures designed to solve the basic NP-Complete problems that are presented in [5]. The design (of the Hamiltonian-path architecture) is based on [2], we add an analysis of the *fan-in, fan-out, efficiency factor*, where the *efficiency factor* is defined as the number of *locations* that light may reach divided by the number of possible solutions, and *depth*, where the *depth* is the maximal number of *locations* a beam traverses from the source to the detectors.

# 3    Architectures for Basic NP-Complete Problems

## 3.1    Hamiltonian-Path

Given as input a directed graph $G = (V, E)$, the objective is to determine whether $G$ contains a Hamiltonian-path.

**Definition 1.** *A path is a sequence of vertices such that from each vertex there is an edge to the next vertex in the sequence. A* Hamiltonian-path *is a path which visits each vertex of the graph exactly once.*

**The architecture.** The configurations in this architecture represent different paths of length zero to $n$, where $n$ is the number of vertices in the graph. The initial configuration is a path of length zero, the final configurations are all possible paths of length $n$.

**The arrangement of the configurations in space.** The *locations* are ordered in $n$ columns, $c_1, \ldots, c_n$, one column per each vertex. *Location* in column $c_i$ represents a path that ends in vertex $v_i$. In addition to the arrangement in $n$ columns, the *locations* are arranged in $n$ levels, $level_1, \ldots, level_n$. Configurations in $level_i$ represent paths of length $i$. *Locations* in column $c_j$ and level $level_i$ correspond to all different paths of length $i$ which end in vertex $v_j$.

The following procedure is used to determine which path corresponds to a *location* in column $c_j$ level $level_i$:

• As stated before, the level of the *location* determines the length of the path, the column of the *location* determines the last vertex in the path. This *location* represents a path of size $i$, which ends in vertex $v_j$.

• In order to determine the previous vertex in the path, we divide the array of *locations* in this column and level into $n$, and check in which part (sub-array) our *location* appears. A *location* which appears in part $k$ indicates that the previous vertex is $v_k$.

• We will repeat the former action $i - 1$ times, each time with the sub-array we obtained earlier, until we will determine the entire path.

The column of vertex $v_1$ in a graph with three vertices is shown in Fig. 1. The black and white colors of the *locations* represent the mask that will be defined later. Sub-array$_i$ corresponds to *locations* for which the incoming beams arrive from vertex $v_i$. The *locations* in this column are ordered in three levels – $level_1$, $level_2$, and $level_3$. We will demonstrate how to determine which path corresponds to the *location* marked by the dot. The *location* appears in $level_3$ in the column of $v_1$. Therefore the path is a path of length three which ends in vertex $v_1$.

We divided $level_3$ into three sub-arrays; the *location* appears in the second sub-array, therefore the previous vertex in the path is $v_2$. We continue and divide the second sub-array into three sub-arrays. The *location* appears in the third sub-array, meaning the previous vertex is $v_3$ and the path represented by this *location* is $(v_3 \rightarrow v_2 \rightarrow v_1)$.

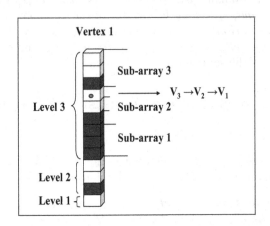

**Fig. 1.** Illustration of the Hamiltonian-path column

**Creating the mask.** We would like to block the light from reaching *locations* which represent paths that contain a repetition of a vertex. In order to do so, we use a *mask*. We view a mask as a screen (e.g., transparency) of material carrying patterns that are either transparent or opaque to the wavelengths used; the mask is created *a priori* regardless of the input graph. The mask will allow further propagation of light to *locations* that represent a feasible path, and will block the light from reaching *locations* which represent paths that contain a repetition of a vertex.

The algorithm of creating the mask is presented in Fig. 2. It is an iterative optical process which occurs level after level. We copy the mask of the previous level $n$ times, then we divide the array of *locations* (in this level) into $n$, and blacken the $i$ part. Although the number of *locations* in a mask is exponential in $n$, the creation of the mask can be done by $n^3$ iterated processes of optical copying in a way similar to the process suggested in [12].

The masked column of vertex $v_1$ is shown in Fig. 1. *Locations* colored in black indicate that the mask is opaque to the wave length used and the light is blocked. *Locations* colored in white indicate that the mask is transparent.

> **Procedure** $CreateMask(c_j)$ :
> 1. **for** $level_i = level_1$ to $level_n$ **do**
> 2.     **if** $level_i = level_1$
> 3.         **then** the mask is transparent
> 4.     **else**
> 5.         **for** $k = 1$ to $n$ **do**
> 6.             copy the mask of $level_{i-1}$ from column $c_j$
> 7.         blacken part $j$ of the mask of $level_i$

**Fig. 2.** Creating the mask

mask is transparent. According to the algorithm, $level_1$ is transparent. The second level is created by copying the transparent mask of the first level three times, then dividing $level_2$ into three, and blackening the first part. The creation of the third level is done by copying the mask of the second level three times, then dividing $level_3$ into three, and blackening the first part.

To create a mask for a certain level we have to copy the former level $n$ times; this takes $n$ iterations. There are $n$ columns, $n$ levels in each column, so the whole process of creating the mask takes $n^3$ iterations.

**The transition relation.** At the initial configuration the beam of light is split into $n$ beams which propagate simultaneously to $level_1$ in $n$ different columns. In this way we get configurations that represent $n$ different paths of length one (the configuration in column $c_i$ represents the path $(v_i)$).

In all *locations* (except the initial one) the beam of light is amplified and split into $(n-1)$ beams. These beams propagate to $(n-1)$ configurations that are located in the other $(n-1)$ columns. Light propagates from a configuration in $level_i$ to $(n-1)$ configurations in $level_{i+1}$. Configuration located in column $c_j$ will propagate light to part $j$ of the next level. The yielding configurations represent the extension of the path (represented by the first *location*) with the vertex each column represents. For example, in the *location* that represents the path $(v_1)$, the beam of light is split into $(n-1)$ beams which propagate simultaneously to $(n-1)$ *locations* in $level_2$ in columns $c_2, \ldots, c_n$; in this way we get configurations that represent $(n-1)$ different paths of length two (for example, the configuration in column $c_i$ represents the path $(v_1 \rightarrow v_i)$).

The transition relation is determined according to the input graph $G$. If the edge $(v_i, v_j) \notin G$ then the transition between a configuration in column $c_i$ to a configuration in column $c_j$ is not legal and should be blocked. In order to block light between configurations we will use $n(n-1)$ barriers, one barrier between every possible pair of columns. If $(v_i, v_j) \notin G$ then a barrier will block the light that propagates from column $c_i$ to column $c_j$.

Detectors in $level_n$ will indicate if a beam arrived to a *location* $l$ in $level_n$ in one of the columns. If the detector sensed a beam then there exists a path of length $n$ with no vertex repetitions, and the transition between any two sequential vertices in the path represented by the *location* $l$ is allowed. Meaning, there exists a Hamiltonian-path in the graph and the output is *accepted*, otherwise the output is *rejected*.

The Hamiltonian-path architecture for a graph with three vertices is illustrated in Fig. 3. Vertices $v_1$ and $v_2$ are not connected, thus a barrier blocks the light between column $c_1$ and column $c_2$. The mask is transparent in *locations* colored in white, and opaque in *locations* colored in black. Light propagates from the initial configuration to $level_1$ in all three columns. To simplify the illustration, we show further propagation from column $c_1$ only. In column $c_1$ the beam is split into two beams. The beam directed to column $c_2$ is blocked by the barrier, the other beam arrives at $level_2$ in column $c_3$, then the beam is split into two beams which propagate simultaneously to $level_3$ in columns $c_1$ and $c_3$. The *location* in column $c_1$ represents the path $(v_1 \rightarrow v_3 \rightarrow v_1)$ which is not feasible; the mask blocks the light from reaching this *location*. The *location* in column

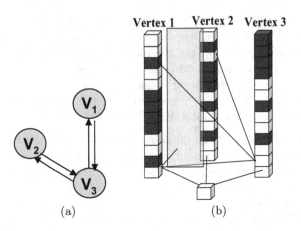

**Fig. 3.** Illustration of the Hamiltonian-path architecture

$c_2$ represents the path $(v_1 \rightarrow v_3 \rightarrow v_2)$ which is a Hamiltonian-path, therefore the output is *accepted*.

A way to obtain a solution to the Traveling Salesman Problem from the Hamiltonian-path architecture is described in [3].

**Lemma 1.** *In the Hamiltonian-path architecture:*
• *max(fan-out) = n.* • *max(fan-in) = one.* • *efficiency factor = n.*
• *depth = n.*

### 3.2   Clique

Given as input an undirected graph $G = (V, E)$, and an integer $k$, the objective is to determine whether $G$ contains a clique of size $k$.

**Definition 2.** *A* clique *is a set of vertices* $S \subseteq V$*, such that* $\forall v_1, v_2 \in S$*,* $(v_1, v_2) \in E$*.*

**The architecture.** The configurations in this architecture represent different subsets of vertices of size zero to $n$, where $n$ is the number of vertices in the graph. The initial configuration is the empty set, the final configurations are all possible subsets.

The architecture is built out of $n$ *vertex architectures*, and an additional column, the AND column. The vertex architecture of $v_i$ will indicate for each set $S$, such that $v_i \in S$, whether all the other vertices in $S$ are connected to $v_i$. A set $S$ is a clique if and only if $\forall v_i \in S$ all the other vertices in $S$ are connected to $v_i$. The AND column will indicate whether the set is a clique.

**The arrangement of the configurations in space**

**Vertex architecture.** The configurations in the vertex architecture are ordered as a tree, with $n + 1$ columns, where configurations in depth $i$ represent all possible subsets of $v_1 \ldots v_i$. The left half of the configurations in depth $i$ correspond to subsets $S$ such that $v_i \in S$, where the right half of the configurations in depth $i$ correspond to subsets which do not contain $v_i$.

The propagation of beams between configurations is done in the following manner. In each *location* the beam is amplified and split into two beams. These beams propagate to two configurations that are located in the next column. Meaning, beam propagates from a configuration in depth $i$ which corresponds to the set $S$ to two configurations in depth $i + 1$, the transition to the right configuration represents the set $S$, and the transition to the left configuration represents the set $\{v_{i+1}\} \cup S$.

For example, in the *location* that represents the initial configuration, meaning the empty set, the beam is split into two beams of light which propagate simultaneously to two *locations* in the first column. The right configuration corresponds to the empty set, the left configuration corresponds to the set $\{v_1\}$.

**The transition relation.** In order to estimate the size of the sets, we will delay the light beams that pass through certain *locations* by $d$ time units. The delay is done in any transition to the right, thus light will reach a configuration in the final column, which corresponds to a set $S$ of size $i$, after $d \cdot (n - i)$ time units (a delay of $d$ time units per each vertex $v_i \notin S$). In addition, we will use barriers to block the light from reaching configurations corresponding to sets which are not a clique.

In each vertex architecture $v_i$: If $(v_i, v_j) \notin E$, we will block the light from reaching configurations that correspond to sets that contain $v_j$, we will place the barriers before the left half of *locations* of depth $j$. In addition, we want to block the light from reaching *locations* corresponding to sets that do not contain $v_i$. In order to achieve this goal we will place a barrier before the right half of the *locations* in depth $i$. In this way, light will reach a leaf configuration corresponding to a set $S$ in the vertex architecture of $v_i$, only if $v_i \in S$ and all the other vertices in $S$ are connected to $v_i$.

We will number the leaf *locations* and the *locations* in the AND column as $l_1 \ldots l_{2^n}$. Light propagates from a leaf *location* $l_j$ in the vertex architecture to *location* $l_j$ in the AND column. In the AND column we will use threshold, just like choosing photographic film sensitivity, to block the light in *locations* where less than $k$ beams of light arrived. In this way, sets which are not clique are blocked. A lens will concentrate the outgoing beams of the AND column to a single detector that will indicate whether a beam arrived after $d \cdot (n - k - 0.5)$ time units (the sensor is activated after $d \cdot (n - k - 0.5)$ time units to ignore sets with size smaller than $k$; light from such sets will arrive at the detector after $d \cdot (n - k)$ time units). If the detector sensed a beam after $d \cdot (n - k - 0.5)$ time units, then the output is *accepted*, otherwise the output is *rejected*.

The clique architecture is illustrated in Fig. 4. *Locations* colored in gray delay the light, *locations* colored in black block the light. Threshold on the number of incoming beams to *locations* in the AND column allow the light to propagate to the detector only if all $k$ incoming beams arrive.

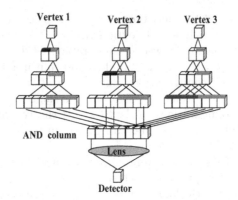

**Fig. 4.** Illustration of the clique architecture

Determining which set $S$ corresponds to a configuration is done by traversing the tree. A turn to the right in depth $i$ in the path from the root to this leaf indicates that $v_{i+1} \notin S$; a left turn indicates that $v_{i+1} \in S$.

A way to obtain a solution to the independent-set problem from the clique architecture is described in [3].

**Lemma 2.** *In the clique architecture:*
• *max(fan-out) = two.* • *max(fan-in) = n.* • *efficiency factor = 2n + 1.*
• *depth = n + 1.*

### 3.3   Vertex Cover

Given as input an undirected graph $G = (V, E)$, and an integer $k$, the objective is to determine whether $G$ contains a vertex cover of size $k$.

**Definition 3.** *A* vertex cover *is a set of vertices $S \subseteq V$, such that $\forall v_i \notin S$, $\exists v_j \in S$ for which $(v_i, v_j) \in E$.*

In order to solve this problem we will use the following observation:

**Observation 1.** *There exists a vertex cover of size $k$ in graph $G$, if and only if there exists a clique of size $n - k$ in the complementary graph $\overline{G}$.*

Given a graph $G$ and an integer $k$, we will use the clique architecture with $\overline{G}$ and $n - k$ as an input. This reduction does not increase the size of the input.

**Lemma 3.** *In the vertex cover architecture:*
• *max(fan-out) = two.* • *max(fan-in) = n.* • *efficiency factor = 2n + 1.*
• *depth = n + 1.*

### 3.4   Partition

Given a set of integers $S = \{a_1, \ldots, a_n\}$, the objective is to determine whether there exists a subset $S_1 \subset S$ such that $\sum_{a_i \in S_1} a_i = \sum_{a_i \in S \setminus S_1} a_i$.

**The architecture.** The configurations in this architecture represent different subsets of length zero to $n$, where $n$ is the size of the set $S$. The initial configuration is the empty set, the final configurations are all possible subsets.

**The arrangement of the configurations in space.** The configurations are ordered as a tree, where configurations in depth $i$ represent all possible subsets of $a_1, \ldots, a_i$. The right half of the configurations in depth $i$ correspond to sets that do not contain $a_i$, where the left half of the configurations correspond to sets that contain $a_i$.

**The transition relation.** In each *location* the beam of light is amplified and split into two beams. These beams propagate into two configurations that are located in the next column. The transition to *locations* that represent sets $S_1$ such that $a_i \in S_1$ delays the light by $a_i$ time units. Meaning, light propagates from a configuration in depth $i$ which corresponds to the set $S_i$ to two configurations in depth $i + 1$. The transition to the right configuration represents the set $S_i$. The transition to the left configuration represents the set $\{a_{i+1}\} \cup S_i$, light is delayed by $a_{i+1}$ time units in this transition.

For example, in the *location* that represents the initial configuration, meaning the empty set, the beam is split into two beams of light which propagate simultaneously to two *locations* in the first column. The right configuration corresponds to the empty set (and does not delay the light), the left configuration corresponds to the set $\{a_1\}$ and the transition to this configuration delays the light by $a_1$ time units.

The partition architecture for a set of three integers $\{a_1, a_2, a_3\}$ is illustrated in Fig. 5. The transition to a configuration in depth $i$ which is colored in gray delays the light in $a_i$ time units. Light reaches the final configuration which correlates to the set $\{a_1, a_2, a_3\}$ after $a_1 + a_2 + a_3$ time units.

A partition of the set $S$ exists if and only if a beam of light reached one of the leaf configurations after $\frac{\sum_{a_i \in S} a_i}{2}$ time units.

We will use a lens to concentrate the light that goes from the leaf configurations to a single detector. The detector will be used to identify arriving beam after $\frac{\sum_{a_i \in S} a_i}{2}$ time units. If a beam arrived after $\frac{\sum_{a_i \in S} a_i}{2}$ time units then the output is *accepted*, otherwise the output is *rejected*.

Determining which subset corresponds to a leaf configuration is done by traversing the tree. A turn to the right in depth $i$ in the path from the root to this leaf indicates that $a_{i+1}$ is in the subset, and a left turn indicates that $a_{i+1}$ is not in the subset.

**Lemma 4.** *In the partition architecture:*
- *max(fan-out) = two.* • *max(fan-in) = one.* • *efficiency factor = two.*
- *depth = n.*

## 3.5   3-SAT

Given a 3-CNF formula $\varphi$ as input, the objective is to determine whether $\varphi$ is satisfiable.

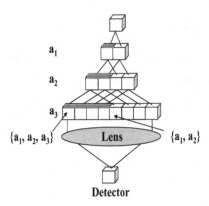

**Fig. 5.** Illustration of the partition architecture

**Definition 4.** *Formula $\varphi$ is* satisfiable *if and only if there exists an assignment for the formula's variables under which $\varphi$ evaluates to* true.

**The architecture.** Configurations in this architecture represent different partial assignments of size zero to $n$, where $n$ is the number of variables. A partial assignment of size $i$ assigns values to the variables $x_1 \ldots x_i$. The initial configuration is the empty assignment, the final configurations are all possible assignments of $x_1 \ldots x_n$.

The architecture is built out of $l$ *clause architectures*, where $l$ is the number of clauses, and an additional column, the AND column. In order to determine whether $\varphi$ is satisfiable, we will check separately whether the assignment satisfies each of the formula's clauses (using the clause architecture). Assignment satisfies $\varphi$ if and only if it satisfies all $\varphi$'s clauses; the AND column will indicate whether the assignment satisfies all clauses.

**The arrangement of the configurations in space**

**Clause architecture.** As demonstrated in Fig. 6, the configurations in the clause architecture are ordered as a tree, where configurations in depth $i$ represent all possible assignments of $x_1 \ldots x_i$. The right half of the configurations in depth $i$ correspond to assignments where $x_i$ evaluates to true. The left half of the configurations in depth $i$ correspond to assignments where $x_i$ evaluates to false.

The propagation of light between configurations is done in the following manner. In each *location* the beam of light is amplified and split into two beams. These beams propagate to two configurations in the column of the next variable. The yielding configurations represent the extension of the assignment represented by the first *location*. In the right transition the value of the next variable is true, and in the left transition it is false. For example, in the *location* that represents the initial configuration, meaning the empty assignment, the beam of light is split into two beams which propagate simultaneously to two *locations* in the first column. In this way we get the right configuration in the first column that represents the partial assignment of size one $x_1 =$ true, and the left configuration in the first column for the partial assignment $x_1 =$ false.

We designed two methods for implementing the transition relation: The first method uses the delay of light, the second method uses a reduction in light intensity. The details of the second method are omitted from this extended abstract and can be found in [3].

**The transition relation (with delay).** In this architecture we will delay the light that goes through certain *locations* by $d$ time units.

In each clause architecture $c$: If $x_i \in c$ then light that goes through the left half of column $i$ is delayed for $d$ time units; if $\overline{x_i} \in c$ the light that goes through the right half of column $i$ is delayed by $d$ time units. The idea is to delay the light by $3d$ time units in assignments where all three literals are assigned to `false`.

We will number the leaf *locations* and the *locations* in the AND column as $l_1 \ldots l_{2^n}$. Light propagates from leaf *location* $l_j$ to *location* $l_j$ in the AND column. In the AND column we will use threshold to block the light in *locations* to which less than $l$ beams of light arrived, where $l$ is the number of clauses in the formula. The idea is to block the assignments where at least one clause evaluates to `false`.

A lens will be used to concentrate all beams from the AND column to one *location* where a detector will indicate whether a beam arrived after $2.5d$ time units. If there exists a *location* in the AND column where all $l$ beams arrived in less than $3d$ time units, then there exists a satisfying assignment. If a beam arrived at the detector after $2.5d$ time units, then the output is *accepted*, otherwise the output is *rejected*.

The 3-SAT architecture for the formula $(x_1 \vee x_2 \vee \overline{x_3}) \wedge (\overline{x_1} \vee \overline{x_2} \vee x_3)$ is illustrated in Fig. 6. A configuration colored in gray delays the light by $d$ time units.

Determining which assignment corresponds to a certain configuration is done by traversing the tree. A turn to the right in depth $i$ of the path from the root to this leaf indicates that the value of $x_{i+1}$ is `true`, and a left turn indicates that it is `false`.

**Lemma 5.** *In the 3-SAT architecture:*
- *$max(fan\text{-}out) = two.$* • *$max(fan\text{-}in) = l.$* • *efficiency factor $= 2l + 1.$*
- *depth $= n + 1.$*

### 3.6   3D Matching

This problem is a generalization of the bipartite matching problem. In the 3D-matching problem, the input is three sets $B$, $G$, and $H$ (boys, girls, and homes), such that $|B| = |G| = |H| = n$, and a set of triples $T$, $T \subseteq B \times G \times H$. The objective is to determine whether there exists a set of $n$ triples in $T$, such that each boy is matched to a different girl, and each couple has home of its own. We will refer to such an arrangement as a perfect match.

**Definition 5.** *A perfect match is a set of triples $S \subseteq T$, such that $\forall o_i \in B \cup G \cup H$, $o_i$ appears exactly in one triple in $S$.*

We will say that a triple $t_i$ is *consistent* with triple $t_j$ if and only if $\forall o_i \in t_i \Rightarrow o_i \notin t_j$. We will construct a graph $G$ where each triple is represented by a vertex. An edge connects $v_i$ and $v_j$ if and only if the triple $v_i$ is consistent with triple $v_j$.

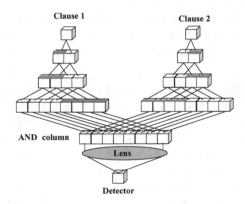

**Fig. 6.** Illustration of the 3-SAT architecture

**Observation 2.** *There exists a perfect match if and only if there exists a clique of size $n$ in the constructed graph $G$.*

Given the input sets $B, G, H$, and $T$ we will construct the graph $G$ and use the clique architecture with $G$ and $n$ as an input. This reduction does not increase the size of the input.

**Lemma 6.** *In the 3D-matching architecture:*
• *max(fan-out) = two.* • *max(fan-in) = t.* • *efficiency factor = $2t + 1$.*
• *depth $= t + 1$.*

## 4    Coordinated Holes in Masks-Made-Blackbox

In the following section we present a different architecture for an optical microprocessor. We will demonstrate how this optical processor is used to solve bounded instances of the Hamiltonian-path problem.

In [12] the authors present an iterative algorithm that produces a binary matrix that represents all possible Hamiltonian-paths using optical copying. Every column in the matrix correlates to a possible edge in the graph, every row in the matrix correlates to a possible Hamiltonian-path. Zero (opaque screen) in the $[i][j]$ entry of the matrix indicates that the edge $e_j$ does not appear in path number $i$, where one (transparent screen) in the $[i][j]$ entry indicates that the edge $e_j$ does appear in path number $i$. We will use a similar technique to produce a matrix where zero indicates the existence of an edge, and vise versa.

In our architecture we will use each column of the binary-matrix mask as a barrier. There are $n(n-1)$ possible barriers, a barrier per each possible directed edge. The barrier of edge $e_i$ blocks the light in paths $p$ where $e_i \in p$, and allows the propagation of light in paths $p$ such that $e_i \notin p$.

Given an input graph $G = (V, E)$, we will select a subset of the barriers. If $e_i \in E$ then we do not use the barrier of edge $e_i$. Otherwise, we will use the barrier of edge $e_i$ which will block the light in paths $p$ where $e_i \in p$.

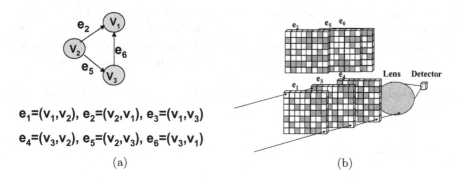

$e_1=(v_1,v_2),\ e_2=(v_2,v_1),\ e_3=(v_1,v_3)$

$e_4=(v_3,v_2),\ e_5=(v_2,v_3),\ e_6=(v_3,v_1)$

(a)                                                    (b)

**Fig. 7.** Illustration of the holes in masks-made-blackbox architecture

The masks that are in use (according to the input) are ordered one behind the other in a way that the entry of the path $p$ in one mask is placed behind the entry of the same path $p$ in the other masks. Note that the masks can be in any 3D shape, for example, nested boxes or balls.

Light propagates from the light source, passes through all the masks, and if possible reaches a detector. A lens will be used to concentrate the light from the final configurations to a single detector. The detector indicates whether light passed all barriers and reached the final configuration. If so, it means that there exists a specific path where the barriers did not block the light from passing. Meaning, all edges in this path exist in the graph. If the detector sensed light then the graph contains a Hamiltonian-path and the output is *accepted*, otherwise the output is *rejected*.

The coordinated holes in masks-made-blackbox architecture for a graph with three vertices is illustrated in Fig. 7. The edges $e_1, e_3$, and $e_4$ do not appear in the graph, thus the correlated barriers are used. To simplify the illustration, we show the propagation of two beams only. The upper left entry correlates to a path that contains the edge $e_3$, thus the mask of $e_3$ in this entry is opaque to the wavelength and the beam is blocked and does not reach the detectors. The lower right entry correlates to a path that does not include edges $e_1, e_3$, and $e_4$, thus this entry is transparent in all three masks, and the beam reaches the detector, which indicates that the graph contains a Hamiltonian-path.

## 5   Concluding Remarks

The advance in optical communication and computing may well serve as a way to cope with the limitations VLSI technologies now face. We suggest ways to use the natural parallelism of wave propagation in space for solving inherent hard problems in the scope of sequential or even parallel electronic computers. The existence of recent industrial attempts to produce optical processing devices (e.g., [8]) as well as the limited implementations in our laboratory (e.g., [12]) encourage us to believe that our new designs will be used in practice for solving combinatorial tasks, at least when there are real-time constraints. Some cryptographic usage of optical processors are suggested in [1,13,14]. We would like to remark that most of our designs solve in fact the #P version of the problem. At last, we view our work as a beginning for further investigations on using beams of light and their location in free space for parallel computations.

**Acknowledgments.** We thank Stephan Messika for discussions in the first steps of this research, and Nati Shaked for fruitful discussions.

# References

1. Dolev, S., Korach, E., Uzan, G.: A Method for Encryption and Decryption of Messages. PCT Patent Application WO 2006/001006 (January 2006)
2. Dolev, S., Yuval, N.: Optical implementation of bounded non-deterministic Turing machines. US Patent 7,130,093 B2, (October 2006)
3. Dolev, S., Fitoussi, H.: The Traveling Beams: Optical Solutions for Bounded NP-Complete Problems. Technical report #07–04, Ben Gurion University of the Negev (January 2007)
4. Feitelson, G.: Optical Computing: A Survey for Computer Scientists. MIT Press, Cambridge (1988)
5. Garey, M.R., Johnson, D.S.: Computers and Intractability, a guide to the theory of NP-completeness. W. H. Freeman and Company, San Francisco (1979)
6. Hyman, A.: Charles Babbage: Pioneer of the Computer. Princeton University Press, Princeton (1982)
7. Kochavi, Z.: Switching and finite automata theory. McGraw-Hill, New York (1978)
8. Lenslet LTD. http://www.hpcwire.com/hpcwire/hpcwireWWW/03/1017/106185.html
9. McAulay, A.D.: Optical computer architectures. John Wiley, Chichester (1991)
10. Rong, H., Liu, A., Jones, R., Cohen, O., Hak, D., Nicolaescu, R., Fang, A., Paniccia, M.: An all-sillicon Raman laser. Nature 433, 292–294 (2005)
11. Reif, J.H., Tygar, D., Yoshida, A.: The Computability and Complexity of Optical Beam Tracing, 31st Annual IEEE Symposium on Foundations of Computer Science, The Computability and Complexity of Ray Tracing. Discrete and Computational Geometry 11, 265–287 (1994)
12. Shaked, N.T., Messika, S., Dolev, S., Rosen, J.: Optical Solution for Bounded NP-Complete Problems. Journal of Applied Optics 46(5), 711–724 (2007)
13. Shamir, A.: Factoring Large Numbers with the TWINKLE device. In: Koç, Ç.K., Paar, C. (eds.) CHES 1999. LNCS, vol. 1717, pp. 2–12. Springer, Heidelberg (1999)
14. Shamir, A., Tromer, E.: Factoring Large Numbers with the TWIRL Device. In: Boneh, D. (ed.) CRYPTO 2003. LNCS, vol. 2729, pp. 1–26. Springer, Heidelberg (2003)

# The Worst Page-Replacement Policy*

Kunal Agrawal[1], Michael A. Bender[2], and Jeremy T. Fineman[1]

[1] MIT, Cambridge, MA 02139, USA
[2] Stony Brook University, Stony Brook, NY 11794-4400, USA

**Abstract.** In this paper, we consider the question: what is the worst possible page-replacement strategy? Our goal is to devise an online strategy that has the highest possible fraction of misses as compared to the worst offline strategy. We show that there is no deterministic, online page-replacement strategy that is competitive with the worst offline strategy. We give a randomized strategy based on the "most-recently-used" heuristic, and show that this is the worst possible online page-replacement strategy.

## 1 Introduction

Since the early days of computer science, thousands of papers have been written on how to optimize various components of the memory hierarchy. In these papers a recurrent question (at least four decades old) is the following: Which page-replacement strategies are the best possible?

The point of this paper is to address the reverse question: *Which page-replacement strategies are the worst possible?* In this paper we explore different ways to formulate this question. In some of our formulations, the worst strategy is a new algorithm that (luckily) has little chance of ever being implemented in software or silicon. In others, the worst strategy may be disturbingly familiar.

We proceed by formalizing the paging problem. We assume a two-level memory hierarchy consisting of a small fast memory, the *cache*, and an arbitrarily large slow memory. Memory is divided into unit-size blocks or *pages*. Exactly $k$ pages fit in fast memory. In order for a program to access a memory location, the page containing that memory location must reside in fast memory. Thus, as a program runs, it makes *page requests*. If a requested page is already in fast memory, then the request is satisfied at no cost. Otherwise, the page must be transferred from slow to fast memory. When the fast memory already holds $k$ pages, one page from fast memory must be evicted to make room for the new page. (The fast memory is initially empty, but once it fills up, it stays full.) The cost of a program is measured in terms of the number of transfers.

The objective of the paging problem is to minimize the number of page transfers by optimizing which pages should be evicted on each page requests. When all

* This research was supported in part by NSF grants CCF 0621439/0621425, CCF 0540897/05414009, CCF 0634793/0632838, CCF 0541209, and CNS 0627645, and by Google Inc.

page requests are known a priori (the offline problem), then the optimal strategy, proposed by Belady, is to replace the page whose next request occurs furthest in the future [2].

More recent work has focused on the online problem, in which the paging algorithm must continually decide which pages to evict without prior knowledge of future page requests. Sleator and Tarjan introduce competitive analysis [10] to analyze online strategies. Let $A(\sigma)$ represent the cost incurred by the algorithm $A$ on the request sequence $\sigma$, and let $\mathrm{OPT}(\sigma)$ be the cost incurred by the optimal offline strategy on the same sequence. For a minimization problem, we say that an online strategy $A$ is *c-competitive* if there exists a constant $\beta$ such that for every input sequence $\sigma$,

$$A(\sigma) \leq c \cdot \mathrm{OPT}(\sigma) + \beta \ .$$

Sleator and Tarjan prove that there is no online strategy for page replacement that is better than $k$-competitive, where $k$ is the memory size. Moreover, they show that the *least-recently-used (LRU)* strategy, in which the page chosen for eviction is always the one requested least recently, is $k$-competitive. If the online strategy operates on a memory that is twice the size of that used by the offline strategy, they show that LRU is 2-competitive. Since this seminal result, many subsequent papers have analyzed paging algorithms using competitive analysis and its variations. Irani [7] gives a good study of many of these approaches.

Page-replacement strategies are used at multiple levels of the memory hierarchy. Between main memory and disk, memory transfers are called *page faults*, and between cache and main memory, they are *cache misses*. There are other differences in between levels besides mere terminology. In particular, because caches must be fast, a cache memory block, called a *cache line*, can only be stored in one or a limited number $x$ of cache locations. The cache is then called *direct mapped* or *x-way-associative*, respectively. There have been several recent algorithmic papers showing how caches with limited associativity can use hashing techniques to acquire the power of caches with unlimited associativity [5, 9].

We are now ready to describe what we mean by the worst page-replacement strategy. First of all, we are interested in "reasonable" paging strategies. When we say that the strategy is *reasonable*, we mean that it is only allowed to evict a page from fast memory when

1. an eviction is necessary to service a new page request, i.e., when the fast memory is full, and
2. the evicted page is *replaced* by the currently requested page.

Without this reasonableness restriction a paging strategy could perform poorly by preemptively evicting all pages from fast memory. In contrast, we explore strategies that somehow try to do the right thing, but just fail miserably.

Thus, our *pessimal-cache problem* is as follows: identify replacement strategies that maximize the number of memory transfers, no matter how efficiently code happens to be optimized for the memory system.

As with traditional paging problems, we use competitive analysis. Since we now have a maximization problem, the definition of competitive is slightly different: An online algorithm $A$ is **c-competitive** if there exists a constant $\beta$ such that for every input sequence $\sigma$,

$$A(\sigma) \geq \frac{1}{c} \cdot \text{OPT}(\sigma) - \beta .$$

An input sequence $\sigma$ consists of a sequence of page requests.[1] The objective of the online algorithm $A$ is to maximize the number of page faults on the input sequence, and OPT is the offline strategy that maximizes the total number of page faults.

Since we are turning the traditional problem on its head, terminology may now seem backwards. Optimal now means optimally bad from a traditional point of view. The adversary is trying to give us an input sequence for which we do not have many more page faults than OPT. Thus, in some sense, the adversary is our friend, who is looking out for our own good, whereas we are trying to indulge in bad behavior.

Note that the pessimal-cache problem still assigns cost in the same way, and thus counts competitiveness in terms of the number of *misses* and not the number of *hits*. We leave the problem in this form because OPT may have no hits, whereas even the best online strategies have infinitely many.

**Results.** In this paper, we present the following results:

- We prove that there is no deterministic, competitive, online algorithm for the pessimal-cache problem (Section 2).
- We show that there is no (randomized) algorithm better than $k$-competitive for the pessimal-cache problem (Section 2).
- We give an algorithm for the pessimal-cache problem that is expected $k$-competitive, and hence optimal (Section 3). Since this strategy exhibits a $1/k$ fraction of the maximum number of page faults on every input sequence, this strategy is the worst page-replacement strategy.
- We next examine page-replacement strategies for caches with limited associativity. We prove that for the direct-mapped caches, the page-replacement is, in fact, worst possible, under the assumption that page locations are random (Section 4).

## 2    Lower Bounds

This section gives lower bounds on the competitiveness of the pessimal-cache problem. We show that no deterministic strategy is competitive. Then we show

---

[1] If a page is requested repeatedly without any other page being interleaved, all strategies have no choice, and there is no page fault on any but the first access. Thus, we consider only sequences in which the same page is repeated only when another page is accessed in between.

that no strategy can be better than expected $k$-competitive, where $k$ is the fast-memory size.

Our first lemma states that there is no deterministic online strategy that is competitive with the offline strategy.

**Lemma 1.** *Consider any deterministic strategy $A$ for the pessimal-cache problem with fast-memory size $k \geq 2$. For any $\varepsilon > 0$ and constant $\beta$, there exists an input sequence $\sigma$ such that $A(\sigma) < \varepsilon \cdot \mathrm{OPT}(\sigma) - \beta$.*

*Proof.* Consider a sequence $\sigma$ that begins by requesting pages $v_1, v_2, \ldots, v_{k+1}$. While the first $k$ pages are requested, all strategies have no choice and have a fast-memory containing pages $v_1, \ldots, v_k$. At the time $v_{k+1}$ is requested, one of the pages must be evicted from fast memory. Suppose that the deterministic strategy chooses to evict page $v_i$. Then for any $j$ with $1 \leq j \leq k$ and $i \neq j$, consider the sequence $\sigma = v_1, v_2, \ldots, v_{k+1}, v_j, v_{k+1}, v_j, v_{k+1}, \ldots$ that alternates between $v_{k+1}$ and $v_j$. Since the deterministic strategy has $v_1, \ldots, v_k$ in fast memory when $v_{k+1}$ is requested, and $v_i \neq v_j$ is evicted, both $v_j$ and $v_{k+1}$ are in fast memory after the request. Thus, all future requests are to pages already in fast memory, and this strategy does not incur any more page faults after the first $k+1$. The offline strategy OPT, on the other hand, still incurs a page fault on every request by evicting page $v_j$ when $v_{k+1}$ is requested and vice versa. Extending the length of the sequence proves the lemma.

Lemma 1 also holds even if we introduce **resource deaugmentation**, that is, the online strategy runs with a *smaller* fast memory of size $k_{on} \geq 2$ and offline optimal strategy runs with a larger fast memory of size $k_{off} \geq k_{on}$.[2] Even so, there is still no competitive deterministic strategy. The same proof still applies with the same sequence–the proof just relies on the fact that there are two particular pages in the fast memory of the online algorithm.

We now turn our attention to randomized strategies with an **oblivious adversary**, meaning that the adversary must choose the entire input sequence before seeing the result of any of the coin tosses used by the randomized algorithm. Note that in the presence of a nonoblivious adversary, randomization does not provide extra power for pessimal-cache problem.

The following lemma states that no randomized strategy is better than expected $k$-competitive when both the online and offline strategies have the same fast-memory size $k$. Moreover, when the offline strategy uses a fast memory of size $k_{off}$ and the online strategy has a fast memory of size $k_{on} \leq k_{off}$, no online strategy is better than $k_{off}/(k_{off} - k_{on} + 1)$.

**Lemma 2.** *Let $k_{off}$ be the fast memory size of the offline strategy and $k_{on}$ (with $1 \leq k_{on} \leq k_{off}$) be the fast memory size of the online strategy. Consider any (randomized) online strategy $A$. For any $c < k_{off}/(k_{off} - k_{on} + 1)$ and constant $\beta$, there exists an input $\sigma$ such that $E[A(\sigma)] < \frac{1}{c} \cdot \mathrm{OPT}(\sigma) - \beta$.*

---

[2] Note that resource deaugmentation in the pessimal-cache problem means is the analog of resource augmentation in the classical problem, in which the online algorithm has a *larger* cache than the offline algorithm.

*Proof.* The proof is similar to that of Lemma 1. After the $(k_{\text{off}}+1)$st page request $v_{k_{\text{off}}+1}$, the online algorithm $A$ has $k_{\text{on}}$ pages in fast memory. Page $v_{k_{\text{off}}+1}$ is definitely in fast memory. Of the remaining $k_{\text{off}}$ pages requested so far, $k_{\text{on}} - 1$ are in $A$'s fast memory.

Now let $v_j$ be a randomly selected page from $v_1, \ldots, v_{k_{\text{off}}}$. Page $v_j$ is in $A$'s fast memory with probability $(k_{\text{on}} - 1)/k_{\text{off}}$. Now consider the sequence $v_1, \ldots, v_{k_{\text{off}}}, v_{k_{\text{off}}+1}, v_j, v_{k_{\text{off}}+1}, v_j, v_{k_{\text{off}}+1}, \ldots$. With probability $(k_{\text{on}} - 1)/k_{\text{off}}$, page $v_j$ is still in fast memory after $v_{k_{\text{off}}+1}$ is requested. In this case, no future page requests cause page faults, giving the online strategy a total of $k_{\text{off}} + 1$ page faults. With probability $(k_{\text{off}} - k_{\text{on}} + 1)/k_{\text{off}}$, $v_j$ is not in memory, and the strategy may be able to attain the optimal $\ell$ page faults, where $\ell$ is the length of the sequence following the first request for $v_{k_{\text{off}}+1}$. Thus, the expected number of page faults is at most $(k_{\text{off}} + 1) + \ell(k_{\text{off}} - k_{\text{on}} + 1)/k_{\text{off}}$, whereas the offline strategy attains $(k + 1) + \ell$. Choosing a long enough sequence proves the lemma.

# 3    Most-Recently Used

This section describes two $k$-competitive strategies for the pessimal-cache problem. The first strategy uses one step of randomization followed by the deterministic "most-recently-used" (MRU) heuristic. The second strategy uses more randomization to achieve the optimal result even when the offline and online strategies have different fast-memory sizes.

Since least-recently-used (LRU) is $k$-competitive and optimal for traditional paging, we explore reverse strategies for the the pessimal-cache problem. The ***most-recently-used*** (***MRU***) heuristic always evicts the page in fast memory that was used most frequently. It might be reasonable to expect MRU to be $k$-competitive for the pessimal-cache problem. MRU, however, is deterministic, and Lemma 1 states that no deterministic strategy can be competitive.

Instead, we consider a natural variation on MRU, which we call ***randomized MRU***. In randomized MRU, the first page evicted is chosen at random. (Recall that this first eviction happens when the $(k+1)$th distinct page is requested.) All subsequent evictions follow the MRU strategy. Randomized MRU gets around the alternating-request strategy used to prove lower bounds in Lemmas 1 and 2. The following lemma shows that MRU keeps a (slightly) random set of pages in fast memory.

**Lemma 3.** *Let $k$ be the size of fast memory (for both online and offline strategies), and consider any request sequence $\sigma$. After the $(k + 1)$st distinct page is requested, randomized MRU guarantees that there are $k$ pages each having probability exactly $1 - 1/k$ of being in fast memory, and there is one page, the most-recently-used page, that has probability $1$ of being in fast memory. All other pages are definitely not in fast memory.*

*Proof.* We prove the claim by induction on the requests over time.

*Base case.* The base case is after the $(k+1)$st distinct page is requested, which is after the first eviction. Since there are $k$ pages in fast memory at the time that the $(k + 1)$st distinct page is requested, and one is chosen to be evicted at random, the claim holds for the base case.

*Inductive step.* Suppose that the claim holds up until the $t$th request. Assume that the next request is for page $v_i$. There are several cases.

*Case 1.* Suppose that $v_i$ is definitely not in fast memory. Then the most-recently-used page $v_j$ is evicted, and hence $v_i$ is definitely in fast memory and $v_j$ is definitely not.

*Case 2.* Suppose that $v_i$ is in fast memory with probability 1. Then none of the probabilities change.

*Case 3.* Suppose that $v_i$ is in fast memory with probability $1 - 1/k$ and that $v_j$ is the most-recently-used page. Then with probability $1/k$ we have $v_i$ not in fast memory, and hence the request for $v_i$ evicts $v_j$. Otherwise, $v_j$ stays in fast memory. Thus, the probability that $v_j$ is in fast memory is $1 - 1/k$, and the probability that the most recently used page $v_i$ is in fast memory is 1.

The probability of any other page (other than the ones mentioned in the appropriate case) being in fast memory is unchanged across the request.

The following theorem states that randomized MRU is $k$-competitive, where $k$ is the size of fast memory.

**Theorem 1.** *Randomized MRU is expected $k$-competitive, where $k$ is the size of fast memory.*

*Proof.* Consider any input sequence $\sigma$. If sequence $\sigma$ contains requests to fewer than $k+1$ distinct pages, then randomized MRU has at the same number of page faults as the offline strategy OPT (Both strategies have page faults only the first time each distinct page is requested.) Consider any request after the first $(k+1)$st distinct page is requested. If the request is for the most-recently-used page, then neither OPT nor randomized MRU have a page fault, since that page must be in fast memory. Otherwise, OPT causes a page fault. By Lemma 3, randomized MRU incurs a page fault with probability at least $1/k$. Specifically, MRU incurs a fault with exactly probability $1/k$ for any of $k$ pages and probability 1 for any of the other pages. Thus, in expectation, randomized MRU incurs at least $1/k$ page faults for each page fault incurred by OPT.

This result for randomized MRU is not quite analogous to the result of Sleator and Tarjan's [10] result for LRU. It is true that LRU is $k$-competitive for the traditional paging problem, and randomized MRU is $k$-competitive for the pessimal-cache problem. However, LRU also has good performance with resource augmentation. Specifically, if LRU has a fast memory of size $k$ and the offline strategy has a fast memory size $(1 - 1/c)k$, then LRU is $c$-competitive. In particular, if the LRU has twice the fast memory of offline, then LRU is 2-competitive. The above result for the pessimal-cache problem does not generalize in the same way–the competitive ratio depends only on the size of randomized MRU's fast memory. If randomized MRU has a size-$k$ fast memory and the offline strategy has a size $2k$ fast memory, then randomized MRU is still only $k$-competitive.

We now give a more powerful MRU algorithm, **reservoir MRU**, that achieves a better competitive ratio for the case of resource deaugmentation. As before, let $k_{off}$ and $k_{on} \le k_{off}$ be the sizes of the offline and online's fast memory, respectively.

The main idea of reservoir MRU is to keep a reservoir of $k_{on} - 1$ pages, where each previously-requested page resides in the reservoir with equal probability. (This technique is based on Vitter's reservoir sampling [11].) Reservoir MRU works as follows. For the first $k_{on}$ distinct requests, the fast memory is not full, and thus there are no evictions. Subsequently, if there is a request for a previously-requested page $v_i$, and the page is not in memory, then the most-recently requested page is evicted. Otherwise, when the $n$th new page is requested, for any $n > k_{on}$, with probability $1 - (k_{on} - 1)/(n - 1)$, the most recently requested page is evicted. Otherwise, the page to evict (other than the most-recently-used page) is chosen uniformly at random.

Reservoir MRU has an invariant that is a generalization of Lemma 3. After any request, the page that was requested most recently has probability 1 of being in fast memory. All other $n - 1$ pages have probability $(k_{on} - 1)/(n - 1)$ probability of being in fast memory.

**Lemma 4.** *Let $k_{off}$ and $k_{on} \le k_{off}$ be the fast memory sizes of the offline strategy and of reservoir MRU, respectively. Consider any page-request sequence $\sigma$ to reservoir MRU. After the $n > k_{on}$th distinct page is requested, there is a single page, the most-recently-used page, that has probability 1 of being in fast memory. All other pages have probability $(k_{on} - 1)/(n - 1)$ of being in fast memory.*

*Proof.* The proof is by induction on the requests, and is reminiscent of the proof of Lemma 3.

*Base case.* After the $(k_{on} + 1)$th distinct request, the $(k_{on} + 1)$th page is definitely in fast memory, and one page randomly chosen has been evicted. Reservoir MRU evicts the most recently used page with probability $1 - (k_{on} - 1)/k_{on} = 1/k_{on}$ and all other page with the same probability. Thus, every page, except the last one has probability $1 - 1/k_{on}$ of being in fast memory, and the lemma holds for the base case.

*Inductive step.* Consider a request for page $v_i$ after the $n$th distinct page has been requested. Assume by induction that the most-recently-used page $v_j$ is definitely in fact memory and that all other $n - 1$ pages are in fast memory with probability $(k_{on} - 1)/(n - 1)$. There are several cases.

*Case 1.* Suppose that page $v_i = v_j$, i.e., $v_j$ is in fast memory with probability 1. Then none of the probabilities change.

*Case 2.* Suppose that page $v_i$ has been previously requested, but $v_i \ne v_j$. If $v_i$ is already in fast memory then nothing is evicted. Otherwise, by the properties of reservoir MRU, page $v_j$ is evicted. Since $v_i$ was in fast memory with probability $(k_{on} - 1)/(n - 1)$, page $v_j$ is evicted with probability $1 - (k_{on} - 1)/(n - 1)$ and remains in fast memory with probability $(k_{on} - 1)/(n - 1)$. None of the probabilities for pages other than $v_i$ and $v_j$ change.

*Case 3.* Suppose that page $v_i$ has never been requested before, that is, $v_i$ is the $(n+1)$st distinct request. By the properties of reservoir MRU, the most-recently-used page $v_j$ (which is definitely in fast memory) is evicted with probability $1 - (k_{on} - 1)/n$ and remains in fast memory with probability $(k_{on} - 1)/n$. Thus, the probability that $v_j$ is in fast memory is at the desired value.

The probability that each additional page is in shared memory now also needs to decrease since the number of distinct pages has increased by one. Since with probability $(k_{on} - 1)/n$, a random page from the other $k_{on} - 1$ pages is evicted from fast memory, each page in fast memory is evicted with probability $1/n$. The probability that any page is in fast memory after this process is the probability that the page was in a fast memory before the $(n + 1)$st distinct page request times the probability that the page was not evicted by this request, which is $(k_{on} - 1)/(n - 1)(1 - 1/n) = (k_{on} - 1)/n$. Since the number of distinct pages requested is now $n + 1$, this probability also matches the lemma statement.

We now use the previous lemma to prove a better competitive ratio for reservoir MRU in the case of resource deaugmentation.

**Theorem 2.** *Reservoir MRU is expected $k_{off}/(k_{off} - k_{on} + 1)$-competitive, where $k_{off}$ is the size of fast memory of the offline strategy, and $k_{on} \leq k_{off}$ is the size of fast memory for reservoir MRU.*

*Proof.* Before $k_{off}$ distinct requests, reservoir MRU has at least as many page faults as the offline strategy. And after this point, each time the offline strategy has a page fault, since $n > k_{off}$, reservoir MRU incurs a page fault with probability at least $1 - (k_{on} - 1)/k_{off}$ from Lemma 4.

This theorem means that when the offline strategy and reservoir MRU have the same fast-memory size $k$, reservoir MRU is $k$-competitive. When reservoir MRU has fast-memory size $k_{on}$ and the offline strategy has fast-memory size $(1 + 1/c)k_{on}$, reservoir MRU is $(c + 1)$-competitive.[3]

Reservoir MRU requires some additional state—in particular, we need one bit per page to indicate whether the page has been requested before. Consequently, if the sequence requests $n$ distinct pages, then we need $O(n)$ extra bits of state. In contrast, Achlioptas et. al.'s [1] optimal randomized algorithm for the page-replacement problem requires only $O(k^2 \log k)$ extra bits of state. The extra state is unavoidable for reservoir MRU, however, because we must know when $n$, the number of distinct pages, increases. Fortunately, these extra bits can be stored in the slow memory, associated with each page—only the more reasonable $O(\log n)$ bits for the counter storing $n$ need be remembered by the algorithm at any given time.

## 4  Direct Mapping

In this section we consider the page-replacement strategy used in direct-mapped caches. We show that for the pessimal-cache problem, direct mapping is

---

[3] In fact, the offline strategy can have a slightly smaller memory—with size $\lceil (1 + 1/c)(k_{on} - 1) \rceil$—and we still attain the $(c + 1)$-competitiveness.

$k$-competitive under some assumptions about the mapping strategy or about the layout in slow memory.

In a direct-mapping strategy (see, e.g., [6]) each page $v_i$ can be stored in only a single location $L(v_i)$ in fast memory. Thus, in a direct-mapped cache, once the function $L(v_i)$ is chosen, there are no algorithmic decisions to make: whenever a page $v_i$ is requested, we must evict the page that is currently stored in location $L(v_i)$ and store $v_i$ there instead.

In the following, we show that if $L(v_i)$ is randomly chosen for each $v_i$, then direct mapping is $k$-competitive with the optimal offline strategy (with no direct-mapped restrictions).

In fact, typically in real caches, the function $L(v_i)$ is determined by the low-order bits in the address of $v_i$ in slow memory; it is not random. However, if each page $v_i$ is stored in a random memory address in slow memory then our theorem still applies. While it is often unrealistic to assume that each page $v_i$ is randomly stored, this approach was also used in [5, 9] to enable direct-mapped caches to simulate caches with no restrictions on associativity.

Observe that direct mapping is not a reasonable strategy when compared with the optimal off-line strategy with no mapping restrictions. In particular, a direct-mapped fast memory may evict a page before the rest of the fast memory is full. However, since caches with limited associativity are so common, it is of interest to explore this special case.

The following theorem states that direct mapping is competitive with the optimal offline strategy for the pessimal-cache problem.

**Theorem 3.** *Direct-mapping is $k$-competitive, where $k$ is the fast-memory size of the both be the direct-mapping and offline strategies.*

*Proof.* We claim that a particular page is requested many times and the offline strategy incurs a page fault on $\ell$ of these requests, then direct mapping incurs at least $\ell/k$ page faults on $v_i$ in expectation. We prove this claim by induction on the number of requests to $v_i$.

The first time that $v_i$ is requested, there is a page fault. If $v_i$ is requested again immediately (without any interleaving page requests), then both strategies have the page in fast memory. If $v_i$ is requested again after another page is requested, then the offline strategy may have a page fault. The direct-mapping strategy incurs a page fault with probability at least $1/k$, because at least one page $v_j$ is requested between $v_i$ requests, and this page $v_j$ has a $1/k$ probability of evicting $v_i$ from fast memory.

## 5   Conclusions

For the pessimal-cache problem, randomization is necessary to achieve any competitive ratio, and the best competitive ratio without resource deaugmentation is $k$. In contrast, for the original problem, deterministic strategies can be $k$-competitive [10], and upper [4, 8, 1] and lower [4] bounds of $\Theta(\log k)$ exist for randomized strategies against oblivious adversaries.

In this paper, competitive ratios are $k$ or larger; is there some model in which the competitive ratio is smaller? Essentially, we're trying to get a better definition of reasonable strategies giving the adversary just the right amount of power. This concept is similar to many approaches for the original page-replacement problem—for example, the graph-theoretic approach [3] tries to better model locality. Unfortunately, the traditional approaches seem to have little impact for the pessimal-cache problem. For example, looking at access patterns matching a graph, little can be said even if the graph is just a simple line. Adding power like lookahead to the online strategy, on the other hand, trivializes the problem since the optimal offline strategy can be implemented with a lookahead of 1. It would be nice to come up with a more accurate model that allows us to beat $k$-competitiveness.

It's interesting that direct-mapped cache is optimally bad when the program shows no locality (i.e., as in a multiprogrammed environment). In this model, however, we cannot show anything about the badness of a 2-way (or, more generally, $c$-way) set-associative cache using LRU. In particular, the LRU subcomponent forces the cache to make the "right" choice for eviction, and the sequence ping-ponging between two pages is sufficient to guarantee no future misses.

One way of weakening the adversary is to restrict the definition of "reasonable" strategies by disallowing the eviction of the most (or perhaps the $c$ most) recently used pages. In some sense, we're forcing the cache to model some small amount of locality, since, after all, that is the purpose of the cache. This modification of the problem has the nice property that it allows us to analyze the pessimal-cache problem for a $c$-way set-associative cache. In particular, a 2-way set-associative cache is roughly $k^2$-competitive for the pessimal-cache problem. This result appears to generalize for $c$-way set-associative caches as well.

It would nice to see if anything from this paper applies to other problems, or generalizations of the paging problem, like the $k$-servers on a line problem, for example.

# References

1. Achlioptas, D., Chrobak, M., Noga, J.: Competitive analysis of randomized paging algorithms. In: Díaz, J. (ed.) ESA 1996. LNCS, vol. 1136, pp. 419–430. Springer, Heidelberg (1996)
2. Belady, L.A.: A study of replacement algorithms for virtual storage computers. IBM Systems Journal 5(2), 78–101 (1966)
3. Borodin, A., Irani, S., Raghavan, P., Schieber, B.: Competitive paging with locality of reference. Journal of Computer and System Sciences 50(2), 244–258 (1995)
4. Fiat, A., Karp, R.M., Luby, M., McGeoch, L.A., Sleator, D.D., Young, N.E.: Competitive paging algorithms. Journal of Algorithms 12(4), 685–699 (1991)
5. Frigo, M., Leiserson, C.E., Prokop, H., Ramachandran, S.:Cache-oblivious algorithms. In: 40th Annual Symposium on Foundations of Computer Science, pp. 285–297, New York, October 17–19 (1999)
6. Hennessy, J.L., Patterson, D.A.: Computer Architecture: a Quantitative Approach, 3rd edn. Morgan Kaufmann, San Francisco, CA (2003)

7. Irani, S.: Competitive analysis of paging. In: Fiat, A. (ed.) Developments from a June 1996 Seminar on Online Algorithms. LNCS, vol. 1442, pp. 52–73. Springer, Heidelberg (1998)
8. McGeoch, L.A., Sleator, D.D.: A strongly competitive randomized paging algorithm. Algorithmica 6, 816–825 (1991)
9. Sandeep Sen and Siddhartha Chatterjee. Towards a theory of cache-efficient algorithms. In: Proceedings of the 11th Annual ACM-SIAM Symposium on Discrete Algorithms (SODA), pp. 829–838, San Francisco, California (January 2000)
10. Sleator, D.D., Tarjan, R.E.: Amortized efficiency of list update and paging rules. Communications of the ACM 28(2), 202–208 (1985)
11. Vitter, J.S.: Random sampling with a reservoir. ACM Transactions on Mathematical Software 11(1), 37–57 (1985)

# Die Another Day

Rudolf Fleischer*

Fudan University
Shanghai Key Laboratory of Intelligent Information Processing
Department of Computer Science and Engineering
Shanghai 200433, China
rudolf@fudan.edu.cn

**Abstract.** The Hydra was a many-headed monster from Greek mythology that would immediately replace a head that was cut off by one or two new heads. It was the second task of Hercules to kill this monster. In an abstract sense, a Hydra can be modeled as a tree where the leaves are the heads, and when a head is cut off some subtrees get duplicated. Different Hydra species differ by which subtress can be duplicated in which multiplicity. Using some deep mathematics, it had been shown that two classes of Hydra species must always die, independent of the order in which heads are cut off. In this paper we identify three properties for a Hydra that are necessary and sufficient to make it immortal or force it to die. We also give a simple combinatorial proof for this classification. Now, if Hercules had known this...

## 1  Introduction

According to Greek mythology, the *Hydra* was a many-headed monster living in a marsh near Lerna [20]. If one head was cut off, one or two new heads grew from the Hydra's body. Nevertheless, in his second task (of twelve, ordered by his cousin Eurystheus) the Greek hero Hercules (a.k.a. Herakles), a son of Zeus, defeated the Hydra, although he did not fully play by the rules: while he was happily hacking away at the heads, his nephew Iolaus burnt the Hydra to prevent new heads from growing [23].

Kirby and Paris studied this epic fight from a graph theoretic point of view and showed that Hercules might even have won without employing unfair tactics (although, then, he would probably still be fighting today) [17]. They proposed to model the Hydra as a rooted tree where the heads are the leaves (see Fig. 1). The classical Hydra would grow one or two new heads replacing any head that was cut off. Clearly, this Hydra species cannot die (except by fire). Now, Kirby and Paris suggested to study another Hydra species which can duplicate entire subtrees which contained the head that was cut off. At the first glance, one might expect this to be a more powerful Hydra variety, but they proved that this species must eventually die (and this cannot be shown by a simple proof by induction).

---

* The work described in this paper was partially supported by a grant from the National Natural Science Fund China (grant no. 60573025).

P. Crescenzi, G. Prencipe, and G. Pucci (Eds.): FUN 2007, LNCS 4475, pp. 146–155, 2007.

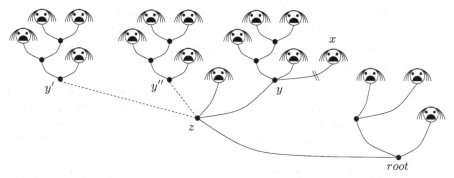

**Fig. 1.** Cutting off $x$ and growing two copies of $T_y^-$ from $z$

To be more precise, they suggested to duplicate subtrees as follows (we call this Hydra species the *i-head Hydra*). For a node $v$, let $T_v$ denote the subtree rooted at $v$. For a leaf $x$, its predecessor $y$ is called the *neck*, and the predecessor's predecessor $z$ is called the *trunk*. The path from the root to $y$ is called the *spine* of $x$. When $x$ is cut off in the $i$-th blow, $i$ new subtrees identical to $T_y$ without $x$, denoted by $T_y^-$, will grow from the trunk $z$. Fig. 1 shows an example of a second blow. We cut off head $x$, and the Hydra grows two copies of $T_y$ out of $z$. If $x$ has no siblings, its neck $y$ becomes a leaf, i.e., a new head, and $i$ new heads (the copies of $T_y$) grow from the trunk $z$. If $y$ is the root, no new subtrees grow. If $x$ is the root, i.e., the root is the only node (and head) of the Hydra, cutting off $x$ will *kill* the Hydra.

A Hydra is *doomed* if it will die in a finite number of steps, for any possible sequence of head cuts. Otherwise, it is *immortal*. In a deep mathematical proof based on transfinite induction (if we add the first ordinal number $\omega$, the smallest infinite number, and arbitrary polynomial expressions of $\omega$ to the set of natural numbers and define the operation $\omega - 1$ as choosing an arbitrary finite number smaller than $\omega$, then the principle of transfinite induction states that we always reach zero in a finite number of steps when counting down from an arbitrary number) Kirby and Paris showed that the $i$-head Hydra is doomed. Later, Luccio and Pagli gave an elementary combinatorial proof based on a potential function defined on the nodes for the special case of the 2-*head Hydra* which can only grow two (or any fixed constant number of) subtree copies in each step. They posed as an open problem to find an elementary combinatorial proof for the $i$-head Hydra [21].

In this paper we give such a proof for the more general class of finite Hydras. As we will see later, the actual number of subtrees grown in each duplication step is not relevant for the fate of a Hydra as long as it is always a finite number, only the locations of the subtrees to be copied do matter.

A *finite Hydra* is a finite tree. If we cut off a head $x$, it can, in any order, copy an arbitrary but finite number of subtrees according to the following three properties (see Fig. 2 for an example).

(P1) The subtree to be copied is rooted at a node on the spine of $x$.

(P2) The subtree copy becomes a child of a node on the spine of $x$.

(P3) The subtree copy is placed at the same level or closer to the root than the original subtree.

In principle, a Hydra may choose not to copy an entire subtree but only part of it (a subtree of the subtree), but obviously it cannot gain anything by doing so, so we may assume w.l.o.g. that a Hydra always copies entire subtrees.

Clearly, the 2-head and $i$-head Hydras are special cases of the finite Hydra. The number of new subtree copies may either be predetermined (e.g., the 2-head Hydra), given as a function of the structure of the tree or the length of the fight (e.g., the $i$-head Hydra), or the Hydra may adapt to the cutting sequence, deciding on the number of tree copies each time a head is cut off (this corresponds to choosing an arbitrary finite number in the operation $\omega - 1$ in the transfinite reduction proof by Kirby and Paris).

Generalizing the results by Luccio *et al.* and Kirby *et al.*, we show that any finite Hydra is doomed.

**Theorem 1 (Hydra Theorem).** *Any finite Hydra is doomed.*

We will give a simple combinatorial proof of the Hydra Theorem in Section 3, after shortly reviewing the mathematical history of this problem in Section 2.

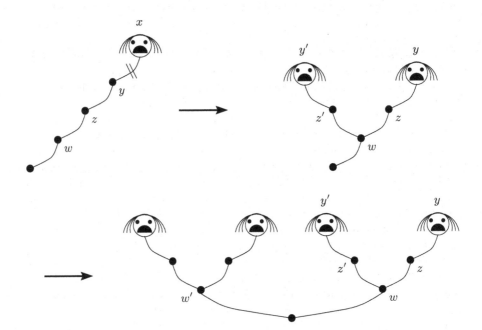

**Fig. 2.** When we cut head $x$ in a finite Hydra, it can first grow a copy of $T_z^-$ at $w$ and then grow a copy of the new $T_w^-$ at the root

In Section 4 we will see that relaxing any one of the three properties (P1)–(P3) can make a Hydra immortal. In Section 5 we shortly discuss worms [13], a one-dimensionally restricted Hydra, and the Buchholz Hydra [3], a truly two-dimensional generalization of the finite Hydra.

## 2   The History of the Hydra Battle

Kurt Tucholsky once noted a speaker should "always start with ancient Rome and mention the historical background of the matter" [27]. We already discussed the pre-Roman story of Hercules and the Hydra, so we may jump directly to the past century (there was not much happening in between; at least, nothing related to this paper).

There is a branch of mathematical logic that is concerned with the relative power of the various axioms of mathematics [9] (the Axiom of Choice (AC), the axioms of Peano Arithmtic (PA), etc.) After Gentzen had shown the consistency of PA [11] (see also [26]) in 1936, Goodstein gave in 1944 a recursive definition of the so-called Goodstein sequence and showed that in PA it cannot be proven to terminate [12] (see also Cichon [5]). Basically, this means there is no classical proof by induction for this theorem. A problem of a similar flavour is the famous $3x + 1$ conjecture (where termination has not been proven yet), also known as Collatz problem, Syracuse problem, Kakutani's problem, Hasse's algorithm, and Ulam's problem [19].

Much later, in 1982, Kirby and Paris gave an alternative proof for the termination of the Goodstein sequence [18] and introduced the Hydra battle as another example to demonstrate their new technique from [22]. They showed that $i$-head Hydras are doomed and this cannot be shown in PA, even if Hercules is required to always cut off the rightmost head of the Hydra (assuming a natural ordering of subtrees by decreasing size). In this case, even if the Hydra has only height two, the length of the battle is not primitive recursive. Another proof was given by Carlucci via reduction from Gentzen's Reduction Strategy [4].

The Kirby-Paris paper led to a flurry of research on the Hydra and related problems. It was quickly observed that their results actually hold for the more general case of *arbitrary-head Hydras* (which decide in every step how many subtree copies are grown). But no generalization to growing copies at arbitrary nodes on the spine (as in our finite Hydra) was suggested, mathematicians only grow new subtrees at the trunk of the cut head.

Since the Kirby-Paris Hydra can only grow in width, the question arose whether there is a generalization of these results to height-growing Hydras. In 1987, Buchholz answered this in the positive with a doomed Hydra species that can grow in width and height [3] (basically, the growth in each dimension is bounded by a Hydra battle). This Hydra is now known as the *Buchholz Hydra*.

Hamano and Okada went the other direction and restricted Hydras to truly one-dimensional objects, so-called worms [13] (see also Beklemishev [2]). Although being recursively defined and of reasonably bounded length, they cannot, in PA, be proven to terminate. Hydras also came to fame in the Scientific

American when Gardner [10] discussed the Hydra battle and Smullyan's Urn Game [24], kind of a simplified version of the arbitrary-head Hydra.

In 2000, Luccio and Pagli discovered the combinatorial beauty of the Hydra battle [21] and asked whether there is a simple combinatorial proof that the $i$-head Hydra is doomed. Luccio's presentation of the problem at FUN 2001 let to lively discussions among the conference participants (futily trying to find proofs) and might be considered the birth hour of the present paper. Here, we give a simple combinatorial proof that finite Hydras are doomed. We use Koenig's Lemma, which pops up here and there in the computer science literature, mainly in proofs in logics and formal languages. Although it is deceivingly simple to state and prove, it is actually a powerful theorem outside of PA, equivalent to AC [6,15,16], so our simple proof does not contradict the Kirby-Paris result.

**Lemma 2 (Koenig's Lemma).** *A tree is finite if and only if every node has finite degree and every simple path from the root is finite.*

*Proof.* If one node has infinite degree or there is an infinite simple path, the tree is infinite. On the other hand, if a tree where all nodes have finite degree is infinite, then one of the subtrees of the root must be infinite. If we follow the edge to that subtree and iterate, we can construct an infinite simple path.  □

Daly also used Koenig's Lemma to give simple proofs for Smullyan's Urn Game and the Goodstein sequence [7]. Weiermann recently showed the exact threshold of the transition between PA-provability (the 2-head Hydra) and non-PA-provability (the $i$-head Hydra) [29]. Readers interested in reading the mathematical papers might first want to read some good introduction into the theory of ordinals (for example, Avigad [1] or Dershowitz [8]).

## 3   Proof of the Hydra Theorem

If not all finite Hydras are doomed, there must be an immortal Hydra $H$ of smallest height. We call the subtrees of the root of $H$ *subhydras*. If we cut a head in a subhydra, $H$ may choose to create a finite number of copies of the subhydra, which we also call subhydras.

Let $\rho$ be an infinite hacking sequence that does not kill $H$. Now we define the *hacking tree $S$*. Initially, $S$ consists of a root with one leaf for each subhydra of $H$. Each time we cut a head in a subhydra, the corresponding leaf of $S$ will become an internal node with $k + 1$ new children leaves, where $k \geq 0$ is the number of copies we create of the subhydra. One of the new leaves corresponds to the subhydra that was copied, the other leaves correspond to the copies. So after each step, each leaf of $S$ corresponds to one of the current subhydras of $H$, and siblings in $S$ represent identical copies of the same subhydra (here we use (P1), that a subtree can only be copied if it contained the head that was cut off).

Since $\rho$ is infinite, $S$ is infinite. Each node has finite degree (here we use that we only create a finite number of subtree copies in each step), thus there must exist an infinite simple path $p$ in $S$ by Koenig's Lemma, corresponding to an infinite subsequence $\sigma$ of $\rho$. For each node $v$ on $p$, we may w.l.o.g. assume that the successor $w$ (the child of $v$ in $S$) is actually the original subhydra that was copied in this cutting step (here we use (P3) in a subtle way; our proof by contradiction is actually a proof by induction on the height of doomed Hydras, where the induction step cuts off the root. Therefore, we cannot allow that subtrees get copied higher up in the tree, because otherwise in the inductive step we could have new subhydras appearing out of the blue sky). Thus, $\sigma$ is an infinite cutting sequence in one of the initial subhydras $G$ of $H$. Note that (P2) implies that none of the cuts in $\rho - \sigma$ can affect $G$. So we have found an immortal Hydra $G$ of smaller height than $H$, a contradiction.

## 4    Immortal Hydras

Fig. 3–5 show examples of immortal Hydras violating exactly one of the properties (P1)–(P3). Note that the Hydras violating (P1) and (P3) are immortal in a strong sense: they can survive *any* cutting sequence.

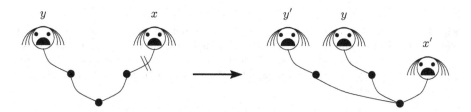

**Fig. 3.** A Hydra that can copy a subtree not containing the head that was cut off, violating (P1). If we cut off $x$ and the Hydra copies the subtree containing $y$, the new Hydra is a supertree of the original one, i.e., it is immortal.

## 5    Worms and the Buchholz Hydra

Hamano and Okada defined a worm as a one-dimensional version of the two-dimensional Buchholz Hydra and showed it to be equivalent to the $i$-head Hydra [13]. Here we give the self-contained definition by Beklemishev [2]. A *worm* is a finite sequence of natural numbers $w = (f(0), f(1), \ldots, f(n))$. $f(n)$ is called the *head* of the worm. The worm battle is defined by the sequence of worms $w_0 = w$ and $w_{n+1} = next(w_n, n + 1)$, defined as follows.

1. If $f(n) = 0$, then $next(w, m) = (f(0), \ldots, f(n - 1))$. That is, in this case we cut off the head of the worm.
2. If $f(n) > 0$, let $k = \max_{i<n}\{f(i) < f(n)\}$. The worm $w$ (with the head decreased by one) is then the concatenation of two parts, the *good* part $r = (f(0), \ldots, f(k))$, and the *bad* part $s = (f(k+1), \ldots, f(n-1), f(n) - 1)$. We define $next(w, m) = r * \underbrace{s * s * \cdots * s}_{m+1 \text{ times}}$.

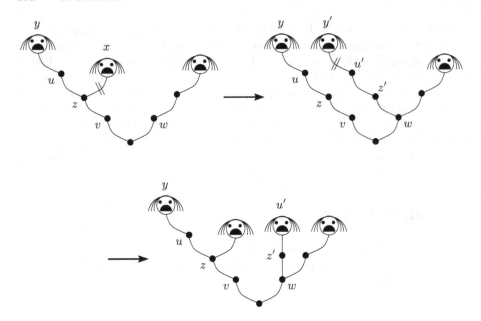

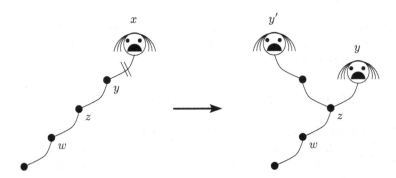

**Fig. 4.** A Hydra that can copy a subtree to nodes not on the spine of the cut head, violating (P2). If we cut off head $x$, the Hydra can grow a copy of $T_v^-$ at node $w$ (which is at the same level as $v$). If we now cut off the new head $y'$, $u'$ will become a head and we can grow a copy of this head at $z$. This Hydra is now a supertree of the original one, so it is immortal.

**Fig. 5.** A Hydra that can place a copy of a subtree higher up on the spine, violating (P3). This is the same Hydra as in Fig. 2, but this time we grow the copy of $T_z$ not at $w$ but at $z$. The new Hydra is a supertree of the original one and thus immortal.

$w_n$ is defined by a primitive recursive function, and the length of a worm is bounded by $|w_n| \leq (n + 2)! \cdot |w_0|$. Still, proving that any worm must eventually die cannot be shown in PA (actually, it is equivalent to the $i$-head Hydra battle).

Buchholz generalized the $i$-head Hydra to a species that can also grow in height [3] by relaxing (P3). He allowed subtree copies to be placed higher up in the tree, but only a bounded number of times. Since this bound is essentially given by the length of a $i$-head Hydra battle, this produces a Hydra that grows as high as it grows wide. To be more precise, a Buchholz Hydra is a finite labeled tree $H$ with the following properties:

1. The root has label $+$.
2. Any other node of $H$ is labeled by some ordinal $v \leq \omega$.
3. All nodes immediately above the root of $H$ have label 0 (zero).

If we cut off head $x$, $H$ will choose an arbitrary number $n \in \mathbb{N}$ and transform itself into a new Hydra $H(x, n)$ as follows. Let $y$ denote that node of $H$ which is immediately below $x$ (the neck), and let $H^-$ denote that part of $H$ which remains after $x$ has been cut off. The definition of $H(x, n)$ depends on the label of $x$.

*Case 1:* label$(x) = 0$. If $y$ is the root of $H$, we set $H(x, n) = H^-$. Otherwise, $H(x, n)$ results from $H^-$ by growing $n$ copies of $H_y^-$ from the node immediately below $y$, i.e., the trunk. Here, $H_y^-$ denotes the subtree of $H^-$ rooted at $y$.

*Case 2:* label$(x) = u + 1$. Let $z$ be the first node below $x$ with label $v \leq u$. Let $G$ be the tree which results from the subtree $H_z$ by changing the label of $z$ to $u$ and the label of $x$ to 0. $H(x, n)$ is obtained from $H$ by replacing $x$ by $G$. In this case, $H(x, n)$ does not depend on $n$.

*Case 3:* label$(x) = \omega$. $H(x, n)$ is obtained from $H$ simply by changing the label of $x$: $\omega$ is replaced by $n + 1$.

Note that a Buchholz Hydra with all labels equal to zero is just the classical $i$-head Hydra. Buchholz showed that every Buchholz Hydra can be killed by repeatedly cutting off the rightmost head, but this cannot be shown in PA (it can be shown in $(\Pi_1^1 - \text{CA}) + \text{BI}$, if you really want to know). Later, Hamano and Okada showed that the Buchholz Hydra is actually doomed with any cutting sequence [14]. Wainer further generalized the underlying mathematics of structured tree-ordinals [28], but it is not clear (to us) what this implies for the Hydra battle.

## 6    Conclusions

We have characterized the elementary Hydras that are doomed or immortal by identifying three properties (P1)–(P3) that are necessary and sufficient for a Hydra to be doomed. We could not find an example of a Hydra only violating (P2) that can survive *any* cutting sequence. We would also like to find a simple combinatorial proof for the Buchholz Hydra.

## Acknowledgments

The particular style to draw the Hydra trees is due to Luccio and Pagli [21]. The literature review was done at the excellent library of Kyoto University during the author's visit in January 2006.

# References

1. Avigad, J.: Ordinal analysis without proofs. In: Sieg, W., Sommer, R., Talcott, C.: (eds.), Reflections on the Foundations of Mathematics: Essays in Honor of Solomon Feferman. Lecture Notes in Logic 15, pp. 1–36. A. K. Peters, Ltd. Wellesley, MA (2002)

2. Beklemishev, L.D.: The worm principle. Technical Report 219, Department of Philosophy, University of Utrecht, Logic Group Preprint Series (2003)

3. Buchholz, W.: An independence result for $(\Pi_1{}^1 - CA) + BI$. Annals of Pure and Applied Logic 33, 131–155 (1987)

4. Carlucci, L.: A new proof-theoretic proof of the independence of the Kirby-Paris' Hydra theorem. Theoretical Computer Science 300(1–3), 365–378 (2003)

5. Cichon, E.A.: A short proof of two recently discovered independence results using recursion theoretic methods. Proceedings of the American Mathematical Society 87(4), 704–706 (1983)

6. Clote, P., McAloon, K.: Two further combinatorial theorems equivalent to the 1-consistency of Peano arithmetic. The. Journal of Symbolic Logic 48(4), 1090–1104 (1983)

7. Daly, B.: (no title). Archived in the Mathematical Atlas at http://www.math.niu.edu/~rusin/known-math/00_incoming/goodstein (17 August 2000)

8. Dershowitz, N.: Trees, ordinals and termination. In: Gaudel, M.-C., Jouannaud, J.-P. (eds.) CAAP 1993, FASE 1993, and TAPSOFT 1993. LNCS, vol. 668, pp. 243–250. Springer, Heidelberg (1993)

9. Feferman, S., Friedman, H.M., Maddy, P., Steel, J.R.: Does mathematics need new axioms? The Bulletin of Symbolic Logic 6(4), 401–446 (2006)

10. Gardner, M.: Mathematical games: Tasks you cannot help finishing no matter how hard you try to block finishing them. Scientific American, pp. 8–13 (1983)

11. Gentzen, G.: Die Widerspruchsfreiheit der reinen Zahlentheorie. Mathematische Annalen, 112:493–565, 1936. Appendix: Galley proof of sections IV and V, Mathematische Annalen received on 11th August 1935. Translated as "The consistency of elementary number theory" in [25], pp. 132–213 (1935)

12. Goodstein, R.J.: On the restricted ordinal theorem. The Journal of Symbolic Logic 9, 33–41 (1944)

13. Hamano, M., Okada, M.: A relationship among Gentzen's proof-reduction, Kirby-Paris' Hydra game and Buchholz's Hydra game. Mathematical Logic Quarterly 43, 103–120 (1997)

14. Hamano, M., Okada, M.: A direct independence proof of Buchholz's Hydra game on finite labeled trees. Archive for Mathematical Logic 37, 67–89 (1998)

15. Ishihara, H.: Weak König's Lemma implies Brouwer's Fan Theorem. Notre Dame Journal of Formal Logic 47(2), 249–252 (2006)

16. Keisler, H.J.: Nonstandard arithmetic and reverse mathematics. The. Bulletin of Symbolic Logic 12(1), 100–125 (2006)

17. Kirby, L., Paris, J.: Accessible independence results for Peano arithmetic. Bulletin of the London Mathematical Society 14, 285–293 (1982)

18. Kirby, L., Paris, J.: Accessible independence results for Peano arithmetic. Bulletin of the London Mathematical Society 14, 285–293 (1982)

19. Lagarias, J.: The $3x + 1$ problem and its generalizations. American Mathematical Monthly 92, 3–23 (1985)

20. Leadbetter, R.: Hydra. In: Encyclopedia Mythica,
    http://www.pantheon.org/articles/h/hydra.html (1999)
21. Luccio, F., Pagli, L.: Death of a monster. ACM SIGACT News 31(4), 130–133 (2000)
22. Paris, J.: Some independence results for Peano arithmetic. The. Journal of Symbolic Logic 43, 725–731 (1978)
23. The Lernean Hydra. In: Crane, G., (ed.), The Perseus Project. Tufts University, Department of the Classics, http://www.perseus.tufts.edu/Herakles/hydra.html (2000)
24. Smullyan, R.M.: Trees and ball games. Annals of the New York Academy of Sciences 321, 86–90 (1979)
25. Szabo, M.E. (ed.): The Collected Papers of Gerhard Gentzen. North Holland, Amsterdam (1969)
26. Tait, W.W.: Gödels reformulation of Gentzen's first consistency proof for arithmetic: the no-counterexample interpretation. The. Bulletin of Symbolic Logic 11(2), 225–238 (2005)
27. Tucholsky, K.: Ratschläge für einen schlechten Redner (Advice for a bad speaker). In: Zwischen gestern und morgen, pp. 95–96. Rowohlt Verlag, Hamburg, 1952. English translation at http://www.nobel133.physto.se/Programme/tucholsky.htm
28. Wainer, S.S.: Accessible recursive functions. The. Bulletin of Symbolic Logic 5(3), 367–388 (1999)
29. Weiermann, A.: Classifying the phase transition of Hydra games and Goodstein sequences, 2006. Manuscript, available at
    http://www.math.uu.nl/people/weierman/goodstein.pdf

# Approximating Rational Numbers by Fractions

Michal Forišek

Department of Informatics,
Faculty of Mathematics, Physics and Informatics,
Comenius University,
Mlynská dolina, 842 48 Bratislava, Slovakia
forisek@dcs.fmph.uniba.sk

**Abstract.** In this paper we show a polynomial-time algorithm to find the best rational approximation of a given rational number within a given interval. As a special case, we show how to find the best rational number that after evaluating and rounding exactly matches the input number. In both results, "best" means "having the smallest possible denominator".

## 1 Motivation

Phillip the physicist is doing an elaborate experiment. He precisely notes each numeric result along with the error estimate. Thus, his results may look as follows: $x = 1.4372 \pm 0.001$.

Larry the lazy physicist is doing similar experiments. However, he just takes the exact value he gets, rounds it to several decimal places and writes down the result. Example of Larry's result: $x \sim 2.3134$.

Cole the computer scientist is well aware of the Occam's Razor principle (in other words, understands what Kolmogorov complexity is). He knows that the simplest answer is often the right one. In the physicists' case, the exact value might very well be a rational number. However, it is not obvious which rational number this might be. Cole's task will be to find the best, simplest one that matches the measured results.

## 2 Previous Results

The sequence of irreducible rational numbers from $[0, 1]$ with denominator not exceeding a given $N$ is known under the name Farey sequence of order $N$. For example $F_5 = \{0/1, 1/5, 1/4, 1/3, 2/5, 1/2, 3/5, 2/3, 3/4, 4/5, 1/1\}$. Several properties of these sequences are known. The one we are most interested in is the following one: Let $p/q$, $p'/q'$, and $p''/q''$ be three successive terms in a Farey sequence. Then

$$p'/q' = (p + p'')/(q + q'') \tag{1}$$

See [3] for Farey's original conjecture of this property, and e.g. [6,1] for a more involved discussion of the history and properties of these sequences.

P. Crescenzi, G. Prencipe, and G. Pucci (Eds.): FUN 2007, LNCS 4475, pp. 156–165, 2007.

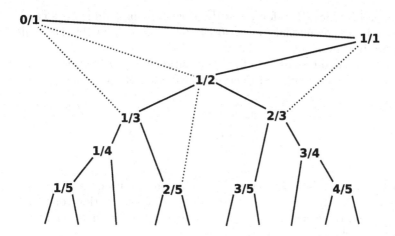

**Fig. 1.** The Stern-Brocot tree. Full lines show the edges of the tree as commonly defined. For select vertices dotted lines show their other "parent" vertex.

Note that the sequence $F_N$ can be created from $F_{N-1}$ by inserting those fractions $a/N$ where $a$ and $N$ are relatively prime. There are $\phi(N)$ such fractions, where $\phi(N)$ is Euler's totient function.

The importance of the property (1) lies in the fact that it gives us more insight into how the sequence is altered with increasing $N$. It tells us exactly where the next elements will appear – or equivalently, what is the next element that will appear between two currently neighboring ones.

When we denote this addition process graphically, we get the Stern-Brocot tree, (see [9,2]) as depicted in Figure 2. The construction and some properties of the tree can be found in [5].

The book [5] also notes the following important fact:

Suppose that $\alpha \in [0, 1)$ is a given real number, and we want to find the best rational approximation of $\alpha$ with denominator not exceeding a given $N$. Clearly, all we have to do is to find $\alpha$'s place in the sequence $F_N$, and consider the closest two elements (or one, if $\alpha$ is an element of $F_N$). This place can be found by an analogy of a "binary search" by descending the Stern-Brocot tree. Each visited vertex of the tree clearly represents either the best lower, or the best upper approximation so far. (Here, "so far" means "with denumerator smaller or equal to the one in the current vertex".)

For example, suppose that $\alpha = 0.39147\ldots$, then the first few vertices visited will be: $1/1$, $1/2$, $1/3$, $2/5$, $3/8$, $5/13$, $7/18$, $\ldots$

This gives us our first algorithm to find a good rational approximation to a given value $\alpha$: Walk down the Stern-Brocot tree until the difference between $\alpha$ and the value in the current vertex is close enough to be acceptable. At this moment, you can be sure that no rational number with a smaller denominator gives a better approximation.

Sadly, this algorithm is far from being polynomial in the input size. As a trivial counterexample, consider $\alpha = 10^{-47}$. The path to $\alpha$ is long and boring, as it contains all fractions $1/x$ for $x \leq 10^{47}$. However, these difficulties can be overcome, and our polynomial-time algorithm is derived from this naive algorithm using two improvements.

Another related tool used to find good rational approximations are continued fractions. A simple continued fraction is a fraction of the form:

$$\alpha = a_0 + \cfrac{1}{a_1 + \cfrac{1}{a_2 + \cfrac{1}{a_3 + \cdots}}} \tag{2}$$

To save space, continued fractions are commonly written as $[a_0, a_1, a_2, a_3, \ldots]$. Every rational number has a finite simple continued fraction (or two of them, depending on the exact definition of terminating). Every irrational number can be represented as an infinite continued fraction.

By truncating the continued fraction of a real number $\alpha$ after a finite number of steps we obtain a rational approximation of $\alpha$. More exactly, the number $c_n$ as defined in (3) is called the $n$-th convergent of $\alpha$. It can be proved that any convergent of $\alpha$ is a best rational approximation of $\alpha$ in the sense that the convergent is nearer to $\alpha$ than any other fraction whose denominator is less than that of the convergent. Moreover, more exact bounds on how good this approximation is are known, but we won't need them in this article.

$$c_n = a_0 + \cfrac{1}{a_1 + \cfrac{1}{a_2 + \cfrac{1}{\cdots + \cfrac{1}{a_n}}}} \tag{3}$$

Note that there may be cases when the desired rational approximation of $\alpha$ is not a convergent of $\alpha$. More exactly, there are fractions $p/q$ such that:

- No fraction with denominator less than $q$ is closer to $\alpha$ than $p/q$.
- The fraction $p/q$ is not a convergent of $\alpha$.

As a simple example, suppose that $\alpha = 0.1$. The fraction $1/7$ satisfies the conditions above.

Furthermore, note that $1/7$ is the first fraction that after evaluating and rounding to one decimal place gives $\alpha$. In other words, $1/7$ is the best rational approximation of $\alpha = 0.1$ with a given tolerance $d = 0.05$.

## 3   Problem Formulation

The exact problem we are trying to solve can be formulated as follows: given is a rational number $\alpha \in (0, 1)$ and a precision $d$. (Both numbers are given in decimal notation, with $N$ digits each.) The goal is to find positive integers $p$, $q$ such that $p/q$ lies within the interval $(\alpha - d, \alpha + d)$, and $q$ is minimal.

Several notes on this definition:

We opted to limit the problem to $\alpha \in (0,1)$ instead of the more general task $\alpha \in \mathbb{R}$. However, note that the general problem is easily reduced to our definition: For $\alpha = 0$ the best answer is always $0/1$. Any $\alpha \in \mathbb{R} \setminus [0,1)$ can be written as $\lfloor \alpha \rfloor + \alpha'$, where $\alpha' \in [0,1)$. The best approximation of $\alpha$ can be computed by finding the best approximation of $\alpha'$ and adding the integer $\lfloor \alpha \rfloor$.

Also, by solving this task for the open interval $(\alpha - d, \alpha + d)$, we can easily obtain solutions for half-closed intervals $[\alpha - d, \alpha + d)$, $(\alpha - d, \alpha + d]$ and for the closed interval $[\alpha - d, \alpha + d]$ simply by checking whether the included bounds correspond to a better approximation than the one found.

**Theorem 1.** *The correct solution of the problem defined above is unique, and* $gcd(p, q) = 1$.

*Proof.* Both parts will be proved by contradiction. First, suppose that there are two best approximations $A = p_1/q$ and $B = p_2/q$ with the same denominator $q > 1$ and $p_1 < p_2 \leq q$. Consider the fraction $C = p_1/(q - 1)$. Clearly we have $C > A$. Moreover, we get:

$$p_2(q - 1) = p_2q - p_2 \geq p_2q - q = (p_2 - 1)q \geq p_1q \qquad (4)$$

From (4) it follows that $p_2/q \geq p_1/(q - 1)$, in other words $B \geq C$. Thus $C$ is also a valid approximation, and it has a smaller denominator.

The coprimeness of $p$ and $q$ in the optimal solution is obvious. Formally, suppose that $gcd(p, q) = g > 1$. Then clearly $p' = p/g$ and $q' = q/g$ is a valid approximation with a smaller denominator.    □

## 4    Mathematica and Similar Software

The software package Mathematica by Wolfram Research, Inc. claims to include a function `Rationalize` that solves the problem as stated above. (And variations thereof, as Mathematica can work both with exact and with approximate values.) Citing from the documentation [11],

- `Rationalize[x]` takes Real numbers in x that are close to rationals, and converts them to exact Rational numbers.
- `Rationalize[x,dx]` performs the conversion whenever the error made is smaller in magnitude than dx
- `Rationalize[x,dx]` yields the rational number with the smallest denominator that lies within dx of x

However, we discovered that this is not the case. We tried to contact the authors of the software package, but to date we received no reply. In Table 1 we demonstrate a set of inputs where Mathematica 5.2 fails to find the optimal solution. The exact command used was `Rationalize[alpha''200,d''200]`, where the ''200 part tells the software to consider the values as arbitrary-precision

**Table 1.** Example inputs where Mathematica fails to find the optimal approximation

| input | correct | Mathematica |
|---|---|---|
| $\alpha = 0.1$, $d = 0.05$ | 1/7 | 1/9 |
| $\alpha = 0.12$, $d = 0.005$ | 2/17 | 3/25 |
| $\alpha = 0.15$, $d = 0.005$ | 2/13 | 3/20 |
| $\alpha = 0.111\,112$, $d = 0.000\,000\,5$ | 888 890/8 000 009 | 1 388 889/12 500 000 |
| $\alpha = 0.111\,125$, $d = 0.000\,000\,5$ | 859/7730 | 889/8000 |

values with 200 digits of accuracy. (See [10] for more details.) The counterexamples in the last two lines of Table 1 seem to scale. (I.e., we can get new inputs by increasing the number of 1s in $\alpha$ and the number of 0s in $d$ by the same amount. Mathematica fails on most of these inputs.)

Furthermore, if the first argument is an exact number, Mathematica leaves it intact, which might be not optimal. E.g., `Rationalize[30/100,4/100]` returns $3/10$ instead of the correct output $1/3$.

We are not aware of any other software that claims to have this functionality. For example, Matlab does include a function `rat`, but the documentation [8] doesn't guarantee optimality of approximation in our sense – on the contrary, it directly claims that `rat` only approximates by computing a truncated continued fraction expansion.

## 5    Outline of Our Algorithm

We will derive the polynomial algorithm in two steps. First, we will show a compressed way of traversing the Stern-Brocot tree. We will show that this approach is related to generating the approximations given by the continued fraction. However, keeping the Stern-Brocot tree in mind will help us in the second step, where we show how to compute the best approximation once we reach a valid approximation using the compressed tree traversal.

Imagine that we start traversing the Stern-Brocot tree in the vertex $1/1$. The path downwards can be written as a sequence of '$L$'s and '$R$'s (going left and right, respectively).

E.g., for $\alpha = 0.112$ we get the sequence $LLLLLLLLLRLLLLLLLLLLLL$. This can be shortened to $L^9RL^{12}$. As noted in [5], there is a direct correspondence between these exponents and the continued fraction representation of $\alpha$. As an example, note that

$$0.112 = 0 + \cfrac{1}{8 + \cfrac{1}{1 + \cfrac{1}{13}}} = 0 + \cfrac{1}{8 + \cfrac{1}{1 + \cfrac{1}{12 + \frac{1}{1}}}} \qquad (5)$$

For a given path in the Stern-Brocot tree let $(a_0, a_1, a_2, \ldots)$ be the sequence of exponents obtained using the path compression as described above. Our algorithm will visit vertices corresponding to paths $L^{a_0}$, $L^{a_0}R^{a_1}$, $L^{a_0}R^{a_1}L^{a_2}$, and so on. In each such vertex we will efficiently compute the next value $a_i$. A detailed description of this computation can be found in Section 6.

For the example above, $\alpha = 0.112$, our algorithm will visit the vertices corresponding to $L^9$, $L^9R$, and $L^9RL^{12}$. These correspond to fractions $1/9$, $2/17$, and $14/125 = \alpha$.

For comparison note that the subsequent approximations of $\alpha = 0.112$ by truncating the continued fractions are: 0, 1/8, 1/9, (13/116 and)[1] 14/125 $= \alpha$.

In Section 7 we will modify the compressed tree traversal algorithm slightly. It can be shown that after the modification the set of examined vertices will be a superset of the convergents, however we omit the proof, as it is not necessary to prove the correctness of our algorithm.

Now, let's assume that the exact input instance is $\alpha = 0.112$ and $d = 0.0006$. For this input, the algorithm as defined above skips the optimal solution $9/80 = 0.1125$.

In Section 7 we will show that the best approximation has to lie on the last compressed branch of the visited path. More generally, it has to lie on the first compressed branch such that its final vertex represents a valid approximation. Also, we will show that this approximation can be computed efficiently.

An example implementation of the algorithm can be found in Section 8.

# 6 Compressed Tree Traversal

We will describe traversal to the left, i.e., to a smaller fraction than the current one. Traversal to the right works in the same way.

Let $p_a/q_a$ and $p_b/q_b$ be the last two vertices visited, with $p_a/q_a < \alpha < p_b/q_b$. At the beginning of the entire algorithm we are in this situation with $p_a = 0$ and $q_a = p_b = q_b = 1$.

We are now in the vertex corresponding to $p_b/q_b$, and we are going to make several steps to the left. Using (1) we get that in our situation each step to the left corresponds to changing the current fraction $p/q$ into a new fraction $(p + p_a)/(q + q_a)$.

Let $x$ be the number of steps performed by the naive algorithm. How to compute the value $x$ without actually simulating the process?

Clearly, $x$ is the smallest such value that the fraction we reach is smaller than or equal to $\alpha$. We get an inequality:

$$\frac{p_b + xp_a}{q_b + xq_a} \leq \alpha \tag{6}$$

This can be simplified to

$$p_b - \alpha q_b \leq x(\alpha q_a - p_a) \tag{7}$$

---

[1] The presence of this continuent depends on the chosen form of the continued fraction, see equation (5).

We assumed that $p_a/q_a < \alpha$, thus $(\alpha q_a - p_a)$ is positive, and we get

$$x \geq \frac{p_b - \alpha q_b}{\alpha q_a - p_a} \tag{8}$$

As $x$ has to be an integer, and we are interested in the smallest possible $x$, the correct $x$ can be computed as follows:

$$x = \left\lceil \frac{p_b - \alpha q_b}{\alpha q_a - p_a} \right\rceil \tag{9}$$

## 7     Locating the Best Approximation

**Theorem 2.** *Suppose that while going in one direction the naive algorithm visited exactly the fractions $p_0/q_0$, $\ldots$, $p_n/q_n$. Consider the fractions $p_0/q_0$, $p_{n-1}/q_{n-1}$ and $p_n/q_n$. If neither of these three fractions represents a valid approximation (i.e., lies within $d$ of $\alpha$), then none of the $p_i/q_i$ represent a valid approximation.*

*Proof.* WLOG let's consider steps to the left. $p_n/q_n$ is the first and only value out of all $p_i/q_i$ such that $p_n/q_n \leq \alpha$. The fact that $p_n/q_n$ is not a valid approximation gives us that in fact $p_n/q_n \leq \alpha - d$. Similarly, $p_{n-1}/q_{n-1} \geq \alpha + d$, and clearly all other $p_i/q_i$ are even greater.                                            □

This gives us a way how to check that our compressed tree traversal algorithm doesn't skip any valid approximations: In each step, after computing the value $x$ from (9), check whether making either $x - 1$ or $x$ steps yields a valid approximation. If not, Theorem 2 tells us that we may make $x$ steps without skipping a valid approximation. If we found a valid approximation, we need to find the smallest one.

Again, for simplicity we will only show the case when the naive algorithm makes steps to the left. In this case, the best approximation simply corresponds to the smallest $k \leq x$ such that the fraction after $k$ steps is a valid approximation – or, equivalently, is less than $\alpha + d$.

We can compute this $k$ in very much the same fashion as when we computed $x$ in the previous section. The exact formula for $k$ for the going-to-left case:

$$k = \left\lfloor \frac{p_b - (\alpha + d)q_b}{(\alpha + d)q_a - p_a} \right\rfloor + 1 \tag{10}$$

## 8     Implementation

A proof-of-concept implementation of our algorithm in the open-source calculator `bc` follows. All computations are done using arbitrary-precision integers. For this reason, the input is given in the variables `alpha_num`, `d_num`, and `denum`. These variables represent the values $\alpha = $ `alpha_num/denum` and

$d =$ d_num/denum. We assume that $0 <$ d_num $<$ alpha_num $<$ denum and that d_num + alpha_num $<$ denum. This exactly covers the cases when $\alpha \in (0,1)$ and the answer is not $0/1$ or $1/1$.

Example: The input $\alpha = 0.33456$ and $d = 0.000005$ can be entered as alpha_num $= 334560$, d_num $= 5$, and denum $= 1000000$.

```
# input variables: alpha_num, d_num, denum
#
# we seek the first fraction that falls into the interval
# (alpha-d, alpha+d) =
#    = ( (alpha_num-d_num)/denum, (alpha_num+d_num)/denum )

# fraction comparison: compare (a/b) and (c/d)
define less(a,b,c,d) { return (a*d < b*c); }
define less_or_equal(a,b,c,d) { return (a*d <= b*c); }

# check whether a/b is a valid approximation
define matches(a,b) {
  if (less_or_equal(a,b,alpha_num-d_num,denum)) return 0;
  if (less(a,b,alpha_num+d_num,denum)) return 1;
  return 0;
}

# set initial bounds for the search:
p_a = 0 ; q_a = 1 ; p_b = 1 ; q_b = 1

define find_exact_solution_left(p_a,q_a,p_b,q_b) {
  k_num = denum * p_b - (alpha_num + d_num) * q_b
  k_denum = (alpha_num + d_num) * q_a - denum * p_a
  k = (k_num / k_denum) + 1
  print (p_b + k*p_a)," ",(q_b + k*q_a),"\n";
}
define find_exact_solution_right(p_a,q_a,p_b,q_b) {
  k_num = - denum * p_b + (alpha_num - d_num) * q_b
  k_denum = - (alpha_num - d_num) * q_a + denum * p_a
  k = (k_num / k_denum) + 1
  print (p_b + k*p_a)," ",(q_b + k*q_a),"\n";
}

while (1) {
  # compute the number of steps to the left
  x_num = denum * p_b - alpha_num * q_b
  x_denum = - denum * p_a + alpha_num * q_a
  x = (x_num + x_denum - 1) / x_denum # = ceil(x_num / x_denum)

  # check whether we have a valid approximation
  aa = matches( p_b + x*p_a, q_b + x*q_a )
  bb = matches( p_b + (x-1)*p_a, q_b + (x-1)*q_a )
  if (aa || bb) { cc = find_exact_solution_left(p_a,q_a,p_b,q_b); break;}
```

```
# update the interval
new_p_a = p_b + (x-1)*p_a ; new_q_a =  q_b + (x-1)*q_a
new_p_b = p_b + x*p_a       ; new_q_b =  q_b + x*q_a

p_a = new_p_a ; q_a = new_q_a
p_b = new_p_b ; q_b = new_q_b

# compute the number of steps to the right
x_num = alpha_num * q_b - denum * p_b
x_denum = - alpha_num * q_a + denum * p_a
x = (x_num + x_denum - 1) / x_denum # = ceil(x_num / x_denum)

# check whether we have a valid approximation
aa = matches( p_b + x*p_a, q_b + x*q_a )
bb = matches( p_b + (x-1)*p_a, q_b + (x-1)*q_a )
if (aa || bb) { cc = find_exact_solution_right(p_a,q_a,p_b,q_b); break;}

# update the interval
new_p_a = p_b + (x-1)*p_a ; new_q_a =  q_b + (x-1)*q_a
new_p_b = p_b + x*p_a       ; new_q_b =  q_b + x*q_a

p_a = new_p_a ; q_a = new_q_a
p_b = new_p_b ; q_b = new_q_b
}
```

## 9    Time Complexity

We claim that after each pass through the main while-loop of our implementation each of the values $q_a$ and $q_b$ at least doubles. To prove this, note the following: After the steps to the left the new two denominators are greater than or equal to $q_b$ and $q_a + q_b$. After the steps to the right the final two denominators are greater than or equal to $q_a + q_b$ and $q_a + 2q_b$.

Now, suppose that the input numbers have at most $N$ digits. The fraction alpha_num/denum is clearly a valid approximation. Therefore the optimal solution has a denominator with at most $N$ digits.

After the while-loop was executed $K$ times, the denominators of the currently examined fractions are greater than or equal to $2^K$. Clearly after $4N$ loops the denominators would exceed $N$ digits in length. Thus the while-loop will be executed $O(N)$ times only.

Each execution of the while-loop involves a constant number of operations with $O(N)$ digits long integers. The required operations are addition, subtraction, multiplication and division. The first two operations can easily be done in $O(N)$. For multiplication and division our implementation uses the naive $O(N^2)$ algorithm. Thus the running time of our solution is $O(N^3)$ – polynomial in the input size.

Clearly the bottleneck are the algorithms for multiplication and division. However, there are faster algorithms for both operations, for example the FFT-based multiplication algorithm running in $O(N \log N)$, and its corresponding division algorithm that uses multiplication and Newton's method to estimate the reciprocal of the denominator. For a discussion of these algorithms, see [7].

## Acknowledgements and Notes

This work has been supported in part by the VEGA grant 1/3106/06.

The author would like to thank three anonymous referees, whose remarks helped to improve the readability of the article.

The author used the main result of this article as a task in the programming competition IPSC, see [4].

## References

1. Beiler, A.H.: Farey Tails. Ch. 16 in Recreations in the Theory of Numbers: The Queen of Mathematics Entertains. Dover, New York (1966)
2. Brocot, A.: Calcul des rouages par approximation, nouvelle methode. Revue Chronométrique 6, 186–194 (1860)
3. Farey, J.: On a Curious Property of Vulgar Fractions. London, Edinburgh and Dublin Phil. Mag. 47, 385 (1816)
4. Forišek, M.: IPSC task Exact? Approximate! http://ipsc.ksp.sk/contests/ipsc2005/real/problems/e.php [accessed January 2007]
5. Graham, R.L., Knuth, D.E., Patashnik, O.: Concrete Mathematics: A Foundation for Computer Science, 2nd edn. Addison-Wesley Professional, London (1994)
6. Hardy, G.H., Wright, E.M.: Farey Series and a Theorem of Minkowski. In: An Introduction to the Theory of Numbers, 5th edn. ch. 3, pp. 23–37. Clarendon Press, Oxford England (1979)
7. Knuth, D.E.: How Fast Can We Multiply? Sec. 4.3.3 in The Art of Computer Programming II. Addison-Wesley Professional, London (1997)
8. The MathWorks, Inc.: MATLAB Function Reference: rat, rats http://www.mathworks.com/access/helpdesk/help/techdoc/ref/rat.html accessed (January 2007)
9. Stern, M.A.: Ueber eine zahlentheoretische Funktion. Journal für die reine und angewandte Mathematik 55, 193–220 (1858)
10. Wolfram Research, Inc.: Mathematica 5.2 Documentation: Numerical Precision http://documents.wolfram.com/mathematica/book/section-3.1.4 accessed (January 2007)
11. Wolfram Research, Inc.: Mathematica 5.2 Documentation: Rationalize http://documents.wolfram.com/mathematica/functions/Rationalize accessed (January 2007)

# Cryptographic and Physical Zero-Knowledge Proof Systems for Solutions of Sudoku Puzzles

Ronen Gradwohl[1,*], Moni Naor[2,**], Benny Pinkas[3,**],
and Guy N. Rothblum[4,* * *]

[1] Department of Computer Science and Applied Math, The Weizmann Institute of
Science, Rehovot 76100, Israel
ronen.gradwohl@weizmann.ac.il
[2] Incumbent of the Judith Kleeman Professorial Chair, Department of Computer
Science and Applied Math, The Weizmann Institute of Science, Rehovot 76100, Israel
moni.naor@weizmann.ac.il
[3] Department of Computer Science, University of Haifa, Haifa, Israel
benny@pinkas.net
[4] CSAIL, MIT, Cambridge, MA 02139, USA
rothblum@csail.mit.edu

**Abstract.** We consider cryptographic and physical zero-knowledge
proof schemes for Sudoku, a popular combinatorial puzzle. We discuss
methods that allow one party, the prover, to convince another party, the
verifier, that the prover has solved a Sudoku puzzle, without revealing
the solution to the verifier. The question of interest is how a prover can
show: (i) that there is a solution to the given puzzle, and (ii) that he
knows the solution, while not giving away any information about the
solution to the verifier.

In this paper we consider several protocols that achieve these goals.
Broadly speaking, the protocols are either cryptographic or physical. By
a cryptographic protocol we mean one in the usual model found in the
foundations of cryptography literature. In this model, two machines exchange messages, and the security of the protocol relies on computational
hardness. By a physical protocol we mean one that is implementable
by humans using common objects, and preferably without the aid of
computers. In particular, our physical protocols utilize scratch-off cards,
similar to those used in lotteries, or even just simple playing cards.

The cryptographic protocols are direct and efficient, and do not involve
a reduction to other problems. The physical protocols are meant to be understood by "lay-people" and implementable without the use of computers.

## 1 Introduction

Sudoku is a combinatorial puzzle that swept the world in 2005 (especially via
newspapers, where it appears next to crossword puzzles), following the lead of

---

* Research supported by US-Israel Binational Science Foundation Grant 2002246.
** Research supported in part by a grant from the Israel Science Foundation.
* * * Research supported by NSF grant CNS-0430450 and NSF grant CFF-0635297.

Japan (see the Wikipedia entry [19] or the American Scientist article [11]). In a Sudoku puzzle the challenge is a 9×9 grid subdivided into nine 3×3 subgrids. Some of the cells are already set with values in the range 1 through 9 and the goal is to fill the remaining cells with numbers 1 through 9 so that each number appears exactly once in each row, column and subgrid. Part of the charm and appeal of Sudoku appears to be the ease of description of the problems, as compared to the time and effort it takes one to solve them.

A natural issue, at least for cryptographers, is how to convince someone else that you have solved a Sudoku puzzle without revealing the solution. In other words, the question of interest here is: how can a prover show (i) that there is a solution to the given puzzle, and (ii) that he knows the solution, while not giving away any information about the solution. In this paper we consider several types of methods for doing just that. Broadly speaking, the methods are either *cryptographic* or *physical*. By a *cryptographic* protocol we mean one in the usual model found in the foundations of cryptography literature. In this model, two machines exchange messages and the security of the protocol relies on computational hardness (see Goldreich [5] for an accessible account and [6] for a detailed one). By a *physical* protocol we mean one that is implementable by humans using common objects, and preferably without the aid of computers. In particular, our protocols utilize scratch-off cards, similar to those used in lotteries.

**This Work:** The general problem of Sudoku (on an $n \times n$ grid) is in the complexity class NP, which means that given a solution it is easy to *verify* that it is correct (In fact, Sudoku is known to be NP-Complete [20], but we are not going to use this fact, at least not explicitly.). Since there are cryptographic zero-knowledge proofs for all problems in NP [7], there exists one for Sudoku, via a reduction to 3-Colorability or some other NP-Complete problem with a known zero-knowledge proof (see definition in Section 2). In this work, however, we are interested in more than the mere existence of such a proof, but rather its efficiency, understandability, and practicality, which we now explain.

First, the benefits of a direct zero-knowledge proof (rather than via a reduction) are clear, as the overhead of the reduction is avoided. Thus, the size of the proof can be smaller, and the computation time shorter. In addition, we wish our proofs to be easy to understand by "non-experts". This is related to the practicality of the proof: the goal is to make the interaction implementable in the real world, perhaps even without the use of a computer. One of the important aspects of this implementability requirement is that the participants have an intuitive understanding of the correctness of the proof, and thus are convinced by it, rather than relying blindly "on the computer". For another example in which this intuitive understanding is important, see the work of Moran and Naor [13] on methods for polling people on sensitive issues.

The contributions of this paper are efficient cryptographic protocols for showing knowledge of a solution of a Sudoku puzzle which do not reveal any other useful information (these are known as zero-knowledge proofs of knowledge) and several transparent physical protocols that achieve the task.

**Organization:** In Section 2 we outline the definition of a zero-knowledge protocol, and the properties of the cryptographic and physical protocols. In section 3 we describe two cryptographic zero-knowledge protocols: the first protocol is very simple and direct, and the second is slightly more involved, but has a lower (better) probability of error. In Section 4 we describe several physical protocols, using envelopes and scratch-off cards. Finally, in Section 5 we discuss further research directions.

## 2    Definitions

**Sudoku:** An instance of Sudoku is defined by the size $n = k^2$ of the $n \times n$ grid, where the subgrids are of size $k \times k$. The indices, values in the filled-in cells and the values to be filled out are all in the range $\{1, \ldots, n\}$. Note that in general the size of an instance is $O(n^2 \log n)$ bits and this is the size of the solution (or witness) as well.

**Cryptographic Functionalities:** We only give rough descriptions of zero-knowledge and commitments. For more details, see the above mentioned books by Goldreich [5] and [6], Chapter 4 or the writeup by Vadhan [18]. In general, a zero-knowledge proof, as defined by Goldwasser, Micali and Rackoff [8], is an interactive-proof between two parties, a *prover* and a *verifier*. They both know an instance of a problem (e.g. a Sudoku puzzle) and the prover knows a solution or a witness. The two parties exchange messages and at the end of the protocol the verifier '*accepts*' or '*rejects*' the execution. The protocol is probabilistic, i.e. the messages that the two parties send to each other are functions of their inputs, the messages sent so far and their private random coins (sequence of random bits that each party is assumed to have in addition to its input). Once the programs of the verifier and prover are fixed, for a given instance the messages sent are a function of the random coins of the prover and verifier only. We will be discussing several properties of such protocols: completeness, soundness, zero-knowledge and proof-of-knowledge.

The *completeness* of the protocol is the probability that an honest verifier accepts a correct proof, i.e. one done by a prover holding a legitimate solution and following the protocol. All our protocols will have *perfect* completeness; a correct proof is *always* accepted (i.e. with probability 1). The probability is over the random coins of the prover and the verifier. The *soundness error* (or soundness) of the protocol is the (upper bound on the) probability that a verifier accepts an incorrect proof, i.e. a proof to a fallacious statement; in our case this is the statement that the prover knows a solution to the given Sudoku puzzle, even though it does not know such a solution.

The goal in designing the protocols is that the verifier should not gain any new knowledge from a *correct* (interactive) proof. I.e. the protocol should be *zero-knowledge* in the following sense: whatever a verifier could learn by interacting with the correct prover, the verifier could learn itself. To formalize this requirement, we require that there is an efficient *simulator* that could have generated the verifier's conversation with the prover without the benefit of the conversation actually occurring, based on knowing the puzzle only, without knowledge of the solution. Since the protocol is probabilistic, we consider the *distribution* of the conversation, the messages sent back and forth between the prover and verifier. We want the two distributions, the one of a conversation between the real prover and verifier, and the one that the simulator produces, to be indistinguishable. Furthermore, we want a simulator for any possible behavior of the verifier, even a verifier that does not follow the prescribed protocol.

Our protocols should also be *proofs-of-knowledge*: if the prover (or anyone impersonating him) can succeed in making the verifier accept, then there is another machine, called the *extractor*, that can communicate with the prover and actually come up with the solution itself. This must involve running the prover several times using the same randomness (which is not possible under normal circumstances), so as not to contradict the zero-knowledge properties.

The only cryptographic tool used by our proofs is a *commitment protocol*. A commitment protocol allows one party, the sender, to commit to a value to another party, the receiver, with the latter not learning anything meaningful about the value. Such a protocol consists of two phases. The first is the *commit* phase, following which the sender is bound to some value $v$, while the receiver cannot determine anything useful about $v$. In particular, this means that the receiver cannot distinguish between the case $v = b$ and $v = b'$ for all $b$ and $b'$. This property is called *hiding*. Later on, the two parties may perform a *decommit* or *reveal* phase, after which the receiver obtains $v$ and is assured that it is the original value; in other words, once the *commit* phase has ended, there is a unique value that the receiver will accept in the *reveal* phase. This property is called *binding*. Bit commitments can be based on any one-way function [14] and are fairly efficient to implement. Both the computational complexity and the communication complexity of such protocols are reasonable and in fact one can amortize the work if there are several simultaneous commitments. In this case, the amortized complexity of committing to a bit is $O(1)$.

Note that in this setting we think of the adversary as controlling one of the parties (prover and verifier) and as being malicious in its actions. The guarantees we make (both against a cheating prover trying to sneak in a fallacious proof and against a cheating verifier trying to learn more than it should) are with respect to *any* behavior.

**Physical Protocols:** While the cryptographic setting is well established and reasonably standard, when discussing 'physical' protocols there are many different options, ranging from a deck of cards [3,17] to a PEZ dispenser [1], a broadsheet newspaper [15], and more (see [12] for a short survey). In our setting we will be using tamper-evident sealed envelopes, as defined by Moran and Naor [12]. It is simplest to think of these as scratch-off cards: each card has a number on it from $\{1, \ldots, n\}$, but that number cannot be determined unless the card is scratched (or the envelope is opened and the seal is broken). Actually for two of our three physical protocols the tamper evident sealed envelopes can be implemented via standard playing cards. These are 'sealed' by turning a card face down and opened by turning the card over. For a demonstration of a zero-knowledge proof for Sudoku using only playing cards, see the web page [10].

We would like our physical protocols to enjoy zero-knowledge properties as well. For this to be meaningful we have to define the power of the physical objects that the protocol assumes as well as the assumptions on the behavior of the humans performing it. In general, the adversarial behavior we combat is more benign than the one in the cryptographic setting. See details in Section 4.

## 3   Cryptographic Protocols

We provide two cryptographic protocols for Sudoku. The setting is that we have a prover and a verifier who both know an instance of an $n \times n$ Sudoku puzzle, i.e. a subset of the cells with predetermined values. The prover knows a solution to the instance and the verifier wants to make sure that (i) a solution exists and (ii) the prover knows the solution.

The protocols presented are in the standard cryptographic setting, as described in Section 2. The structure of the proof is as follows, which is common to many zero-knowledge protocols:

1. The prover commits to several values. These values are functions of the instance, the solution and some randomization known only to the prover.
2. The verifier requests that the prover open some of the committed values – this is called the *challenge*. The verifier chooses the challenge at random from a collection of possible challenges.
3. The prover opens the requested values.
4. The verifier checks the consistency of the opened values with the given instance, and accepts or rejects accordingly.

The only cryptographic primitive we use in both protocols is bit or string *commitment* as described above.

To prove a protocol with the structure above is zero-knowledge we use the so called 'standard' argument, due to [7]: we require that the distribution of the values opened in Step 3 is an efficiently computable function of the Sudoku puzzle and the challenge the verifier sent in Step 2 (but *not* of the puzzle's solution, e.g. it could be a random permutation of $\{1, \ldots, n\}$). If the *number* of possible challenges in Step 2 is polynomial in the size of the Sudoku puzzle, then this property, together with the indistinguishably property of the commitment protocol, implies the existence of an efficient simulator, as described below.

The simulator operates in the following way: it picks at random a challenge that the verifier might send in Step 2 (i.e. it guesses what the verifier's challenge will be), and computes commitments for Step 1 that will satisfy this challenge. The simulator simulates sending these commitments to the verifier, then it runs the verifier's algorithm with the puzzle as its input, a fresh set of random bits and these commitments being the first message it receives. It then obtains the challenge the verifier sends in Step 2. If this challenge is indeed the value it guessed, then the simulator can open the commitments it sent and the verifier should accept; the simulator can continue simulating the protocol and output the transcript of the simulated protocol execution. Otherwise, the simulator resets the simulation and starts it all over again.

If the number of possible challenges is polynomial, then each time the simulator "guesses" the verifier's challenge, it is correct with some 'reasonably high' probability (i.e. at least an inverse polynomial). Therefore within a polynomial number of tries the simulator is expected to guess the verifier's challenge correctly and the simulation process succeeds. This procedure guarantees that the protocol is zero knowledge because the output of the simulator looks very much like a successful execution of the proof protocol. I.e., the output of the simulator is indistinguishable from what the verifier would see when interacting with the prover, but is computed without ever talking with the prover!

The two protocols we provide are based on two classic zero-knowledge protocols for NP problems: for 3-Colorability and Graph Hamiltonicity. We find it interesting that while the original protocols seem to fit different types of problems, we could efficiently adapt both of them for the same problem.

### 3.1    A Protocol Based on Coloring

The following protocol is an adaptation of the famed GMW zero-knowledge proof of 3-Colorability of a graph [7] (see [6]) for Sudoku puzzles. The idea there was for the prover to randomly permute the colors and then commit to the (permuted) color of each vertex. The verifier picks a random edge and checks that its two end points are colored differently. To apply this idea in the context of Sudoku it helps to think of the graph as being partially colored to begin with, so one should also check consistency with the partial coloring. The resulting protocol consists of the prover randomly permuting the numbers and committing to the resulting solution. What the verifier checks is either the correctness of the values of one of the rows, columns or subgrids, or consistency with the filled-in values. The protocol operates in the following way:

**Protocol 1.** *A cryptographic protocol with* $1 - \frac{1}{3n+1}$ *soundness error* **Prover**

1. *Prover chooses a random permutation* $\sigma : \{1, \ldots, n\} \mapsto \{1, \ldots, n\}$.
2. *For each cell* $(i, j)$ *with value* $v$, *prover sends to verifier a commitment for the value* $\sigma(v)$.

**Verifier:** *Chooses at random one of the following* $3n + 1$ *possibilities: a row, column or subgrid* (*3n possibilities*), *or 'filled-in cells', and asks the prover to open the corresponding commitments. After the prover responds, in case the verifier chose a row, column or subgrid, the verifier checks that all values are indeed different. In case the verifier chose the filled-in cells option, it checks that cells that originally had the same value still have the same value (although the value may be different), and that cells with different values are still different, i.e. that* $\sigma$ *is indeed a permutation over the values in the filled-in cells.*

*Proof sketch for the required properties:* The perfect *completeness* of the protocol is straightforward. *Soundness* follows from the fact that any cheating prover must cheat either in his commitments for a row, column, subgrid, or the filled-in cells (namely, there is at least one question of the verifier for which the prover cannot provide a correct answer). Thus, the verifier catches a cheating prover with probability at least $1/(3n+1)$. Note also that the protocol is a *proof-of-knowledge*, since if the prover convinces the verifier with high probability this means that *all* the $3n+1$ queries can be answered properly, and then it is possible to find a solution to the original puzzle (simply find a reverse permutation $\sigma^{-1}$ mapping the filled-in values). The distribution on the values of the answer when the challenge is a row, column or subgrid is simply a random permutation of $\{1, \ldots, n\}$. The distribution in case the challenge is filled-in cells is a random injection of the values appearing in those cells to $\{1, \ldots, n\}$. Therefore the zero-knowledge property of the protocol follows the standard arguments. The witness/solution size, as well as the number of bits committed, are both $O(n^2 \log n)$ bits.

### 3.2   An Efficient Cryptographic Protocol with Constant Soundness Error

Below is a more efficient zero-knowledge protocol for the solution of a Sudoku puzzle. It is closest in nature to Blum's protocol for proving the existence of a Hamiltonian Cycle [2]. The protocol described has constant (2/3) soundness error for an $n \times n$ Sudoku problem, and its complexity in terms of the number of bits committed to is $O(n^2 \log n)$, which is also the witness/solution size.

The idea of the protocol is to triplicate each cell, creating a version of the cell for the row, column and subgrid in which it participates. The triplicated cells are then randomly permuted and the prover's job is to demonstrate that the following properties hold:

a. The cells corresponding to the rows, columns and subgrids have all possible values.
b. The three copies of each cell have the same value.
c. The cells corresponding to the predetermined values indeed contain them.

If all three conditions are met, then, as we show below, there is a solution and the prover knows it. The following protocol implements this idea:

**Protocol 2. *A cryptographic protocol with 2/3 soundness error***
**Prover**

1. *Commit to $3n^2$ values $v_1, v_2, \ldots, v_{3n^2}$ where each cell of the grid corresponds to three randomly located indices $(i_1, i_2, i_3)$. The values of $v_{i_1}, v_{i_2}$ and $v_{i_3}$ should be the value $v$ of the cell in the solution.*
2. *Commit to $n^2$ triples of locations in the range $\{1, \ldots, 3n^2\}$, where each triple $(i_1, i_2, i_3)$ corresponds to the locations of a cell of the grid in the list of commitments of Item 1.*
3. *Commit to the names of the grid cells of each triple from Item 2.*

4. *Commit to $3n$ sets of locations from Item 1, corresponding to the rows, columns and subgrids, where each set is of size $n$ and no two cells intersect.*

**Verifier:** *Ask one of the following three options at random:*

a. *Open all $3n^2$ commitments of Item 1 and the commitments of Item 4. When the answer is received, verify that each set contains $n$ different numbers.*

b. *Open all $3n^2$ commitments of Item 1 and the commitments of Item 2. When the answer is received, verify that each triple contains the same numbers.*

c. *Open the commitments of Items 2, 3 and 4 as well as the commitments of Item 1 corresponding to filled-in cells in the Sudoku puzzle. When the answer is received, verify the consistency of the commitments with (i) the predetermined values, (ii) the set partitions of Item 4 and (iii) the naming of the triples.*

Each option for the verifier's query checks a corresponding property from the list of properties that the prover must prove. Option (a) checks the constraint that all values should appear in each row, column and subgrid (item (a) in the list of properties above). Option (b) makes sure that the value of the cell is consistent in its three appearances. Option (c) makes sure that the filled-in cells have the correct value and that the partitioning of the cells to rows, columns and subgrids is as it should be. Therefore, if all three challenges (a, b and c) are met, then we have a solution to the given Sudoku puzzle. This is a proof-of-knowledge as well, since the answers to all the options of the verifier's queries reveal the solution to the puzzle. As for soundness, a cheating prover cannot successfully answer all three of the possible challenges, and thus with probability at least $1/3$ the verifier rejects. The probability of cheating is at most $2/3$. As before, perfect completeness of the protocol is straightforward. Regarding the zero-knowledge property, note that for each challenge it is easy to describe the distribution on the desired response, and so the zero-knowledge of the protocol follows from standard arguments, as outlined in the beginning of the section.

**Overhead of our protocols:** The communication complexity and computation time of both protocols presented here is similar (assuming efficient commitments), and is $O(n^2 \log n)$. However, the first protocol allows the prover to cheat (without being caught) with relatively high probability, $(1 - 1/(3n+1))$, while the second protocol has a constant probability of catching a cheater. In both cases the soundness can be decreased by repeating the protocols several times, either sequentially or in parallel (for parallel repetition more involved protocols have to be applied, see [6], to preserve the zero-knowledge property). Therefore, to reduce the cheating probability to $\varepsilon$, the first protocol has to be repeated $O(n \log(1/\varepsilon))$ times and the resulting communication complexity is $O(n^3 \log n \log 1/\varepsilon)$ bits, while the second protocol should be repeated only $O(\log 1/\varepsilon)$ times, and the resulting communication complexity is $O(n^2 \log n \log 1/\varepsilon)$ bits.

# 4  Physical Protocols

The protocols described in Section 3 can both have a physical analog, given some physical way to implement the commitments. The problematic point is that tests such as checking that the set partitions and the naming of the triples are consistent (needed in challenge (c) of the protocol in Section 3.2) are not easy for humans to perform. In this section we describe protocols that are designed with human execution in mind, taking into account the strengths and weaknesses of such beings.

**Tamper evidence as a physical cryptographic primitive:** A locked box is a common metaphoric description of bit (or string) commitment, where the commiter puts the hidden secret inside the box, locks it, keeps the key but gives the box to the receiver. At the *reveal* stage he gives the key to the receiver who opens it. The problem with this description is that the assumption is that the receiver can *never* open the box without the key. It is difficult to imagine a physical box with such a guarantee that is also readily available, and its operation transparent to humans.[1] A different physical metaphor was proposed by Moran and Naor [12], who suggested concentrating on the *tamper-evident* properties of sealed envelopes and scratch-off cards. That is, anyone holding the envelopes can open them and see the value inside, but this act is not reversible and it will be transparent to anyone examining the envelope in the future. Another property we require from our envelopes is that they be indistinguishable, i.e. it should be *impossible to tell two envelopes apart*, at least by the party that did not create them (this is a little weaker than the indistinguishable envelope model formalized in [12]).

Another distinction between our physical model and the cryptographic one has to do with the way in which we regard the adversary. Specifically, the adversary we combat in the physical model is more benign than the one considered in the cryptographic setting or the one in [12,13]. We can think of our parties as not wanting to be labelled 'cheaters', and so the assurance we provide is that either the protocol achieves its goal or the (cheating) party is labelled a cheater.

We think of the prover and verifier as being present in the same room, and in particular the protocols we describe are *not* appropriate for execution over the postal system (see Section 5). The presence of the two parties in the same room is required since the protocols use such operations as shuffling a given set of envelopes - one party wants to make sure that the shuffle is appropriate, while the other party wants to make sure that the original set of envelopes is indeed the one being shuffled.

We also need two of additional functionalities that are not included in the vanilla model of sealed envelopes ([12,13]): *shuffle* and *triplicate*. The *shuffle* functionality is essentially an indistinguishable shuffle of a set of seals. Suppose some party has a sequence of seals $L_1, \ldots, L_i$ in his possession. Invoking the *shuffle* functionality on this sequence is equivalent to picking $\sigma \in_R S_i$, i.e. a

---

[1] Perhaps quantum cryptography can yield an approximation to such a box, but not a perfect one. See the discussion in [12].

random permutation on $i$ elements, to yield the sequence $L_{\sigma(1)}, \ldots, L_{\sigma(i)}$. The *triplicate* functionality is used only in our last protocol, so we defer its description to Section 4.2.

It is easy to apply in the physical setting described above, the same definitions of completeness and soundness as in the cryptographic setting. The definition of zero-knowledge in the physical setting can be made rigorous: as in the cryptographic case, we need to come up with a simulator that can emulate the interaction between the prover and verifier. The simulator interacts with a cheating verifier, runs in probabilistic polynomial time, and produces an interaction that is indistinguishable from the verifier's interaction with the prover. The simulator does not have a correct solution to the Sudoku instance, but it does have an advantage over the prover: at any point in time it is allowed to swap an unscratched off card with another. This advantage replaces the ability of simulators to "rewind" the verifier in cryptographic zero-knowledge protocols. The appropriate analogy is editing a movie, as first suggested in [16]. When making a movie of the proof one can swap the cards and edit the movie so it is unnoticeable. The result is indistinguishable from what one would see in a real execution. We will describe such simulators in Sections 4.1 and 4.2.

Finally, since we want protocols that are also proofs-of-knowledge, we will describe *extractors* that interact with honest provers in the physical setting and extract a correct solution for the Sudoku instance.

**An implementation without using any scratch-off cards:** Given that the setting we consider involves the prover and receiver being in the same room there is a very simple implementation for sealed envelopes, without scratch-off cards or envelopes: standard playing cards. Sealing a value means that a card with this value is placed faced down. The equivalent of scratching off or opening the value is simply turning the card over so that it is face up. Tamper evidence is achieved by making sure that no card is turned over before it should be. The prevalence of playing cards and the experience people have in shuffling such cards makes this implementation very attractive. This implementation is relevant for the first two protocols. A demonstration of running the first protocol using only playing cards is documented in the web page [10].

### 4.1   A Physical Zero-Knowledge Protocol with Constant Soundness Error

In the following protocol, the probability that a cheating prover will be caught is at least 8/9. The main idea is that each cell should have three (identical) cards; instead of running a subprotocol to check that the values of each triple are indeed identical we let the verifier make the assignment of the three cards to the corresponding row, column and subgrid at random. The protocol operates in the following way:

**Protocol 3.** *A physical protocol with 1/9 soundness error*

- *The prover places three scratch-off cards on each cell. On filled-in cells, he places three cards with the correct value, which are already open (scratched).*

- *For each row/column/subgrid, the verifier chooses (at random) one of the three cards of each cell in the corresponding row/column/subgrid.*
- *The prover makes packets of the verifier's requested cards (i.e. for every row/column/subgrid, he assembles the requested cards). He then shuffles each of the 3n packets separately (using the* shuffle *functionality), and hands the shuffled packets to the verifier.*
- *The verifier scratches off all the cards in each packet and verifies that each packet contains all of the numbers.*

**An implementation with playing cards:** As mentioned above, this protocol can be implemented using standard playing cards, without any scratch-off layer. In the first step the prover puts all cards face down, except for those cards in filled-in cells, which are put face up. In the following steps the verifier chooses cards, and the prover makes packets and shuffles them, without turning over the cards. Only in the last step do the parties turn the cards over and examine their values.

Consider the typical 9×9 case. The total number of cards needed is $3 \cdot 81 = 243$ cards, 27 cards of each type. We want to use standard packs of playing cards, (it is important that they have identical backs). Using only the cards numbered 1 to 9, discarding all other cards, requires 7 packs (if all the cards are used, 5 packs suffice). So the equipment needed to execute the protocol for any puzzle is a large sheet with the $9 \times 9$ grid marked on it and several packs of cards. A demonstration of running the protocol in this manner is documented in the web page [10].

**Completeness:** Perfect completeness of the protocol is straightforward.

**Soundness:** We claim that the soundness error of the protocol is $1/9$. We describe a simple argument showing that the soundness error is $1/3$ and provide a more involved analysis showing that it is indeed $1/9$. Assume that the prover does not know a valid solution for the puzzle. Then he is always caught by the protocol as a liar if he places the cards such that each cell has three cards of identical value. The only way a cheating prover can cheat is by placing three cards that are not all of the same value on a cell, say cell $a$. This means that in this cell at least one value $y$ must be different from all others. Suppose that for all other cells the verifier has already assigned the cards to the rows, columns and subgrids. A necessary condition for the (cheating) prover to succeed is that given the assignments of all cells except $a$ there is exactly one row, column or subgrid that needs $y$ to complete the values in $\{1, \ldots, n\}$. The probability that for cell $a$ the verifier assigns $y$ to the row, column or subgrid that needs it is $1/3$.

We now sketch a more involved argument that shows that the soundness error is actually $1/9$. We know that there is a cell where not all three values are the same. Also, the total number of cards of each value must be correct, otherwise the prover will be caught with probability 1. Thus, there must be at least two cells on which the prover cheats, say $a$ and $b$. We will consider different ways in which a prover can cheat on these cells, and show that his success probability is bounded above by $1/9$.

First suppose the prover cheats on exactly two cells, say $a$ and $b$, and suppose the values are $(x, x, y)$ for cell $a$ and $(y, y, x)$ for cell $b$. Note that this is the only way he can cheat on exactly two cells without being caught with probability 1. There are three possibilities for the location of cells $a$ and $b$: it may be that they are not in the same row, column or sub-grid, they may be same row, column or subgrid (exactly one of them), and they may be both in the same row or column and in the same subgrid. We have to analyze the cheating prover's probability of being caught for each of these cases. This analysis (as well as the case where there are more than two cells on which the prover cheats) is given in the full version [9].

**Zero-Knowledge:** To show that Protocol 3 is zero-knowledge, we have to describe an efficient simulator that interacts with a cheating verifier, and produces an interaction that is indistinguishable from the verifier's interaction with the prover. The simulator does not have a correct solution to the Sudoku instance, but it does have an advantage over the prover: before handing the shuffled packets to the verifier, it is allowed to swap the packets for different ones (see the discussion above). The simulator acts as follows:

- The simulator places three *arbitrary* scratch-off cards on each cell.
- After the verifier chooses the cards for the corresponding packets, the simulator takes them and shuffles them (just as the prover does).
- Before handing the packets to the verifier, the simulator swaps each packet with a randomly shuffled packet of scratch-off cards, in which each card appears once. If there is a scratched card in the original packet, there is one in the new packet as well.

Note that the final packets, and therefore the entire execution, are indistinguishable from those provided by an honest prover, since the *shuffle* functionality guarantees that the packets each contain a randomly shuffled set of scratch-off cards.

**Knowledge extraction:** To show that the protocol constitutes a proof-of-knowledge, we describe the extractor for this protocol, which interacts with the prover to extract a solution to the Sudoku instance: After the prover places the cards on the cells, the extractor simply scratches all the cards. If the proof convinces the verifier with high probability, then the scratched-cards give a solution.

**Overhead:** Finally, in terms of the complexity of the protocol, we utilize $3n^2$ scratch-off cards, and $3n$ shuffles by the prover. However, recall that we are interested in making the protocols accessible to humans. For a standard $9 \times 9$ Sudoku grid, this protocol requires 27 shuffles by the prover, which seems a bit much. Thus, we now give a variant of this protocol that reduces the number of shuffles to one.

**Reducing the Number of Shuffles.** We now discuss a variant of the previous protocol, where the number of required shuffles is $c - 1$, at the expense of each shuffle being applied to a larger set of envelopes (expected size $3n^2/c$) and with worse soundness $(1 - \frac{8}{9}\frac{c-1}{c})$. The idea is to run the protocol as above, but then pick a random subset of the rows, columns and subgrids and perform the shuffle on all of them simultaneously. Note that the special case of only one shuffle has soundness error $4/9$.

**Protocol 4. *A physical protocol with $c-1$ shuffles and $1 - \frac{8(c-1)}{9c}$ soundness error***

- *The prover places three scratch-off cards on each cell. On filled-in cells, he places three scratched cards with the correct value.*
- *For each row/column/subgrid, the verifier chooses (at random) one of the three cards for each cell in the corresponding row/column/subgrid.*
- *The prover makes packets of the verifier's requested cards (i.e. for every row/column/subgrid, he assembles the requested cards into a packet).*
- *The verifier marks each packet with a number chosen uniformly at random from $0, \ldots, c-1$, where $0$ corresponds to leaving the packet unmarked.*
- *For $i = 1, \ldots, c-1$:*
  *The prover takes all packets marked with $i$, shuffles them all together, and hands them to the verifier.*
- *The verifier scratches off all the cards and verifies that in each packet, each number appears the correct number of times (namely, if $t$ packets were marked $i$, each number must appear $t$ times in the packet corresponding to $i$).*

As before, the protocol is perfectly complete, since an honest prover will always succeed. For analyzing the soundness, note that if the prover is cheating, then with probability $8/9$ (as above) there is at least one packet which is unbalanced. If this packet is marked (i.e. by a number $i$ from $1$ to $c-1$), and no other unbalanced packet is marked by $i$, then the final count of values is unbalanced and the prover fails. However, we have to be a bit careful here, since there may be two or more unbalanced packets that, when marked together, balance each other out.

By a more careful analysis we will show that the cheating probability is at most $(1 - \frac{8}{9}\frac{c-1}{c})$: With probability $8/9$, some packet, say $a$, is unbalanced. Now suppose the verifier has already gone through all other packets and marked them. Thus far, each marked packet is either balanced or unbalanced. If they are all balanced, then with probability $(c-1)/c$ the verifier will mark packet $a$ with one of $1, \ldots, c-1$, and the final mix will be unbalanced. If one marked packet is unbalanced, then with probability $(c-1)/c$ the verifier will **not** mark the packet $a$ with the correct number, and again the final mix will be unbalanced. Finally, if more than one marked packet is unbalanced, then with probability $1$ the final mix will be unbalanced. Thus, with probability $(c-1)/c$, the final mix will be unbalanced, and the verifier will be caught. Note that this was conditioned on the fact that some packet is unbalanced, so overall, the probability that a cheating prover will be caught is $8/9 \cdot (c-1)/c$ as claimed.

The zero-knowledge and proof-of-knowledge properties can be proved in the same way as they were proved for Protocol 3.

## 4.2   A Physical Zero-Knowledge Protocol with No Soundness Error

In this section we describe another physical zero-knowledge protocol, this time with the optimal soundness error of 0. This comes at the expense of a slightly stronger model, as we also make use of the *triplicate* functionality of the tamper-evident seals. This functionality generates three identical copies of a card, without revealing its value. We show here two possible methods of implementing the triplicate functionality:

**Triplicate using a trusted setup:** It is simplest to view this functionality as using some supplementary "material" that a trusted party provides to the parties. For instance, if the Sudoku puzzles are published in a newspaper, the newspaper could provide this material to its readers. The material consists of a bunch of scratch-off cards with the numbers $\{1, \ldots, n\}$ ($3n$ of each value). The cards come in triples that are connected together with an open title card on top that announces the value. The title card can be torn off (see figure below). It is crucial that the three unscratched cards hide the same value, and that it is impossible to forge such triples in which the hidden numbers vary.

**Fig. 1.** A scratch-off card with the triplicate functionality

**Triplicate without trusted setup:** It is preferable to be able to implement the triplicate functionality in the absence of a trusted party preparing the cards in advance. To do so we utilize a property of the human visual system: it can easily distinguish between a uniformly colored patch and one which has more than one color. We will use scratch-off cards as before, but the underlying numbers are replaced by colors, in a straightforward encoding, e.g. '1' is encoded by yellow, '2' by red etc. The idea is that the prover prepares a scratch-off card which is (or at least should be) uniformly colored. The verifier partitions (cuts) the card at random to three parts of equal shape and size. When it is time to peel off the top layer, if the color in one of the parts is not uniform then it is evident the prover was cheating and the verifier will summarily reject. Concretely, let the prover use a circular scratch card. When the prover wishes to triplicate a card, he asks the verifier to cut the card into three equally shaped parts (if it is easier to perform, he could ask the verifier to partition into four parts, one of which will be thrown away or shuffled and checked separately). The point is that the partitioning should be *random*.

If this task is performed by humans (which is the objective of this procedure), then slight variations in shapes will most likely go unnoticed by the human eye. A cheating prover may cheat by coloring some third a different color from the rest. However, assuming the cards are circles, there are (infinitely) many places in which the verifier can cut the cards. Thus, the probability that he cuts along the border separating two different colors (which is the only way the prover will not be caught) is nearly zero (the exact value depends on assumptions on resolution and on the model random partition).

Using the tamper-evident seals with the additional *shuffle* and *triplicate* functionalities, the protocol is as follows:

**Protocol 5. *A physical protocol with 0 soundness error, using triplicate***

- *The prover lays out the seals corresponding to the solution in the appropriate place. The seals placed on the filled-in squares are scratched off; they and must be the correct value (otherwise the verifier rejects).*
- *The verifier triplicates the seals (using the* triplicate *functionality).*
- *For each seal, each third is taken to be in its corresponding row / column / subgrid packet, and the packets are shuffled by the prover (using the* shuffle *functionality). The prover hands the packets to the verifier.*
- *The verifier scratches off the cards of each packet, and verifies that in each packet all numbers in $\{1, \ldots, n\}$ appear.*

Note that the *triplicate* functionality solves the problem of the first physical protocol, by preventing the prover from assigning different values to the same cell. Therefore the prover has no way of cheating. Thus, the soundness error of the protocol is 0 (assuming that the triplicate functionality is perfect, i.e., that the prover can never generate different copies of the same card).

The simulator for this protocol is nearly identical to that of Protocol 1, with the exception that the cards in the swapped packets are also formed using the *triplicate* functionality. Since we are assuming that triplicated cards are indistinguishable by the verifier, the packets swapped by the simulator will look the same to the verifier as the original packets. The protocol will therefore be zero-knowledge and be a proof-of-knowledge.

## 5    Open Problems and Discussion

Is there an implementable physical protocol that can be executed by (snail) mail, i.e. without assuming that the prover and the verifier are in the same room? In principle we know that such protocols exist, based on the scratch-off functionality, since in [12] it was shown how to construct commitments from this functionality and hence the cryptographic protocols of Section 3 can be used. However, since there is an amplification step in the construction of commitments from the tamper-evident envelopes of [12], involving a large number of repetitions, the result is not really human implementable.

One of the major applications of zero-knowledge proofs in the cryptographic setting is as a mechanism for converting a protocol that is resilient to semihonest behavior of the participants into one that is resilient to *any* malicious behavior. This conversion is not necessarily always possible with physical protocols. It would be interesting to see whether it is possible to do so for the Sudoku protocols.

**Acknowledgments.** We are grateful to Tal Moran for helpful discussions and comments. We also thank Tobias Barthel and Yoni Halpern for providing the initial motivation for this work. We thank Efrat Naor for helping to implement the protocol with a deck of playing cards and Yael Naor for diligently reading the paper.

# References

1. Balogh, J., Csirik, J.A., Ishai, Y., Kushilevitz, E.: Private computation using a PEZ dispenser. Theoretical Computer Science 306(1-3), 69–84 (2003)
2. Blum, M.: How to Prove a Theorem So No One Else Can Claim It. In: Proc. of the International Congress of Mathematicians, Berkeley, California, USA, pp. 1444–1451 (1986)
3. Crépeau, C., Kilian, J.: Discreet Solitary Games, Advances in Cryptology - CRYPTO'93. In: Stinson, D.R. (ed.) CRYPTO 1993. LNCS, vol. 773, pp. 319–330. Springer, Heidelberg (1994)
4. Fagin, R., Naor, M., Winkler, P.: Comparing Information Without Leaking It. Comm. of the ACM 39, 77–85 (1996)
5. Oded Goldreich, Modern Cryptography, Probabilistic Proofs and Pseudorandomness, Springer, Algorithms and Combinatorics, Vol 17 (1998)
6. Goldreich, O.: Foundations of Cryptography: Basic Tools. Cambridge University Press, Cambridge (2001)
7. Goldreich, O., Micali, S., Wigderson, A.: Proofs that Yield Nothing But their Validity, and a Methodology of Cryptographic Protocol Design. J. of the ACM 38, 691–729 (1991)
8. Goldwasser, S., Micali, S., Rackoff, C.: The knowledge complexity of interactive proof systems. SIAM J. Computing 18(1), 186–208 (1989)
9. Gradwohl, R., Naor, M., Pinkas, B., Rothblum, G.N.: Cryptographic and Physical Zero-Knowledge Proof Systems for Solutions of Sudoku Puzzles. http://www.wisdom.weizmann.ac.il/~naor/PAPERS/sudoku_abs.html
10. Gradwohl, R., Naor, E., Naor, M., Pinkas, B., Rothblum, G. N.: Proving Sudoku in Zero-Knowledge with a Deck of Cards, http://www.wisdom.weizmann.ac.il/~naor/PAPERS/SUDOKU_DEMO/ (January 2007)

11. Hayes, B.: Unwed Numbers. American Scientist Vol. 94(1), http://www.americanscientist.org/template/AssetDetail/assetid/48550 (January-February 2006)
12. Moran, T., Naor, M.: Basing Cryptographic Protocols on Tamper-Evident Seals. In: Caires, L., Italiano, G.F., Monteiro, L., Palamidessi, C., Yung, M. (eds.) ICALP 2005. LNCS, vol. 3580, pp. 285–297. Springer, Heidelberg (2005)
13. Moran, T., Naor, M.: Polling With Physical Envelopes: A Rigorous Analysis of a Human Centric Protocol. In: Vaudenay, S. (ed.) EUROCRYPT 2006. LNCS, vol. 4004, pp. 88–108. Springer, Heidelberg (2006)
14. Naor, M.: Bit Commitment Using Pseudo-Randomness. Journal of Cryptology 4, 151–158 (1991)
15. Naor, M., Naor, Y., Reingold, O.: Applied kid cryptography or how to convince your children you are not cheating http://www.wisdom.weizmann.ac.il/~naor/PAPERS/waldo.ps (March 1999)
16. Jean-Jacques, Q., Myriam, Q., Muriel, Q., Michaël, Q., Louis, G., Marie Annick, G., Gaïd, G., Anna, G., Gwenolé, G., Soazig, G., Berson, T.: How to explain zero-knowledge protocols to your children. In: Brassard, G. (ed.) CRYPTO 1989. LNCS, vol. 435, pp. 628–631. Springer, Heidelberg (1990)
17. Schneier, B.: The solitaire encryption algorithm, 1999. http://www.schneier.com/solitaire.html
18. Vadhan, S.P.: Interactive Proofs & Zero-Knowledge Proofs, lectures for the IAS/Park City Math Institute Graduate Summer School on Computational Complexity http://www.eecs.harvard.edu/~salil/papers/pcmi-abs.html
19. Sudoku, Wikipedia, the free encyclopedia, (based on Oct 19th 2005 version) http://en.wikipedia.org/wiki/Sudoku
20. Yato, T.: Complexity and Completeness of Finding Another Solution and its Application to Puzzles, Masters thesis, Univ. of Tokyo, Dept. of Information Science. Available: http://www-imai.is.s.u-tokyo.ac.jp/~yato/data2/MasterThesis.ps (January 2003)

# Sorting the Slow Way: An Analysis of Perversely Awful Randomized Sorting Algorithms

Hermann Gruber[1], Markus Holzer[2], and Oliver Ruepp[2]

[1] Institut für Informatik, Ludwig-Maximilians-Universität München,
Oettingenstraße 67, D-80538 München, Germany
`gruberh@tcs.ifi.lmu.de`
[2] Institut für Informatik, Technische Universität München,
Boltzmannstraße 3, D-85748 Garching bei München, Germany
`{holzer,ruepp}@in.tum.de`

**Abstract.** This paper is devoted to the "Discovery of Slowness." The archetypical perversely awful algorithm bogo-sort, which is sometimes referred to as Monkey-sort, is analyzed with elementary methods. Moreover, practical experiments are performed.

## 1 Introduction

To our knowledge, the analysis of perversely awful algorithms can be tracked back at least to the seminal paper on pessimal algorithm design in 1984 [2]. But what's a perversely awful algorithm? In the "The New Hacker's Dictionary" [7] one finds the following entry:

> **bogo-sort:** /boh'goh-sort'/ /n./ (var. 'stupid-sort') The archetypical perversely awful algorithm (as opposed to → **bubble sort**, which is merely the generic *bad* algorithm). Bogo-sort is equivalent to repeatedly throwing a deck of cards in the air, picking them up at random, and then testing whether they are in order. It serves as a sort of canonical example of awfulness. Looking at a program and seeing a dumb algorithm, one might say "Oh, I see, this program uses bogo-sort." Compare → **bogus**, → **brute force**, → **Lasherism**.

Among other solutions, the formerly mentioned work contains a remarkably slow sorting algorithm named *slowsort* achieving running time $\Omega\big(n^{\log n/(2+\epsilon)}\big)$ even in the best case. But the running time, still being sub-exponential, does not improve (i.e., increase) in the average case, and not even in the worst case. On the contrary, the analysis of bogo-sort carried out here shows that this algorithm, while having best-case expected running time as low as $O(n)$, achieves an asymptotic expected running time as high as $\Omega(n \cdot n!)$ already in the average case. The pseudo code of bogo-sort reads as follows:

---
**Algorithm 1.** Bogo-sort

---
1: Input array $a[1 \ldots n]$
2: **while** $a[1 \ldots n]$ is not sorted **do**
3:    randomly permute $a[1 \ldots n]$
4: **end while**

---

P. Crescenzi, G. Prencipe, and G. Pucci (Eds.): FUN 2007, LNCS 4475, pp. 183–197, 2007.
© Springer-Verlag Berlin Heidelberg 2007

The test whether the array is sorted as well as the permutation of the array have to be programmed with some care:

```
1: procedure sorted: {returns
       true if the array is sorted and
       false otherwise}
2: for i = 1 to n − 1 do
3:     if a[i] > a[i + 1] then
4:         return false
5:     end if
6: end for
7: return true
8: end procedure
```

```
1: procedure randomly permute:
       {permutes the array}
2: for i = 1 to n − 1 do
3:     j := rand[i . . . n]
4:     swap a[i] and a[j]
5: end for
6: end procedure
```

The second algorithm is found, e.g., in [5, p.139]. Hence the random permutation is done quickly by a single loop, where **rand** gives a random value in the specified range. And the test for sortedness is carried out from left and right.

In this work we present a detailed analysis, including the exact determination of the expected number of comparisons and swaps in the best, worst and average case. Although there are some subtleties in the analysis, our proofs require only a basic knowledge of probability and can be readily understood by non-specialists. This makes the analysis well-suited to be included as motivating example in courses on randomized algorithms. Admittedly, this example does not motivate coursework on *efficient* randomized algorithms. But the techniques used in our analysis cover a wide range of mathematical tools as contained in textbooks such as [4].

We will analyze the expected running time for bogo-sort under the usual assumption that we are given an array $\bar{x} = x_1 x_2 \ldots x_n$ containing a permutation of the set of numbers $\{1, 2, \ldots, n\}$. In a more abstract fashion, we are given a list containing all elements of a finite set $S$ with $|S| = n$ and an irreflexive, transitive and antisymmetric relation $\sqsubset$. To analyze the running time of the algorithm, which is a comparison-based sorting algorithm, we follow the usual convention of counting on one hand the number of comparisons, and on the other hand the number of swaps. An immediate observation is that the algorithm isn't guaranteed to terminate at all. However, as we will prove that the *expectation* of the running time $T$ is finite, we see by Markov's inequality

$$\mathbb{P}[T \geq t] \leq \frac{\mathbb{E}[T]}{t} \quad , \text{ for } t > 0,$$

that the probability of this event equals 0. There are essentially two different initial configurations: Either the list $\bar{x}$ is initially sorted, or it is not sorted. We have to make this distinction as the algorithm is smart enough to detect if the given list is initially sorted, and has much better running time in this case. This nice built-in feature also makes the running time analysis in this case very easy: The number of total comparisons equals $n - 1$, and the total number of swaps equals zero, since the while-loop is never entered.

We come to the case where the array is not initially sorted. Note that the first shuffle yields a randomly ordered list, so the behavior of the algorithm does no longer depend on the initial order; but the number of comparisons before the first shuffle depends on the structure of the original input.

# 2  How Long Does It Take to Check an Array for Sortedness?

## 2.1  The Basic Case

We make the following important

Observation 1 *If the kth element in the list is the first one which is out of order, the algorithm makes exactly $k - 1$ comparisons (from left to right) to detect that the list is out of order.*

This motivates us to study the running time of the subroutine for detecting if the list is sorted on the average:

**Theorem 2.** *Assume $\bar{x}$ is a random permutation of $\{1, 2, \ldots, n\}$, and let $C$ denote the random variable counting the number of comparisons carried out in the test whether $\bar{x}$ is sorted. Then*

$$\mathbb{E}[C] = \sum_{i=1}^{n-1} \frac{1}{i!} \sim e - 1.$$

*Proof.* For $1 \le k < n$, let $I_k$ be the random variable indicating that (at least) the first $k$ elements in $\bar{x}$ are in order. A first observation is that $I_k = 1 \Leftrightarrow C \ge k$. For on one hand, if the first $k$ elements are in order, then at least $k$ comparisons are carried out before the for-loop is left. On the other hand, if the routine makes a minimum of $k$ comparisons, the $k$th comparison involves the elements $x_k$ and $x_{k+1}$, and we can deduce that $x_1 < x_2 < \cdots < x_{k-1} < x_k$.

Thus, we have also $\mathbb{P}[C \ge k] = \mathbb{P}[I_k]$. This probability computes as

$$\mathbb{P}[I_k] = \frac{\binom{n}{k} \cdot (n-k)!}{n!}.$$

The numerator is the product of the number of possibilities to choose $k$ first elements to be in correct order and the number of possibilities to arrange the remaining $n - k$ elements at the end of the array, and the denominator is just the total number of arrays of length $n$. Reducing this fraction, we obtain $\mathbb{P}[I_k] = \frac{1}{k!}$. As the range of $C$ is nonnegative, we obtain for the expected value of $C$:

$$\mathbb{E}[C] = \sum_{k>0} \mathbb{P}[C \ge k] = \sum_{k>0} \mathbb{P}[I_k] = \sum_{k=1}^{n-1} \frac{1}{k!} = \sum_{k=0}^{n-1} \frac{1}{k!} - \frac{1}{0!}.$$

And it is a well-known fact from calculus that the last sum appearing in the above computation is the partial Taylor series expansion for $e^x$ at $x = 1$.  □

Wasn't that marvelous? Theorem 2 tells us that we need only a constant number of comparisons on the average to check if a large array is sorted, and for $n$ large enough, this number is about $e - 1 \approx 1.72$. Compare to the worst case, where we have to compare $n - 1$ times.

## 2.2   A Detour: Random Arrays with Repeated Entries

In a short digression, we explore what happens if the array is filled not with $n$ distinct numbers. At first glance we consider the case when $n$ numbers in different multiplicities are allowed. Then we have a look at the case with only two distinct numbers, say 0 and 1. In the former case the expected number of comparisons remains asymptotically the same as in the previous theorem, while in the latter the expected number of comparisons jumps up dramatically.

**Theorem 3.** *Assume $\bar{x}$ is an array chosen from $\{1, 2, \ldots, n\}^n$ uniformly at random, and let $C$ denote the random variable counting the number of comparisons carried out in the test whether $\bar{x}$ is sorted. Then*

$$\mathbb{E}[C] = \sum_{k=1}^{n-1} \binom{n-1+k}{k} \left(\frac{1}{n}\right)^k \sim e - 1.$$

*Proof.* The random variable $C$ takes on a value of at least $k$, for $1 \le k \le n - 1$, if the algorithms detects that the array is out of order after the $k$th comparison. In this case $\bar{x}$ is of the form that it starts with an increasing sequence of numbers chosen from $\{1, 2, \ldots, n\}$ of length $k$, and the rest of the array can be filled up arbitrarily. Thus, the form of $\bar{x}$ can be illustrated as follows:

$$\underbrace{1^{t_1} 2^{t_2} \ldots n^{t_n}}_{k} \underbrace{* \ldots *}_{n-k} \quad \text{with } t_1 + t_2 + \ldots + t_n = k \text{ and } t_i \ge 0, \text{ for } 1 \le i \le n.$$

Hence we have to determine how many ways an integer $k$ can be expressed as sum of $n$ non-negative integers. Image that there are $k$ pebbles lined up in a row. Then if we put $n - 1$ sticks between them we will have partitioned them into $n$ groups of pebbles each with a non-negative amount of marbles. So we have basically $n - 1 + k$ spots, and we are choosing $n - 1$ of them to be the sticks—this is equivalent to choosing $k$ marbles. Therefore the number of arrays of this form is $\binom{n-1+k}{k} n^{n-k}$, and $\mathbb{P}[C \ge k] = \binom{n-1+k}{k} \left(\frac{1}{n}\right)^k$, as there is a total of $n^n$ arrays in $\{1, 2, \ldots, n\}^n$. But then

$$\mathbb{E}[C] = \sum_{k=1}^{n-1} \mathbb{P}[C \ge k] = \sum_{k=1}^{n-1} \binom{n-1+k}{k} \left(\frac{1}{n}\right)^k \tag{1}$$

$$= \left(\sum_{k=0}^{\infty} \binom{n-1+k}{k} \cdot x^k\right)_{x=\frac{1}{n}} - \left(\sum_{k=n}^{\infty} \binom{n-1+k}{k} \cdot x^k\right)_{x=\frac{1}{n}} - 1. \tag{2}$$

Next let us consider both infinite sums in more detail. By elementary calculus on generating functions we have for the first sum

$$\sum_{k=0}^{\infty} \binom{n-1+k}{k} \cdot x^k = \frac{1}{(1-x)^n}, \tag{3}$$

which in turn gives $\left(\frac{n}{n-1}\right)^n$ because $x = \frac{1}{n}$ and by juggling around with double fractions. It remains to consider the second sum. Shifting the index $n$ places left gives us a more convenient form for the second sum:

$$\sum_{k=0}^{\infty}\binom{2n-1+k}{k+n}\cdot x^{k+n} = x^n\sum_{k=0}^{\infty}\binom{2n-1+k}{k+n}\cdot x^k \qquad (4)$$

Doesn't look that bad. As the coefficients of this power series are binomial coefficients, there might be quite a good chance that this sum can be expressed as a (generalized) hypergeometric function. In general, a hypergeometric function is a power series in $x$ with $r + s$ parameters, and it is defined as follows in terms of rising factorial powers:

$$F\left[\begin{array}{c} a_1, a_2, \ldots, a_r \\ b_1, b_2, \ldots, b_s \end{array}\middle|\ x\right] = \sum_{k\geq 0}\frac{a_1^{\overline{k}} a_2^{\overline{k}} \ldots a_r^{\overline{k}}}{b_1^{\overline{k}} b_2^{\overline{k}} \ldots b_s^{\overline{k}}}\cdot\frac{x^k}{k!}.$$

In order to answer this question we have to look at the ratio between consecutive terms—so let the notation of the series be $\sum_{k\geq 0} t_k \cdot \frac{x^k}{k!}$ with $t_0 \neq 0$. If the term ratio $t_{k+1}/t_k$ is a rational function in $k$, that is, a quotient of polynomials in $k$ of the form

$$\frac{(k + a_1)(k + a_2)\ldots(k + a_r)}{(k + b_1)(k + b_2)\ldots(k + b_s)}$$

then we can use the ansatz

$$\sum_{k\geq 0} t_k\cdot\frac{x^k}{k!} = t_0\cdot F\left[\begin{array}{c} a_1, a_2, \ldots, a_r \\ b_1, b_2, \ldots, b_s \end{array}\middle|\ x\right].$$

So let's see whether we are lucky with our calculations. As $t_k = \binom{2n-1+k}{k+n}\cdot k!$, the first term of our sum is $t_0 = \binom{2n-1}{n}$, and the other terms have the ratios given by

$$\frac{t_{k+1}}{t_k} = \frac{(2n+k)!(k+n)!(n-1)!(k+1)!}{(n+k+1)!(n-1)!(2n-1+k)!k!} = \frac{(k+2n)(k+1)}{(k+n+1)},$$

which are rational functions of $k$, yeah .... Thus, the second sum equals

$$\sum_{k=0}^{\infty}\binom{2n-1+k}{k+n}\cdot x^k = \binom{2n-1}{n}\cdot F\left[\begin{array}{c} 2n, 1 \\ n+1 \end{array}\middle|\ x\right]$$

$$= \frac{1}{2}\cdot\binom{2n}{n}\cdot F\left[\begin{array}{c} 2n, 1 \\ n+1 \end{array}\middle|\ x\right], \qquad (5)$$

because $\binom{2n-1}{n} = \frac{(2n-1)!}{n!(n-1)!} = \frac{n}{2n}\cdot\frac{(2n)!}{n!n!} = \frac{1}{2}\cdot\binom{2n}{n}$. This looks much nicer, and it's even a Gaussian hypergeometric function, i.e., $r = 2$ and $s = 1$. What about

a closed form for $F(1, 2n; n + 1 \mid x)$? Supercalifragilisticexpialidoceous[1] ....
That's fresh meat for the Gosper-Zeilberger algorithm. Next the fact

$$2S_x(n)x(x - 1)(2n - 1) + nS_x(n - 1) = 0,$$

where $S_x(n)$ is the indefinite sum $\sum\limits_{k=-\infty}^{\infty} \frac{(2n)^{\overline{k}}(1)^{\overline{k}}}{(n+1)^{\overline{k}}} \cdot \frac{x^k}{k!}$, can be easily verified with a
symbolic computation software at at hand.[2] Hence the sum is Gosper-Zeilberger
summable. Ah, ... Maybe it's worth a try to check whether the original sum
given in Equation (1) is Gosper-Zeilberger summable as well. Indeed, with a
similar calculation as above we obtain

$$(x - 1)S_x(n) + S_x(n - 1) = 0,$$

where $S_x(n)$ now equals $\sum\limits_{k=-\infty}^{\infty} \binom{n-1+k}{k} \cdot x^k$. That's even nicer than above. Since
we don't remember all details of the Gosper-Zeilberger algorithm by heart, we
peek into a standard book like, e.g., [4]. Wow, ... our sum (with slight modifications) from Equation (1) is already "solved"—[4, page 236]: The recurrence for
the definite sum $s_x(n) = \sum_{k=0}^{n-1} \binom{n-1+k}{k} \cdot x^k$—note that $\mathbb{E}[C] = s_{1/n}(n) - 1$—reads
as

$$s_x(n) = \frac{1}{1 - x}\left(s_x(n - 1) + (1 - 2x)\binom{2n - 3}{n - 2} \cdot x^{n-1}\right).$$

Because $s_x(1) = 1$, we can solve the recurrence and obtain

$$s_x(n) = \frac{1}{(1 - x)^{n-1}} + (1 - 2x)\sum_{k=1}^{n-1}\binom{2k - 1}{k - 1} \cdot \frac{x^k}{(1 - x)^{n-k}}. \tag{6}$$

---

[1] According to Pamela L. Travers' "Mary Poppins" it is a very important word every-
body should know—see, e.g., [6]:

Jane: Good morning, father. Mary Poppins taught us the most wonderful word.
Michael: Supercalifragilisticexpialidocious.
George W. Banks: What on Earth are you talking about? Superca - Super - or what-
ever the infernal thing is.
Jane: It's something to say when you don't know what to say.
George W. Banks: Yes, well, I always know what to say.

[2] The actual computation is done by Maple's hsum-package as follows:

```
> read "hsum10.mpl";
     Package "Hypergeometric Summation", Maple V – Maple 10
     Copyright 1998-2006, Wolfram Koepf, University of Kassel
> sumrecursion(hyperterm([1, 2*n], [n + 1], x, k), k, S(n));
     2 (2 n + 1) (x - 1) x S(n + 1) + (n + 1) S(n) = 0
```

Here S(n) plays the role of $S_x(n)$. Moreover, we have shifted the index $n$ one to the
right to obtain the above mentioned recurrence.

Unfortunately this "closed form" is more complicated than the original sum.[3] So we are happier with Equation (1) as a solution.

What about the asymptotic behaviour for $x = \frac{1}{n}$ and growing $n$. For both Equations (5) and (6) taking limits is no fun at all, in particular for the respective second terms! But still we are lucky, because it is not too hard to give an estimate for $x^n \sum_{k=0}^{\infty} \binom{2n-1+k}{k+n} \cdot x^k$ from Equation (4) by noting that $\binom{2n-1+k}{k+n} \leq 2^{2n-1+k}$. So this sum is upper-bounded by a geometric series: $x^n 2^{2n-1} \sum_{k=0}^{\infty} (2x)^k = x^n 2^{2n-1} \frac{1}{1-2x}$, which is valid for $x < 1/2$. For $n > 2$, we can plug in $x = 1/n$, and get $\sum_{k=n}^{\infty} \binom{2n-1+k}{k+n}(1/n)^{n+k} \leq \frac{1}{2} \left(\frac{4}{n}\right)^n$, and this even holds for $n \geq 2$. Thus we have

$$\left(\frac{n}{n-1}\right)^n - \frac{1}{2}\left(\frac{4}{n}\right)^n - 1 \leq \mathbb{E}[C] \leq \left(\frac{n}{n-1}\right)^n - 1. \qquad (7)$$

Since $\left(\frac{4}{n}\right)^n$ tends to $0$ as $n$ grows and $\left(\frac{n}{n-1}\right)^n \sim e$, we see that $\mathbb{E}[C]$, the expectation of $C$, is asymptotically $e - 1$. $\qquad \square$

The behavior of (the analytic continuations of) these two functions is compared in Figure 1. We turn to the binary case, which again turns out to be easier.

**Theorem 4.** *Assume $\bar{x}$ is an array chosen from $\{0,1\}^n$ uniformly at random, and let $C$ denote the random variable counting the number of comparisons carried out in the test whether $\bar{x}$ is sorted. Then*

$$\mathbb{E}[C] = 3 - (2n+4)2^{-n} \sim 3.$$

*Proof.* Assume $k \in \{1, 2, \ldots, n-2\}$. If $C$ takes on the value $k$, then the algorithm detects with the $k$th comparison that the array is out of order. Thus $\bar{x}$ must be of a special form: it starts with a number of 0s, then follows a nonempty sequence of 1s, which is again followed by a 0 at index $k+1$. The rest of the array can be filled up arbitrarily with zeroes and ones. This can be illustrated as follows:

$$\underbrace{0\ldots0}_{\ell}\underbrace{1\ldots1}_{k-\ell>0}0\underbrace{*\ldots*}_{n-k-1}.$$

Counting the number of arrays of this form, we obtain: $\sum_{\ell=0}^{k-1} 2^{n-k-1} = k2^{n-k-1}$, and $\mathbb{P}[C = k] = k\left(\frac{1}{2}\right)^{k+1}$, as there is a total of $2^n$ arrays in $\{0,1\}^n$.

The remaining case is that the number of comparisons equals $n - 1$. In this case, either the array is sorted, or $x$ has the following form:

$$\underbrace{0\ldots0}_{\ell}\underbrace{1\ldots1}_{n-1-\ell>0}0.$$

---

[3] Our detour on hypergeometric functions was not useless because by combining Equations (2), (5), and (6) and evaluating at $x = \frac{1}{2}$ results in the quaint hypergeometric identity $\binom{2n}{n}F\left[\begin{matrix} 2n, 1 \\ n+1 \end{matrix} \middle| \frac{1}{2}\right] = 2^{2n}$, for integers $n \geq 2$.

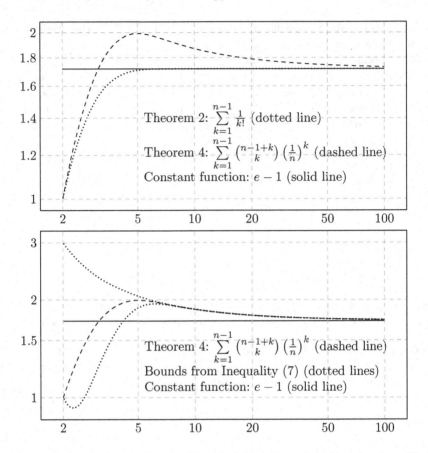

**Fig. 1.** The functions on the number of expected number of comparisons from Theorems 2 and 3 compared with the constant $e - 1$

The set $\{0, 1\}^n$ contains exactly $n + 1$ sorted arrays, and the number of arrays of the second form clearly equals $n - 1$. Thus we have $\mathbb{P}[C = n - 1] = 2n2^{-n}$. Now we are ready to compute the expected value as

$$\mathbb{E}[C] = \sum_{k=1}^{n-2} k^2 \left(\frac{1}{2}\right)^{k+1} + (n-1)\mathbb{P}[C = n-1] = \frac{1}{2}\left(\sum_{k=1}^{n-2} k^2 x^k\right) + (2n^2 - 2n)2^{-n},$$

for $x = \frac{1}{2}$. Next, the fact

$$(x-1)^3 \cdot \sum_{k=1}^{m} k^2 x^k = m^2 x^{m+3} - 2m(m-1)x^{m+2} + (m+1)^2 x^{m+1} - x^2 - x$$

can be easily verified if we have a symbolic computation software at hand. Then we briskly compute $\sum_{k=1}^{n-2} k^2 \left(\frac{1}{2}\right)^k = 6 - (4n^2 + 8)2^{-n}$, and finally get it: $\mathbb{E}[C] = 3 - (2n + 4)2^{-n}$. $\qquad \square$

We can use a similar approach to determine the expected value in the setup where the array is drawn uniformly at random from all arrays with a fixed number of zeroes, but it apparently cannot be expressed in a neat form as above. As we feel that ugly expressions are outside the scope of this conference, we refuse to further report on this here.

## 2.3   The Expected Number of Swaps in Bogo-Sort

When computing the expected number of iterations in bogo-sort, we concentrate on the case where the input $\bar{x}$ is not sorted; for the other case it equals 0, because of the intelligent design of the algorithm. In each iteration, the array is permuted uniformly at random, and we iterate until we hit the ordered sequence for the first time. As the ordered sequence is hit with probability $\frac{1}{n!}$ in each trial, the number of iterations $I$ is a random variable with

$$\mathbb{P}[I = i] = \left( \frac{n! - 1}{n!} \right)^i \cdot \frac{1}{n!}.$$

That is, $I$ is a geometrically distributed random variable with hitting probability $p = \frac{1}{n!}$, and $\mathbb{E}[I] = p^{-1} = n!$

In each iteration, the array is shuffled; and a shuffle costs $n - 1$ swaps. As the algorithm operates kind of economically with respect to the number of swaps, these are *the only* swaps carried out while running the algorithm. If $S$ denotes the random variable counting the number of swaps, we have $S = (n - 1) \cdot I$. By linearity of expectation, we derive:

**Theorem 5.** *If $S$ denotes the total number of swaps carried out for an input $\bar{x}$ of length $n$, we have*

$$\mathbb{E}[S] = \begin{cases} 0 & \textit{if } \bar{x} \textit{ is sorted} \\ (n - 1)n! & \textit{otherwise.} \end{cases}$$

**Corollary 6.** *Let $S$ denote the number of swaps carried out by bogo-sort on a given input $\bar{x}$ of length $n$. Then*

$$\mathbb{E}[S] = \begin{cases} 0 & \textit{in the best case,} \\ (n - 1)n! & \textit{in the worst and average case.} \end{cases}$$

## 2.4   The Expected Number of Comparisons in Bogo-Sort

Now suppose that, on input $\bar{x}$, we iterate the process of checking for sortedness and shuffling eternally, that is we do not stop after the array is eventually sorted. We associate a sequence of random variables $(C_i)_{i \geq 0}$ with the phases of this process, where $C_i$ counts the number of comparisons before the $i$th shuffle. Recall the random variable $I$ denotes the number of iterations in bogo-sort. Then the total number of comparisons $C$ in bogo-sort on input $\bar{x}$ is given by the sum

$$C = \sum_{i=0}^{I} C_i.$$

Wait ... This is a sum of random variables, where the summation is eventually stopped, and the time of stopping is again a random variable, no? No problem. We can deal with this rigorously.

**Definition 7.** *Let $(X_i)_{i \geq 1}$ be a sequence of random variables with $\mathbb{E}[X_i] < \infty$ for all $i \geq 1$. The random variable $N$ is called a* stopping time *for the sequence $(X_i)_{i \geq 1}$, if $\mathbb{E}[N] < \infty$ and, $\mathbb{1}_{(N \leq n)}$ is stochastically independent from $(X_i)_{i > n}$, for all $n$.*

For the concept of stopping times, one can derive a useful (classical) theorem, termed Wald's Equation. For the convenience of the reader, we include a proof of this elementary fact.

**Theorem 8 (Wald's Equation).** *Assume $(X_i)_{i \geq 1}$ is a sequence of independent, identically distributed random variables with $\mathbb{E}[X_1] < \infty$, and assume $N$ is a stopping time for this sequence. If $S(n)$ denotes the sum $\sum_{i=0}^{n} X_i$, then*

$$\mathbb{E}[S(N)] = \mathbb{E}[X_1] \cdot \mathbb{E}[N].$$

*Proof.* We can write $S(n)$ equivalently as $\sum_{i=1}^{\infty} X_i \cdot \mathbb{1}_{(N \geq i)}$ for the terms with $i > N$ are equal to zero, and the terms with $i \leq N$ are equal to $X_i$. By linearity of expectation, we may write $\mathbb{E}[S(n)]$ as $\sum_{i=1}^{\infty} \mathbb{E}[X_i \cdot \mathbb{1}_{(N \geq i)}]$. Next, observe that $X_i$ and $\mathbb{1}_{(N \geq i)}$ are stochastically independent: Since $N$ is a stopping time, $X_i$ and $\mathbb{1}_{(N \leq i-1)}$ are independent. But the latter is precisely $1 - \mathbb{1}_{(N \geq i)}$. Thus we can express the expectation of $X_i \cdot \mathbb{1}_{(N \geq i)}$ as product of expectations, namely as $\mathbb{E}[X_i] \cdot \mathbb{E}[\mathbb{1}_{(N \geq i)}]$. And finally, as the $X_i$ are identically distributed, we have $\mathbb{E}[X_i] = \mathbb{E}[X_1]$. Putting these together, we get

$$\mathbb{E}[S(n)] = \sum_{i=1}^{\infty} \mathbb{E}[X_1] \cdot \mathbb{E}[\mathbb{1}_{(N \geq i)}] = \sum_{i=1}^{\infty} \mathbb{E}[X_1] \mathbb{P}[N \geq i] = \mathbb{E}[X_1] \cdot \mathbb{E}[N],$$

as $\mathbb{E}[\mathbb{1}_{(N \geq i)}] = \mathbb{P}[N \geq i]$ and $\mathbb{E}[N] = \sum_{i=1}^{\infty} \mathbb{P}[N \geq i]$.    □

Now we have developed the tools to compute the expected number of comparisons:

**Theorem 9.** *Let $C$ denote the number of comparisons carried out by bogo-sort on an input $\bar{x}$ of length $n$, and let $c(\bar{x})$ denote the number of comparisons needed by the algorithm to check $\bar{x}$ for being sorted. Then*

$$\mathbb{E}[C] = \begin{cases} c(\bar{x}) = n - 1 & \text{if } \bar{x} \text{ is sorted} \\ c(\bar{x}) + (e-1)n! - O(1) & \text{otherwise.} \end{cases}$$

*Proof.* The random variable $C_0$ has a probability distribution which differs from that of $C_i$ for $i \geq 1$, but its value is determined by $\bar{x}$, that is $\mathbb{P}[C_0 = c(\bar{x})] = 1$. By linearity of expectation, $\mathbb{E}[C] = c(\bar{x}) + \mathbb{E}[\sum_{i=1}^{I} C_i]$. For the latter sum, the random variables $(C_i)_{i \geq 1}$ are independent and identically distributed. And $I$ is indeed a stopping time for this sequence because the time when the algorithm stops does not depend on future events. Thus we can apply Wald's equation and get $\mathbb{E}[\sum_{i=1}^{I} C_i] = \mathbb{E}[C_1] \cdot \mathbb{E}[I]$. After the first shuffle, we check a random array for being sorted, so with Theorem 2 and the following remark holds $\mathbb{E}[C_1] = e - 1 - O(\frac{1}{n!})$. The left inequality follows by an easy induction. And recall from Section 2.3 that $\mathbb{E}[I] = n!$.    $\square$

**Corollary 10.** *Let $C$ denote the number of comparisons carried out by bogo-sort on a given input $\bar{x}$ of length $n$. Then*

$$\mathbb{E}[C] = \begin{cases} n - 1 & \text{in the best case,} \\ (e-1)n! + n - O(1) & \text{in the worst case, and} \\ (e-1)n! + O(1) & \text{in the average case.} \end{cases}$$

*Proof.* In the best case, the input array $\bar{x}$ is already sorted, and thus the total number of comparisons equals $n - 1$. In the worst case, $\bar{x}$ is not initially sorted, but we need $n - 1$ comparisons to detect this. Putting this into Theorem 9, we obtain $\mathbb{E}[C] = \left(e - 1 - O(\frac{1}{n!})\right) n! + n - 1$. For the average case, recall in addition that $c(\bar{x}) = e - 1 - O(\frac{1}{n!})$ holds for an average input $\bar{x}$ by Theorem 2.    $\square$

# 3    Variations and Optimizations

## 3.1    A Variation: Bozo-Sort

We can generalize the template of repeated testing and shuffling by using other shuffling procedures than the standard shuffle. For instance, the set of transpositions, or swaps, generates the symmetric group $S_n$. Thus one can think of the following variation of bogo-sort, named bozo-sort: After each test if the array is ordered, two elements in the array are picked uniformly at random, and swapped. The procedure is iterated until the algorithm eventually detects if the array is sorted.

---

**Algorithm 2.** Bozo-sort

---

1: Input array $a[1 \ldots n]$
2: **while** $a[1 \ldots n]$ is not sorted **do**
3:     randomly transpose $a[1 \ldots n]$
4: **end while**

---

We note that this specification is ambiguous, and two possible interpretations are presented in pseudo-code:

```
1: procedure rand. transpose:
   {swaps two elements chosen
   independently}
2: i := rand[1 ... n]
3: j := rand[1 ... n]
4: swap a[i] and a[j]
5: end procedure
```

```
1: procedure rand. transpose:
   {swaps a random pair }
2: i := rand[1 ... n]
3: j := rand[1 ... i - 1, i + 1, ... n]
4: swap a[i] and a[j]
5: end procedure
```

We refer to the variant on the left as bozo-sort and to the right variant as bozo-sort$^+$. Note the apparent difference to bogo-sort: This time there are permutations of $\bar{x}$ which are not reachable from $\bar{x}$ with a single exchange, and indeed there are inputs for which the algorithm needs at least $n - 1$ swaps, no matter how luckily the random elements are chosen.

We conclude that the respective process is not stateless. But it can be suitably modeled as a finite Markov chain having $n!$ states. There each state corresponds to a permutation of $\bar{x}$. For bozo-sort$^+$, transition between a pair of states happens with probability $1/\binom{n}{2}$ if the corresponding permutations are related by a transposition. The expected hitting time of the sorted array on $n$ elements for this Markov chain was determined using quite some machinery in [3]. Translated to our setup, the relevant result reads as:

**Theorem 11 (Flatto/Odlyzko/Wales).** *Let $S$ denote the number of swaps carried out by bozo-sort$^+$ on an input $\bar{x}$ of length $n$. Then*

$$\mathbb{E}[S] = n! + 2(n - 2)! + o((n - 2)!)$$

*in the average case.*

The expected number of swaps in the best case is clearly 0, but we do not know it in the worst case currently. The expected number of comparisons is still more difficult to analyze, though it is easy to come up with preliminary upper and lower bounds:

**Theorem 12.** *Let $C$ denote the number of comparisons carried out by bozo-sort$^+$ on an input $\bar{x}$ of length $n$. Then*

$$n! + 2(n - 2)! + o((n - 2)!) \leq \mathbb{E}[C] \leq (n - 1)n! + 2(n - 1)! + o((n - 1)!)$$

*in the average case.*

*Proof.* We can express the number of comparisons as a sum of random variables as in Section 2.4: If $I$ denotes the number of iterations on an input $\bar{x}$ chosen uniformly at random, and $C_i$ the number of iterations before the $i$th swap, then the total number $C$ of comparisons equals $C = \sum_{i=0}^{I} C_i$. Obviously $1 \leq C_i \leq n - 1$, and thus $\mathbb{E}[S] \leq \mathbb{E}[C] \leq (n - 1)\mathbb{E}[S]$ by linearity of expectation.     □

The results obtained in Section 2.4 even suggest that the expected total number of comparisons on the average is as low as $O(n!)$. This would mean that the running time of bogo-sort outperforms (i.e. is higher than) the one of bozo-sort on the average. In particular, we believe that bozo-sort has the poor expected running time of only $O(n!)$ in the average case. Compare to bogo-sort, which achieves $\Omega(n \cdot n!)$.

*Conjecture 13.* For arrays with $n$ elements, the expected number of comparisons carried out by bozo-sort$^+$ is in $\Theta(n!)$ in the average case, as $n$ tends to infinity.

## 3.2   Comments on Optimized Variants of Bogo-Sort

Though optimizing the running time seems somewhat out of place in the field of *pessimal* algorithm design, it can be quite revealing for beginners in both fields of optimal *and* pessimal algorithm design to see how a single optimization step can yield a dramatic speed-up. The very first obvious optimization step in all aforementioned algorithms is to swap two elements only if this makes sense. That is, before swapping a pair, we check if it is an inversion: A pair of positions $(i, j)$ in the array $a[1 \ldots n]$ is an *inversion* if $i < j$ and $a[i] > a[j]$. This leads to optimized variants of bogo-sort and its variations, which we refer to as bogo-sort$_{\mathrm{opt}}$, bozo-sort$_{\mathrm{opt}}$, and bozo-sort$^{+}_{\mathrm{opt}}$, resp. As there can be at most $\binom{n}{2}$ inversions, this number gives an immediate upper bound on the number of swaps for these variants—compare, e.g., to $\Omega(n \cdot n!)$ swaps carried out by bogo-sort. It is not much harder to give a similar upper bound on the expected number of iterations. As the number of comparisons during a single iteration is in $O(n)$, we also obtain an upper bound on the expected total number of comparisons:

**Lemma 14.** *The expected number of iterations (resp. comparisons) carried out by the algorithms bogo-sort$_{\mathrm{opt}}$, bozo-sort$_{\mathrm{opt}}$, and bozo-sort$^{+}_{\mathrm{opt}}$ on a worst-case input $\overline{x}$ of length $n$ is at most $O\bigl(n^2 \log n\bigr)$ (resp. $O\bigl(n^3 \log n\bigr)$).*

Thus a single optimization step yields *polynomial* running time for all of these variants. The proof of the above lemma, which is based on the coupon collectors' problem, is elementary and well-suited for education. A very similar fact is shown in [1], so the details are omitted. Besides, variations of bozo-sort based on this optimization have been studied in [1]: A further optimization step is to run the procedure sorted only after every $n$th iteration, which results in the algorithm *guess-sort*, designed and analyzed in the mentioned work.

## 4   Experimental Results

We have implemented the considered algorithms in C and have performed some experiments. The source code as well as the test scripts are available on request by email to one of the authors. The experiments were conducted on our lab pool, roughly 10 PCs AMD Athlon XP 2400+ and Intel Pentium 4 CPU 3.20 GHz with 3 to 4 GB RAM. It took quite some time to collect our results, but this was no problem, since the lab courses start in late February and the PCs were idle anyway. The results are shown in Figure 2, for the number swaps and comparisons for the bogo-sort and both bozo-sort variants. For the values $n = 2, 3, \ldots, 6$ all $n!$ permutations were sorted more than 1000 times. For the remaining cases $n = 7, 8, 9, 10$ only $6! \cdot 1000$ randomly generated permutations were sorted. The average values depicted in the diagrams nicely fit the theoretical results. Moreover, our conjecture on the number of comparisons carried out by bozo-sort$^{+}$ is supported by the given data. We can also conclude from the data that in practice bogo-sort outperforms, i.e., is slower than, the bozo-sort variants w.r.t. the number of swaps by a linear factor, whereas all variants perform equally

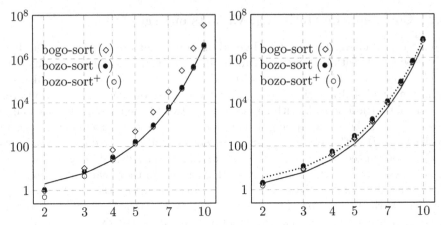

**Fig. 2.** Expected number of swaps (left) and comparisons (right) for the three considered randomized sorting algorithms—both axes are logarithmically scaled. The factorial function is drawn as a solid line, while the factorial times $(e - 1)$ is drawn as dotted line.

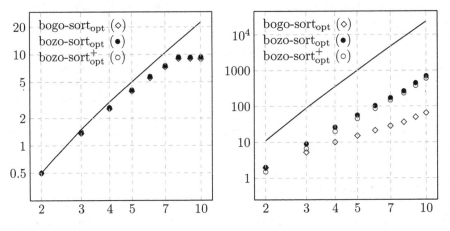

**Fig. 3.** Expected number of swaps (left) and comparisons (right) for the three considered optimized randomized sorting algorithms—both axes are logarithmically scaled. Both, the function $\frac{1}{4}n(n - 1)$ (left), which is the number of expected inversions, and $n^4 \log n$ (right) are drawn as solid lines.

good w.r.t. the number of comparisons. This is somehow counter-intuitive since one may expect at first glance that the bozo-sorts are slower.

Finally, we have evaluated the performance of optimized variants of bogo-sort and bozo-sort empirically on the same data-set as described above. The data in Figure 3 suggests that the upper bounds on the expected running time we obtained are probably not sharp and can be improved. In particular, we experience that the optimized variant of bogo-sort performs considerably less comparisons than the appropriate counterparts bozo-sort$_{\text{opt}}$ and bozo-sort$_{\text{opt}}^+$.

# 5 Conclusions

We contributed to the field of pessimal algorithm design with a theoretical and experimental study of the archetypical perversely awful algorithm, namely bogo-sort. Remarkably, the expected running time in terms of the number of swaps and comparisons can be determined exactly using only elementary methods in probability and combinatorics. We also explored some variations on the theme: In Section 2.2, we determined the number of comparisons needed to detect sort-edness on the average in different setups. And in Section 3, we introduced two variants of bogo-sort which are based on random transpositions. The analysis of these variants seems to bear far more difficulties. There our results essentially rely on a technical paper on random walks on finite groups. Quite opposed, we showed that the expected running time becomes polynomial for all variants by a simple optimization. We contrasted our theoretical study with computer experiments, which nicely fit the asymptotic results already on a small scale.

## Acknowledgments

Thanks to Silke Rolles and Jan Johannsen for some helpful discussion on random shuffling of cards, and on the structure of random arrays of Booleans, resp.

## References

1. Biedl, T., Chan, T., Demaine, E.D., Fleischer, R., Golin, M., King, J.A., Munro, J.I.: Fun-sort—or the chaos of unordered binary search. Discrete Applied Mathematics 144(3), 231–236 (2004)
2. Broder, A., Stolfi, J.: Pessimal algorithms and simplexity analysis. SIGACT News 16(3), 49–53 (1984)
3. Flatto, L., Odlyzko, A.M., Wales, D.B.: Random shuffles and group presentations. Annals of Probability 13(1), 154–178 (1985)
4. Graham, R.L., Knuth, D.E., Patashnik, O.: Concrete Mathematics. Addison-Wesley, London (1994)
5. Knuth, D.E.: Seminumerical Algorithms, The Art of Computer Programming, vol. 2. Addison-Wesley, London (1981)
6. Walt Disney's Mary Poppins. VHS Video Cassette, [Walt Disney Home Video 023-5](1980)
7. Raymond, E.S.: The New Hacker's Dictionary. MIT Press, Cambridge (1996)

# The Troubles of Interior Design–A Complexity Analysis of the Game Heyawake

Markus Holzer and Oliver Ruepp

Institut für Informatik, Technische Universität München,
Boltzmannstrasse 3, D-85748 Garching bei München, Germany
{holzer,ruepp}@in.tum.de

**Abstract.** HEYAWAKE is one of many recently popular Japanese pencil puzzles. We investigate the computational complexity of the problem of deciding whether a given puzzle instance has a solution or not. We show that Boolean gates can be emulated via HEYAWAKE puzzles, and that it is possible to reduce the Boolean Satisfiability problem to HEYAWAKE. It follows that the problem in question is N P-complete.

## 1 Introduction

HEYAWAKE is one of many pencil puzzles published by the Japanese company *Nikoli Inc.* which specializes in logic games. Pencil puzzles have gained considerable popularity during recent years. The arguably most prominent example is the game of Number Place (jap. Sudoku), which first appeared as early as 1979 in an American magazine, but did not receive much attention until *Nikoli Inc.* published their version of the puzzle on the Japanese market. Being a big hit in Japan, the puzzle later became very popular around the whole world, and now the interest in other pencil puzzles is also rising.

As most other pencil puzzles, HEYAWAKE (engl. "divided rooms") is played on a finite, two-dimensional rectangular grid. Compared to most other pencil puzzles however, HEYAWAKE seems to be substantially more complicated due to its many rules. The grid is sub-divided into smaller rectangles (which are also called rooms, hence the name), and each of these rectangles may or may not contain a number. The sub-rectangles must form a disjoint partition of the whole grid. The goal of the game is to paint the cells of the board either white or black, according to the following rules:

1. Black cells are never horizontally or vertically adjacent.
2. All white cells must be interconnected. Diagonal connections do not count.
3. If a sub-rectangle contains a number, it must contain exactly that many black fields. Otherwise, any number of black cells is allowed.
4. Any horizontal or vertical straight line of white cells must not pass through more than 2 sub-rectangles.

Figure 1 shows an example HEYAWAKE puzzle and its solution. The reader is encouraged to verify that the solution is unique. Lots of puzzles are available on

P. Crescenzi, G. Prencipe, and G. Pucci (Eds.): FUN 2007, LNCS 4475, pp. 198–212, 2007.
© Springer-Verlag Berlin Heidelberg 2007

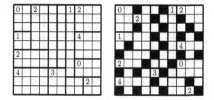

**Fig. 1.** HEYAWAKE example puzzle (left) and its solution (right)

the internet, e.g., on `http://www.nikoli.com` or on `http://www.janko.at` (in German).

For a computer scientist, pencil puzzles are especially interesting from the computational complexity point of view. The probably most basic problem is finding a solution for a given puzzle, and in most cases, the corresponding decision problem ("is there a solution?") turns out to be NP-complete. Here NP denotes the class of problems solvable in polynomial time on a nondeterministic Turing machine. To our knowledge, the first result on pencil puzzles is due to Ueda and Nagao [8], who showed that Nonogram is NP-complete. Since then, a number of other pencil puzzles have been found to be NP-complete, e.g., Corral [2], Pearl [3], Spiral Galaxies [4], Nurikabe [6], Cross Sum (Jap. Kakkuro) [7], Slither Link [10], Number Place (Jap. Sudoku) [10], and Fillomino [10]. We contribute to this list by showing that HEYAWAKE is NP-complete, too, proving the following theorem:

**Theorem 1.** *Solving a* HEYAWAKE *puzzle is* NP-*complete.*

To this end, we show how to emulate Boolean circuits via HEYAWAKE puzzles. We assume the reader to be familiar with the basics of complexity theory as contained in [5]. Hardness and completeness are always meant with respect to deterministic many-one log-space reducibilities.

## 2    Heyawake Is Intractable

To prove Theorem 1, we have to show that the problem in question is contained in NP, and that it is NP-hard. The containment in NP is immediate, since it is obvious that a nondeterministic Turing machine can guess a black and white pattern and check if that pattern constitutes a valid solution in polynomial time. It remains to prove the NP-hardness of the problem. We achieve this by showing how to reduce a 3SAT formula to HEYAWAKE. We define the problem 3SAT as follows:

**Instance:** A finite set of Boolean variables $X = \{x_1, x_2, \ldots, x_n\}$ and a finite set of clauses $C = \{c_1, c_2, \ldots, c_m\}$, where each clause consists of 3 literals.

**Question:** If the input is interpreted in the obvious way as a 3CNF formula, is there an assignment for the variables such that the formula evaluates to true?

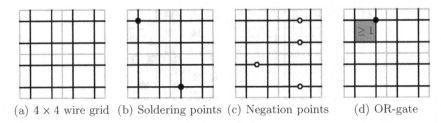

(a) 4 × 4 wire grid     (b) Soldering points     (c) Negation points     (d) OR-gate

**Fig. 2.** Basic devices, simplified representation

It is well known that this problem is NP-complete [5]. In our construction, we will use a variant of it where clauses contain 4 literals instead of three. That variant is obviously NP-complete as well, and using it simplifies the construction.

Now we are ready to present our reduction. We use the following gadgets to emulate Boolean formulas:

- A two-dimensional grid of wires: Each wire in the grid will carry a Boolean value. The gadget is designed such that crossing wires will not affect each other (unless connected by a soldering point).
- Soldering points: They are used to synchronize the values between two crossing wires.
- NOT-gates: Used to invert a Boolean signal.
- OR-gates: Each of these has 4 inputs, and works in a slightly "nonstandard" way. Instead of producing another Boolean output value, it won't allow that all of the input values are "false."

Before we start to explain in detail how to emulate these gadgets through HEYAWAKE puzzles, we want to provide the reader with a more schematic overview of our reduction. Figure 2 shows some simplified symbolic drawings for our gadgets. The basic size of our devices is chosen such that 2 horizontal and 2 vertical wires fit, device boundaries are indicated by grey bounding boxes. This means that some space is wasted, but as we will see later, it also helps to greatly simplify and clarify the actual HEYAWAKE construction. The 4 inputs to the OR-gate are indicated by the soldering points on the corners of the gadget. Note that these soldering points are not merely decoration, but they actually function as soldering points. This means that the horizontal and vertical wires that meet at one input have their values synchronized. However, the 4 input values are of course decoupled through the OR-gate, so, e.g., the value at the lower left input may be different from the value at the lower right input and so forth.

There will be one OR-gadget for each clause, placed at a certain position in the grid, and by using the soldering point gadgets, the appropriate Boolean values can be transported to the OR-gate. Note that we don't need an AND-gate, because its functionality is already implicitly present: The only thing that the AND-operator in a 3CNF formula does is that it forces the value of **every** clause to "true," and as we explained above, our OR-gate already does that. Figure 3 shows an example construction.

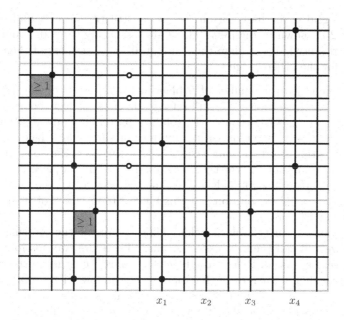

**Fig. 3.** Schematic construction for example formula $(x_1 \vee x_2 \vee x_3 \vee \overline{x_4}) \wedge (\overline{x_1} \vee \overline{x_2} \vee \overline{x_3} \vee x_4)$

It is quite easy to see how this example construction can be extended to represent arbitrary formulae. For every variable that occurs in a formula, we add two vertical wires, one carrying the signal for the variable and the other is just a spacer which carries an arbitrary signal.

Another two unused vertical wires are added left of the variable wires. They are added so we can negate values on the horizontal wires that cross these vertical wires. That way, we can easily determine the original content of any clause in the input formula by just reading off the negation points in that vertical layer of the construction.

Finally, we add an OR-gadget for each clause, and we transport the Boolean variable signals to the gadget using 6 horizontal and two vertical wires. The first two gadgets are placed as shown in the example, and any additional gadget is then added two wires to the left and 6 wires above the last.

We will now present the HEYAWAKE sub-puzzles used in our construction. Rooms that have a single unique solution have their solution pattern entered. Cells are painted light green if we know that they cannot possibly be colored black because of any of our 4 rules. Whenever we talk about coordinates of form $(x, y)$, they have the following meaning: The upper left corner has coordinates $(1, 1)$, and the values denote first the horizontal and then the vertical position relative to that spot. Sizes of gadgets are specified as $h \times v$ with $h$ and $v$ denoting horizontal and vertical extent. We leave a border of undetermined cells around the sub-puzzles to indicate that they must never be regarded as stand-alone puzzles, and there will always be some kind of surrounding, like, e.g., another gadget.

The wire grid is the basis for our whole construction. An example wire grid of size $2 \times 2$ is shown in Figure 4. There are essentially three different kinds of sub-rectangles used. We explain these in detail, because most of them are also important for the other gadgets.

- $3 \times 3$ rooms with a number of 5 black fields. There is obviously only one solution for these rooms. We will refer to them as "spacer rooms," and they are basically used to decouple the solution patterns in any adjacent rooms.
- $3 \times 4$ (resp. $4 \times 3$) rooms with a number of three black fields. These are the rooms that represent our wires, thus we will also refer to them as "wire rooms." Because of the vertically (resp. horizontally) adjacent $3 \times 3$ rooms, only $3 \times 2$ (resp. $2 \times 3$) fields are left for the black cells, and thus there are exactly two possible solutions for this kind of room. Each of these solutions represents a Boolean value.
- $4 \times 4$ rooms with two black fields. These represent the crossing of wires, hence we will call them "crossing rooms." There are two possible solutions for these: Because of Rule 4, the black cells must be either in the upper left and lower right corner, or in the upper right and lower left corner. Choosing a solution for one of these rooms also determines the solution for all other rooms of this kind (because of Rule 4), they must all be filled with the same solution.

To propagate the Boolean values, we use only Rule 4. If a certain solution is chosen for a horizontal resp. vertical wire room, we have to choose the other solution for the next horizontal resp. vertical wire room. Figure 4 also shows that choosing the same solution leads to an invalid configuration, the red fields indicate a white line that spans over three rooms and thus violates Rule 4. So the solution pattern representing a certain Boolean value alternates from one wire room to the next. We still have to define which solution represents which value, but since the interpretation depends on the actual implementation of the OR-gadget, we will get to this later.

To combine several wire gadgets to form a larger grid, we can just copy the gadget such that the spacer rooms on the border coincide. If we want, e.g., to create a $4 \times 2$ wire grid from the example shown in Figure 4 (which has a size of $17 \times 17$ cells), we would need to take the $17 \times 17$ block of cells at $(3,3)$ and copy it to position $(17,3)$. Vertical extension works analogous.

The HEYAWAKE puzzle corresponding to a soldering point is shown in Figure 5. The actual gadget has size $10 \times 10$ and can be found at position $(10, 10)$ in the diagram. Whenever we copy that $10 \times 10$ block over a wire crossing in a wire grid (again such that the $3 \times 3$ spacers coincide), the corresponding vertical and horizontal wires have their values synchronized. Note that we also have to modify the four adjacent crossing rooms: They are extended by one cell towards the middle of the gadget. If we leave the crossing rooms unchanged, the two black fields of the those rooms could no longer be placed arbitrarily (because of Rule 2), but a certain solution would be enforced. This solution depends on the Boolean input value of the gadget and it thus would conflict with any other soldering point gadget that has the other Boolean value as input.

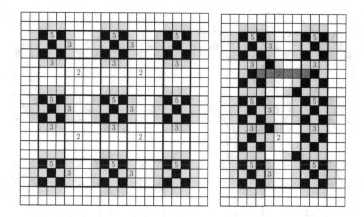

**Fig. 4.** Wire construction (left) and propagating values with Rule 4 (right), showing an inconsistent solution (top right) and a consistent solution (bottom right)

Note that because we need to modify the adjacent crossing rooms, combining soldering points is difficult: If we place a soldering point on a wire crossing, we cannot place another soldering point on any of the 8 neighbouring wire crossings. Also, if two soldering points are placed on the same wire with only one wire crossing in between, then the crossing room between these soldering points has to grow in both directions.

Again, Rule 4 is used to propagate information: There are four rooms of sizes $2 \times 4$ resp. $4 \times 2$ and one room of size $2 \times 2$ in the middle. Each of these rooms effectively contains a $2 \times 2$ region of cells that has to be filled with two black cells each. The information is transferred through the rooms marked with 0. All of the $2 \times 2$ regions must be filled with the same pattern, because otherwise Rule 4 is violated.

Figure 5 also contains a negation point in the right part of the diagram. The basic idea is to shrink one of the wire rooms horizontally, thus one of the adjacent crossing rooms must grow, as in the soldering point gadget. The alternating pattern of solutions in the wire rooms is interrupted by this, and the signal is inverted. The principle can also be applied on a vertical wire, but we won't make use of that. The shrinking of the right crossing room also leads to a modification in the crossing room on the left: Two black fields are enforced there due to Rule 4. This means that we need to change the number of black fields from 2 to 3 in the left crossing room. The neighboring crossing rooms on the top, left and bottom assure that the remaining black field is placed in the lower left or upper left corner of the room, which assures that the rest of the construction is not influenced in any way. Since the crossing rooms are modified, we also cannot combine this gadget arbitrarily with any other gadgets that involve modified crossing rooms. However, we can easily avoid any conflicts by simply making our construction large enough.

The final and most important gadget in our construction is the OR-gate, shown in Figure 6. Again, the gadget requires a modification to the adjacent

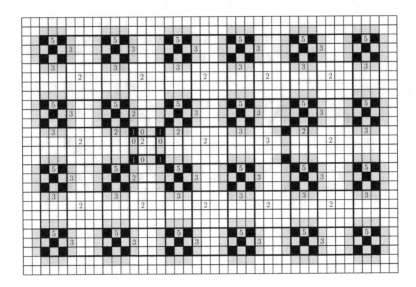

**Fig. 5.** HEYAWAKE construction for soldering points (left) and negation gadget (right)

crossing rooms, and we have to be cautious about placing other such gadgets too close. The input values determine the solution of the $2 \times 2$ blocks on the corners of the gadget. A certain pattern in these rooms would lead to an invalid configuration, so the corresponding combination of input values is forbidden. Figure 7 shows that configuration: The red highlighted cells (which are actually black cells) enclose a certain area, and the white cells inside that area are disconnected from the white cells outside, which means that Rule 2 is not satisfied.

It is also easy to see that this sub-puzzle has a solution if any of the input values are changed, i.e., if not all of the corners are "closed," since the black and white pattern that has been chosen for the inner rooms remains valid independently from the input values, and all of the inner white fields get connected to the outer white fields as soon as one of the four corners "opens."

Now it also becomes clear how we have to interpret the solutions in the wire gadgets as Boolean values: The solution that leads to a "closed" corner has to be identified with "false," the other solution corresponds to "true." Looking again at Figure 3, we would like to be able to "read off" a satisfying variable assignment from a solved puzzle. It would probably be most convenient if we could just interpret the patterns that occur at the lower end of the variable wires as Boolean values. So let us examine all four input corners of the OR-gadget and find out which pattern is induced in the wire gadgets by an input value of "false." Figure 8 shows the example circuit with every variable set to "false." The signal shapes are similar to the solution shapes in the underlying HEYAWAKE puzzle, so this diagram illustrates the following argumentation.

Since the solution pattern alternates in every wire room, the pattern that appears at the lower end of the variable wire depends on the number of wire crossings the signal passes on its way to the input corner of the OR-gate, so all

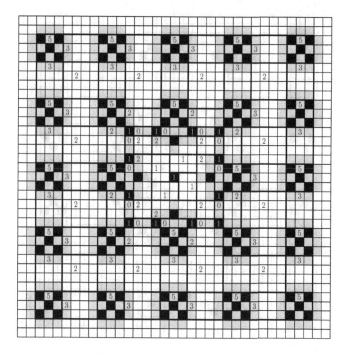

**Fig. 6.** The OR-gate

we need to do is count these crossings. In our example, the lower left input node of the OR-gadget is connected to the variable wire that represents $x_1$, and the corresponding signal passes over 6 wire crossings, the edge of the OR-gadget is not counted. From that observation, we can derive that the pattern that plays the role of "false" is the pattern that resembles an array pointing to the left. Now if we would want to connect any of the other variables to the lower left input of the OR-gate, we would have to move the corresponding soldering point from the $x_1$ wire horizontally to some other variable wire. Since the parity of the number of crossings is left unchanged by this process, we can conclude that the same pattern plays the role of "false" for all variables.

The next input we want to check is the lower right input. In the example, it is connected to the variable wire carrying $x_2$. The number of wire crossings is now odd, but also the "closing" pattern for the lower right input is inverse to that for the lower left input. This means that the same pattern as before represents the value "false" at the lower end of the variable wire. With the same argument as above, this also holds true if we connect a different variable than $x_2$ to the lower right input.

With analogous reasoning, we can conclude that the pattern representing "false" at the end of a variable wire is always the "left arrow" pattern, and the pattern for "true" is given by the inverse solution.

This is where the construction gets easier because of the $2 \times 2$ wire size of our gadgets. If, e.g., the variable wires were placed directly next to each other

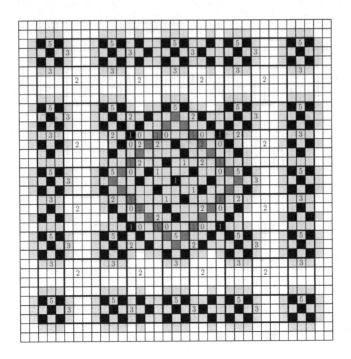

**Fig. 7.** The forbidden configuration

(without the unused spacing wire in between), the pattern for "false" would alternate for each variable. And if we tried to save some more space, then the interpretation of the patterns would certainly get more complicated.

The description of our construction is almost complete, but one last thing remains to be explained. We never talked about how to handle the border of the game board. If we place, e.g., a $3 \times 3$ room with five black cells directly into the upper left corner of the game grid, there is a problem: Two of the white cells (the upper and the left white cell) get isolated, Rule 2 is not satisfied. The simplest way to fix this problem is to add some border rooms to the construction. To add these rooms, we will have to increase the size of the construction by two cells on each side, so if the original construction has size $h \times v$, the size after adding the border rooms will be $(h+4) \times (v+4)$, and the original construction will be placed at $(3,3)$. The border is made up of two rooms with a size of $(h+2) \times 2$ and two rooms of size $2 \times (v+2)$. Both rooms need not contain a number. However, if we wished to abandon the feature of using blank rooms (see Rule 3), we could mark the border rooms with the number of vertical/horizontal wires used in the construction. The additional rooms are arranged such that the $2 \times (v+2)$ rooms are placed at $(1,1)$ and $(h+3,3)$, and the $(h+2) \times 2$ rooms are placed at $(3,1)$ and $(1,v+3)$. Note that they cannot possibly interfere with the rest of the construction, because all the mechanisms used in our gadgets are independent from the surroundings.

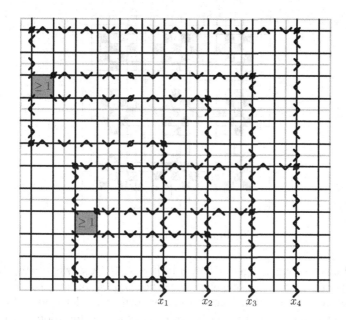

**Fig. 8.** Example circuit from Figure 3, with the relevant patterns of the underlying HEYAWAKE puzzle indicated

## 3  A Heyawake Variant

Now that the proof of NP-completeness for the regular HEYAWAKE ruleset is finished, it would be interesting to know if there are rules that are not essential for the NP-completeness. We will look at the variant where rule 2 is absent, which indeed turns out to be NP-complete as well:

**Theorem 2.** *Solving a* HEYAWAKE *puzzle, when played with Rules 1, 3 and 4, but not with Rule 2, is* NP-*complete.*

To show NP-completeness in this new situation, we can basically reuse the construction that has been developed for the original HEYAWAKE puzzle. All of our gadgets can easily be adapted to the new ruleset, and only slight modifications are necessary.

Figure 9 shows the new wire grid construction. It works completely analogous to the normal wire grid. The crossing rooms remain unchanged, and because of Rule 4, the two black fields will be placed in the corners of the room, just as before. Note however that choosing a solution for one crossing room no longer determines the solution for other crossing rooms, because now there is one black field between the corners. The $3 \times 2$ resp. $2 \times 3$ rooms now play the role of the wire rooms. It is obvious that these rooms have exactly two possible solution patterns, and the pattern propagates as before due to Rule 4. The new solution pattern of the wire rooms also has the convenient property that the patterns no longer alternate between wire rooms.

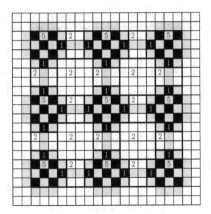

**Fig. 9.** The modified wire grid

The soldering point gadget is shown in Figure 10, it works exactly as before: The patterns of the two crossing wires are synchronized, because otherwise Rule 4 is not satisfied. The construction is even somewhat simpler than before: We no longer have to make modifications to the adjacent crossing rooms, because this was only necessary due to Rule 2.

The negation gadget, in contrast, is a little bit more complicated to construct under the new ruleset. Because of the modifications that have been made to the wire construction, it is not as easy as in the original construction to resize a wire room, and neither would this suffice to achieve the desired negation effect. We could use a wire room containing 3 black cells instead of 2 black cells to invert the signal, but in that case, we would need to introduce two blank rooms. Although this yields a rather simple negation gadget, we still want to get by without using the mentioned feature, so we discard the idea.

It seems substantially more complicated to emulate a negation gadget without using unspecified rooms. The best device we could find for this job is also shown in the right half of Figure 10. It is quite different from the negation gadgets discussed so far because it doesn't simply negate the signal that is carried on the affected wire. The gadget rather operates like a soldering point gadget that, instead of forcing the values on the horizontal and vertical crossing wires to be the same, forces them to be inverse to each other.

The OR-device also requires extensive modifications. In the original construction, its functionality relied entirely on Rule 2, which is not available any more. But fortunately, a simple counting mechanism can also be used to achieve the desired effect, and Figure 11 shows the resulting gadget. The idea is as follows: In the middle of the device, there is a big room that is marked with the number 9. A certain input value forces two black fields into this room, while the other value forces only one black field. The pattern that forces two fields corresponds to the value "false," and the other one corresponds to "true." If all input values are "false," we would need to put at least 10 fields into the big room, but there

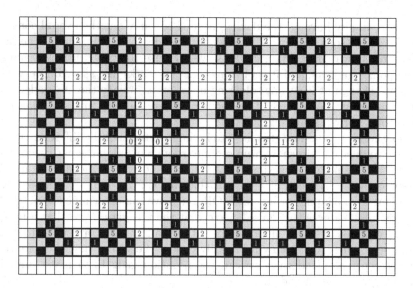

**Fig. 10.** Modified soldering point (left) and "negating soldering point" (right)

are only up to 9 allowed. If any of the other input values are "true" however, a solution exists. So all in all, the gadget works just like its counterpart in the original construction did.

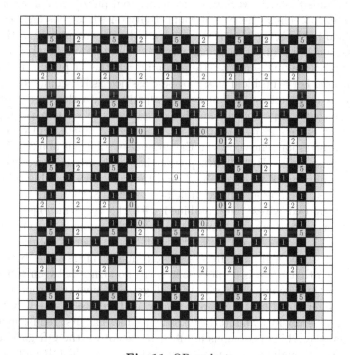

**Fig. 11.** OR-gadget

Now we finally have all the devices needed to carry out the overall construction. There are some minor differences, but the idea basically remains the same. Instead of dedicated negation gadgets, we now use the new "negation and soldering point" gadgets, and it is obvious that this works just as well: Every value is routed through at least one soldering point on its way to the OR-gadget, and if the value must be negated, then we simply use the negating soldering point instead of a normal soldering point.

Also, the wire room patterns have to be interpreted slightly different: Because the pattern does not alternate from room to room any more, the parity argument used in the original construction is not valid anymore. This means that it is not as easy as before to "read off" the variable values from a HEYAWAKE solution. By looking at the input corners of all OR-gates, we can determine the pattern in the wires that induces the "false" pattern at the OR-gate. Note that a pattern that corresponds to "false" in one corner of the gadget may correspond to "true" at another corner. So we may arbitrarily choose the values that the wire room patterns represent, and then we can use negation gadgets to make sure that everything works as intended.

## 4    Conclusion

In this paper we have shown that the game of HEYAWAKE is NP-complete by reducing the Boolean Satisfiability problem to the problem under consideration. We also examined a slight variation of the original puzzle and showed it to be NP-complete, too. Thus, the rule of connectivity on white cells has been shown to be artifical in the sense that it does not add to the complexity of the game.

There is another problem that has not been discussed so far, but commonly arises in puzzle making practice: Determining whether a puzzle has a unique solution or not. For a problem that can be classified as NP-complete, it is quite common that its counting variant can be classified into the class #P(see [9]). It is obvious that computing the number of solutions to a HEYAWAKE puzzle is #P-complete: The counting variant of 3SAT is #P-complete, and we can easily determine a correspondence between the number of solutions of a 3SAT instance and the corresponding HEYAWAKE puzzle. For each unused wire, we have to multiply the number of solutions by 2, and there are two possible solutions for the wire rooms. Looking at our example from Figure 8, there are 6 vertical and 4 horizontal unused wires, and combined with the two solutions that can be chosen for the wire rooms, we have to multiply the number of solutions of the underlying SAT instance with $2 \cdot 2^{6+4}$ to get the number of solutions that the HEYAWAKE instance will have. It is possible to make things easier by improving our result such that there is a one-to-one correspondence between solutions.

To achieve this, we have to verify that all gadgets have the property of being completely determined by their input values, and we need some way to fix the solutions in unused wires, and in the crossing rooms. The first requirement is met all of our gadget, which is obvious in all cases except for the OR-gate. But as it turns out, even though we have not mentioned it before, the OR-gadget actually

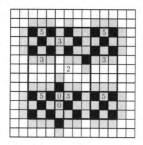

**Fig. 12.** Modified wire room that forces a wire value and a crossing room solution at the same time

does have that property. This can be verified by trial and error as follows: It is clear that the the inputs are completely determined, and thus the only possibly problematic part is the interior, i.e., the four rooms that are arranged around the 1 in the middle, and the four adjacent rooms that contain a 2. Each of the rooms containing a 2 has only 4 possible solutions, and if we try to solve the gadget starting with one of these solutions, a conflict will arise in all cases except for one solution.

Forcing a value in an unused wire is also quite easy: Let us look, e.g., at the lowest wire room of a vertical wire. If we simply split up that room, such that it is partitioned into one room of size 3 × 3 and one of size 1 × 3. We mark the latter room with a 0 (just because we do not want to use blank rooms), and with this we will have forced a certain pattern into the other room, due to Rule 4. The technique works completely analogous for horizontal wires.

To force a specific solution for a crossing room, we further modify the wire room at the bottom of an unused wire that has already been modified as described. We split the 1 × 3 room again to get a 1 × 1 and a 1 × 2 room, both marked with a 0. Because of Rule 4, this will force black fields above and below the two rooms. But this means that a black field is enforced in one corner of a crossing room, and so the whole room is determined, and in turn all other crossing rooms. The other black field will be forced in one of the border rooms, but that is no problem if we increase the number contained in that room by 1. Figure 12 shows a diagram illustrating the overall idea.

Using this modified construction, we can also derive a somewhat different, more adequate result: Checking whether a HEYAWAKE puzzle has a unique solution is complete for the class US, which is the class of sets of type $\{ x \mid f(x) = 1 \}$, for some $f \in \#P$, that has been introduced and studied in [1]. The uniqueness result is more adequate since we are not really interested in knowing the number of solutions of a HEYAWAKE instance, as long as we know whether the solution is unique or not.

There remain some open questions: There are other rulesets for which it is unknown whether the problem remains NP-complete or not. Furthermore, it would be interesting to know if HEYAWAKE remains NP-complete if there is a restriction on the values of the numbers in the rooms. Our original construction can be done using only numbers from 0 to 3, if the spacer rooms are divided into several single cell rooms containing the numbers 0 and 1 to imitate the spacer pattern. We don't know if this result is optimal, or if the problem remains NP-complete if the range of used numbers is reduced further.

# References

1. Blass, A., Gurevich, Y.: On the unique satisfiability problem. Information and Control 55(1-3), 80–88 (1982)
2. Friedman, E.: Corral puzzles are NP-complete. Technical report, Stetson University, DeLand, FL 32723 (2002)
3. Friedman, E.: Pearl puzzles are NP-complete. Technical report, Stetson University, DeLand, FL 32723 (2002)
4. Friedman, E.: Spiral galaxies puzzles are NP-complete. Technical report, Stetson University, DeLand, FL 32723 (2002)
5. Garey, M.R., Johnson, D.S.: Computers and Intractability; A Guide to the Theory of NP-Completeness. W. H. Freeman & Co, New York, NY, USA (1990)
6. Holzer, M., Klein, A., Kutrib, M.: On the NP-completeness of the NURIKABE pencil puzzle and variants thereof. In: Ferragnia, P., Grossi, R.: (eds.). In: Proceedings of the 3rd International Conference on FUN with Algorithms, pp. 77–89, Island of Elba, Italy, Edizioni Plus, Università di Pisa (May 2004)
7. Seta, T.: The complexities of puzzles, cross sum and their another solution problems (asp). Senior thesis, Univ. of Tokyo, Dept. of Information Science, Faculty of Science, Hongo 7-3-1, Bunkyo-ku, Tokyo 113, Japan (February 2001)
8. Ueda, N., Nagao, T.: NP-completeness results for NONOGRAM via parsimonious reductions, Technical Report TR96-008, Dept. of Computer Science, Tokyo Institute of Technology (1996)
9. Valiant, L.G.: The complexity of computing the permanent. Theoretical Computer Science 8, 189–201 (1979)
10. Yato, T.: Complexity and completeness of finding another solution and its application to puzzles. Master's thesis, Univ. of Tokyo, Dept. of Information Science, Faculty of Science, Hongo 7-3-1, Bunkyo-ku, Tokyo 113, Japan (January 2003)

# Drawing Borders Efficiently

Kazuo Iwama[1,*], Eiji Miyano[2,**], and Hirotaka Ono[3,***]

[1] School of Informatics, Kyoto University, Kyoto 606-8501, Japan
iwama@kuis.kyoto-u.ac.jp
[2] Department of Systems Innovation and Informatics, Kyushu Institute of
Technology, Fukuoka 820-8502, Japan
miyano@ces.kyutech.ac.jp
[3] Department of Computer Science and Communication Engineering,
Kyushu University, Fukuoka 819-0395, Japan
ono@csce.kyushu-u.ac.jp

**Abstract.** A spreadsheet, especially MS Excel, is probably one of the
most popular software applications for personal-computer users and gives
us convenient and user-friendly tools for drawing tables. Using spreadsheets, we often wish to draw several vertical and horizontal black lines
on selective gridlines to enhance the readability of our spreadsheet. Such
situations we frequently encounter are formulated as the Border Drawing
Problem (BDP). Given a layout of black line segments, we study how
to draw it efficiently from an algorithmic view point, by using a set of
border styles and investigate its complexity. (i) We first define a formal
model based on MS Excel, under which the drawability and the efficiency
of border styles are discussed, and then (ii) show that unfortunately the
problem is $\mathcal{NP}$-hard for the set of the Excel border styles and for any
reasonable subset of the styles. Moreover, in order to provide potentially
more efficient drawing, (iii) we propose a new compact set of border
styles and show a necessary and sufficient condition of its drawability.

## 1  Introduction

MS Excel is probably one of the most popular software applications for personal-
computer users. Among other nice features, it gives us a convenient and user-
friendly tool for drawing tables. Suppose, for example, we wish to draw a table
as shown in Figure 1. Other than characters, we have to draw several black
lines called *borders*. To do so, we click "Border Style" button and then there
appears the drop-down menu as shown in Figure 2. This includes 12 different
styles, style (1) through style (12) in the order of top-left, top-second, through
bottom-right. To draw the top horizontal border of the table, for example, we

---

\* Supported in part by Scientific Research Grant, Ministry of Japan, 16092101,
16092215, and 16300002.
\*\* Supported in part by Scientific Research Grant, Ministry of Japan, 16092223 and
17700022.
\*\*\* Supported in part by Scientific Research Grant, Ministry of Japan, 18300004 and
18700014.

P. Crescenzi, G. Prencipe, and G. Pucci (Eds.): FUN 2007, LNCS 4475, pp. 213–226, 2007.
© Springer-Verlag Berlin Heidelberg 2007

| | | Tokyo | | |
| --- | --- | --- | --- | --- |
| 9, am | 12, noon | 3, pm | 6, pm | 9, pm |
| S | S | PC | PC | PC |
| | | Paris | | |
| 9, am | 12, noon | 3, pm | 6, pm | 9, pm |
| PC | C | C | R | R |
| | | London | | |
| 9, am | 12, noon | 3, pm | 6, pm | 9, pm |
| C | C | PC | PC | PC |

**Fig. 1.** Borders

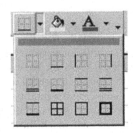

**Fig. 2.** Excel border styles

select the five horizontal cells just above this border and click style (2). Since the table includes 21 line segments, it is easy to draw it in 21 steps by using only style (2) and style (3). However, it turns out that the same table can be drawn in much less steps by using other styles, in as few as four steps!

Thus, there can be a big difference in the efficiency between naive users and highly trained users. It should be noted that such a mechanism as above, namely applying ready-made templates sequentially to do something, is an important paradigm in many different systems, including in theoretical models. One of the best known examples is the *PQ-tree* [2], which was introduced for checking the consecutive-one property of a Boolean matrix and has also been studied recently for application to bioinformatics (e.g., [6,10]). Data structures are also a nice example, where clever use of basic operations plays a key role for efficient programs. However, such a rigorous research from an algorithmic point of view has not extended to more practical systems like MS Excel, Tgif [12] and Xfig [14] (See Previous Work).

**Our Contribution.** In this paper, we concentrate ourselves on MS Excel and investigate the complexity of the Border Drawing Problem (BDP), which is basically the same as drawing a table described above. Our model has been carefully designed, which we believe does not lose the basic nature of Excel and at the same time can be used for more general discussion such as the completeness of the style set.

As for the complexity of BDP, our results are somewhat negative. Namely, the problem is $\mathcal{NP}$-hard for the style set of Excel and is also $\mathcal{NP}$-hard for any reasonable subset of styles. We also give some observations on which styles are important for several kinds of instances. Furthermore, we consider the possibility of designing a style set which is better than the Excel set. More concretely, we give an interesting set of styles which is natural, compact, and more efficient than Excel by up to a factor of $n$ for some instances, but unfortunately is not complete. It is apparently important to give approximation algorithms and/or heuristic algorithms, but in this paper, we only give a few basic observations.

**Previous Work.** The most related problem is probably the rectilinear polygon covering problem [11] (also known as the rectilinear picture compression one), which is, given a Boolean matrix, to cover (or to draw) all the 1's with as few rectangles as possible. The problem has a number of important practical applications, such as in data mining [4], and in the VLSI fabrication process [8]. Thus, it has received a considerable amount of attention and there are a lot of its variants [1,3,6,7]. In [13] (on p.433), the time complexities for various polygon covering problems are listed; almost all variations are $\mathcal{NP}$-hard. The difference is that our problem to draw (and also to delete) lines by using several different border styles, which provide numerous varieties for drawing a picture; this certainly makes the problem harder but more attractive than just rectangles.

## 2   Models

We first give a formal definition of the terminology (basically we follow that of Excel). A *spreadsheet* (or *worksheet*) is delineated by $n + 1$ horizontal and $n + 1$ vertical *gridlines* of length $n$, which are illustrated by dotted lines in this paper. Note that the gridlines are always viewable on the screen, however, any gridline will not be actually drawn or not printed on a spreadsheet. A single addressable unit surrounded by two consecutive horizontal gridlines and two consecutive vertical lines is called a *cell*. Let $c(i, j)$ be a cell on the intersection of the $i$th row from the top and the $j$th column from the left for $1 \leq i, j \leq n$. For example, reading left-to-right across the spreadsheet on the top row, we encounter $c(1, 1)$ through $c(1, n)$. The intersection of the $k$th horizontal and the $\ell$th vertical gridlines forms a *vertex*, $(k, \ell)$, for $0 \leq k, \ell \leq n$. That is, there are $(n + 1)^2$ vertices, $(0, 0)$ through $(n, n)$. Throughout the paper, we assume $n$ is not too small, for example $n \geq 4$, to avoid trivial cases.

A rectangle surrounded by two (not necessarily consecutive) horizontal gridlines and two vertical gridlines is called an *extended cell* or an *e-cell* in short. See Figure 3. An e-cell is specified by an ordered pair of its upper-left cell and lower-right one with a colon, for example, $c(1, 2) : c(3, 3)$ for $3 \times 2$ cells, whose four corners are $(0, 1)$, $(0, 3)$, $(3, 1)$, and $(3, 3)$. Also, as a special case, $c(i, j) : c(i, j)$ means a single cell $c(i, j)$. A portion of a (single) gridline is called a *line segment*, which is denoted by its two endpoints, $[(x, y), (x + u, y)]$, if it is vertical and by $[(x, y), (x, y + v)]$ if it is horizontal. Two horizontal line segments that

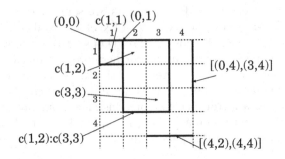

**Fig. 3.** Extended cell (e-cell)

**Fig. 4.** Border style

are touching, namely $[(x, y_1), (x, y_2)]$ and $[(x, y_2), (x, y_3)]$, are equivalent to the single line segment $[(x, y_1), (x, y_3)]$. Similarly for vertical line segments.

In many situations, we may wish to draw several vertical and horizontal black lines on selective gridlines to enhance the readability of our spreadsheet, or enclose a selected range of cells with four black lines to highlight data in the range. The Border Drawing Problem (BDP) is formulated by such situations we frequently encounter in spreadsheet applications. An instance of BDP, called a *pattern*, is given as a set of $N$ black line segments, each of which is called a *border*. Given a pattern as an input, we study how to draw it by using a set of border styles defined as follows.

According to the Excel border styles shown in Figure 2, a *border style* (or *style*) is defined as a mapping from $\{1, 2, 3, a, b, c\}$ into $\{B, W, T\}$. It is convenient to use an illustration as in Figure 4 to represent a style, where three horizontal lines correspond to 1, 2 and 3 from top to bottom and three vertical lines to $a$, $b$ and $c$ from left to right. $B$, $W$, and $T$ stand for black, white and transparency, respectively. In the figure, the left-side vertical line is given as a thick straight line, which means $a$ is mapped to $B$ in this style. Similarly, 1, 2, $b$ and $c$ are thin dotted lines, which means those are mapped to $T$. 3 is a thick dotted line, meaning it is mapped to $W$.

A pattern is drawn by a sequence of operations. A single operation is given by a pair of an e-cell and a style. For example, see Figures 5-(1) and (2). Here we selected the e-cell whose four corners are $(2, 1)$, $(2, 5)$, $(5, 1)$ and $(5, 5)$. Thus this e-cell includes four horizontal line segments and five vertical ones, each of which is represented by a symbol in $\{1, 2, 3, a, b, c\}$, namely, 1 shows the uppermost horizontal line segment, 3 the bottom horizontal one, 2 the remaining

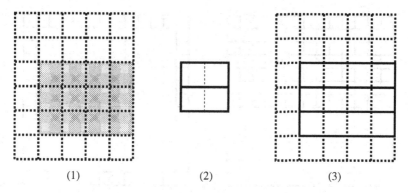

**Fig. 5.** (1) Original e-cell   (2) Style   (3) New e-cell

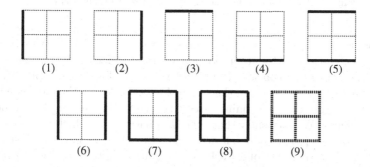

**Fig. 6.** Excel border styles: (1) $\ell$-style   (2) $r$-style   (3) $t$-style   (4) $b$-style   (5) $tb$-style   (6) $\ell r$-style (7) $o$-style   (8) $\theta$-style (9) $\phi$-style

(intermediate) horizontal ones, $a$ the left most vertical one, $b$ the intermediate vertical ones and $c$ the right most vertical one. Now suppose that our style is the one illustrated in Figure 5-(2) then the "colors" of the nine line segments of this e-cell will change as shown in Figure 5-(3) if the original colors of them are all white. Note that $B$ ($W$, respectively,) requires that the corresponding line segments become black (white, respectively,) regardless of their original colors and $T$ does not change the original colors. We assume that all the gridlines are white at the beginning and the drawing is completed if the colors of all the borders have become black and all the others remain white.

MS Excel basically allows us to use nine different styles which are given in Figure 6. Styles (1) through (9) are referred to by $\ell$, $r$, $t$, $b$, $tb$, $\ell r$, $o$, $\theta$, $\phi$, respectively. A set of styles is said to be *complete* if we can draw any pattern by using only styles in the set. It is easy to see that $\{\ell, r, t, b\}$, denoted by $S_4$, is complete (and therefore any set including $S_4$ is also complete).

**Theorem 1.** $S_4$, $\{\ell r, t, b, \phi\}$, $\{\ell, r, tb, \phi\}$, $\{\ell r, tb, \phi\}$ *are all of the minimal complete style sets. (Proof is straightforward and omitted.)*

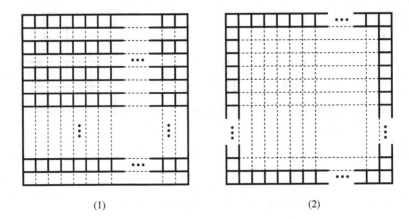

(1)                                    (2)

**Fig. 7.** (1) Proof of Theorem 2-(i)   (2) Proof of Theorem 2-(ii)

Thus, just to draw every pattern, for example, we need only four styles $\{\ell, r, t, b\}$. However, some other styles are important when considering the efficiency of the drawing. For example, consider the set $\{\ell, r, t, b, \phi\}$. This set, $S_4$ plus the style which makes all line segments of the e-cell white, is probably the most convenient for beginners. Note that $\phi$ is mainly used to correct mistakes, but it is important for the efficiency. We now show that there is a pattern for which $S_4$ needs $\Omega(n^2)$ steps, but $O(n)$ steps are enough for $S_4 \cup \{\phi\}$ : Consider the pattern illustrated in Figure 7-(1). For simplicity of exposition, we assume that $n$ is divided by 2. It has $\frac{n}{2}$ ladder-shaped tables. Since there exist disjoint $\frac{n}{2} \times (n-1)$ vertical segments, $S_4$ obviously requires $\Omega(n^2)$ steps. For $S_4 \cup \{\phi\}$, one can see that the following $O(n)$ sequence of operations draws the pattern: (i) Using the first $n-1$ steps, we place $n-1$ vertical lines segments, $[(0,1),(n,1)]$ through $[(0,n-1),(n,n-1)]$. (ii) In the next $\frac{n}{2}$ steps, all borders of $\frac{n}{2} - 1$ e-cells, $c(2,1) : c(2,n)$, $c(4,1) : c(4,n)$, through $c(n-2,1) : c(n-2,n)$ are deleted by the sequence of the $\phi$ styles. Here, each $\phi$ style can disconnect $(n-1)$ segments at a time. (iii) $n+1$ horizontal line segments of length $n$ are added. (iv) Finally, the leftmost and the rightmost line segments of length $n$ are placed.

As observed above, the deletion operation by using the $\phi$-style gives us efficient drawing sequences. Also, the $\theta$-style sometimes helps: See Figure 7-(2). Since there are $(n-3) \times 2 + 4$ vertical and $(n-3) \times 2 + 4$ horizontal borders, $S_4 \cup \{\phi\}$ obviously needs $\Omega(n)$, but only three steps suffice for $S_4 \cup \{o, \theta, \phi\}$ to draw those borders: $(c(1,1) : c(n,n), \theta), (c(2,2) : c(n-1,n-1), \phi), (c(2,2) : c(n-1,n-1), o)$ in this order.

**Theorem 2.** *(i) There is a pattern for which $S_4$ takes $\Omega(n^2)$ steps, but $S_4 \cup \{\phi\}$ does $O(n)$ steps. (ii) There is a pattern for which $S_4 \cup \{\phi\}$ takes $\Omega(n)$ steps, but $S_4 \cup \{o, \theta, \phi\}$ does $O(1)$ steps.*

To quantify the efficiency of drawing a pattern $P$ by using style sets $A$ or $B$, we introduce an *acceleration factor* of $A$ for $B$ to draw $P$, as $\alpha_{(A,B)}(P) =$

**Table 1.** The acceleration factors $\alpha_{(A,B)}(n)$ of $A$ (row) for $B$ (column)

| | $S_4$ | $S_4 \cup \{\phi\}$ | $S_4 \cup \{\theta\}$ | $S_4 \cup \{\phi,\theta\}$ | $\{\ell r, tb, \phi\}$ | $\{\ell r, tb, \phi\} \cup \{\theta\}$ |
|---|---|---|---|---|---|---|
| $S_4$ | – | subset | subset | subset | $\Omega(1)$ | $\Omega(1)$ |
| $S_4 \cup \{\phi\}$ | $\Omega(n)$ | – | $\Omega(n)$ | subset | $\Omega(1)$ | $\Omega(1)$ |
| $S_4 \cup \{\theta\}$ | $\Omega(n)$ | $\Omega(n)$ | – | subset | $\Omega(n)$ | $\Omega(1)$ |
| $S_4 \cup \{\phi,\theta\}$ | $\Omega(n)$ | $\Omega(n)$ | $\Omega(n)$ | – | $\Omega(n)$ | $\Omega(1)$ |
| $\{\ell r, tb, \phi\}$ | $\Omega(n)$ | $\Omega(1)$ | $\Omega(n)$ | $\Omega(1)$ | – | subset |
| $\{\ell r, tb, \phi\} \cup \{\theta\}$ | $\Omega(n)$ | $\Omega(n)$ | $\Omega(n)$ | $\Omega(1)$ | $\Omega(n)$ | – |

$step_B(P)/step_A(P)$, where $step_A(P)$ and $step_B(P)$ are the *minimum* numbers of the steps to draw $P$ by using style sets $A$ and $B$, respectively. For the size $n$ of the spreadsheet, we define the acceleration factor of $A$ for $B$ as

$$\alpha_{(A,B)}(n) = \max\{\alpha_{(A,B)}(P) \mid P \in \mathcal{P}_n\},$$

where $\mathcal{P}_n$ is the set of all possible patterns in the spreadsheet with size $n$. Table 1 summarizes the acceleration factors between representative styles that we found.

For some patterns, $S_4 \cup \{\phi, \theta\}$ can be more efficient than $S_4$ by up to a factor of $n$, and similarly for the full set of the Excel border styles, denoted by $S_{\text{Excel}}$, and $S_4$. One might ask whether there is a pattern that this factor is significantly more than $n$, such as $\Omega(n^2)$ steps for $S_4$ and $O(\sqrt{n})$ steps for $S_{\text{Excel}}$. The answer is NO:

**Theorem 3.** $S_{\text{Excel}}$ *can be simulated by* $S_4$ *in an overhead factor of* $O(n)$. *That is,* $\alpha_{(S_{\text{Excel}}, S_4)}(n) = \Theta(n)$.

*Proof.* We show that $S_{\text{Excel}} \setminus S_4$ can be simulated by $S_4$ in $O(n)$ steps. (1) A single use of the $tb$-style (resp. $\ell r$-style) in $S_{\text{Excel}}$ can be achieved only by using a pair of the $t$- and $b$-styles (resp. $\ell$- and $r$-styles) in $S_4$. (2) The $o$-style in $S_{\text{Excel}}$ is equal to be a sequence of four styles in $S_4$. (3) The $\theta$-style can be simulated in $O(n)$ steps because it includes at most $n$ horizontal and at most $n$ vertical line segments. (4) The remaining is the $\phi$-style. Suppose that the $\phi$ style is now used. Then, it divides one horizontal (resp. vertical) line segment into at most two pieces, which means that a single operation of the $\phi$-style can be simulated by at most two operations of the $t$-style (resp. $\ell$-style) per horizontal (resp. vertical) line segment. Since the $\phi$-style cuts at most $2n$ line segments at a time, it can be simulated by $S_4$ in $O(n)$ steps. ☐

## 3    Complexity of Border Drawing Problem

The *border drawing problem with a style set* $S$, BDP($S$), is to find a drawing sequence of minimum size for a given pattern where every style is in $S$. Restating this optimization problem as a decision problem, BDP($S, k$), we wish to determine whether a pattern has a drawing sequence with $S$ of a given size $k$. As mentioned in the previous section, this problem is obviously in $P$ for the set $S_4$. In this section we show that the problem becomes intractable if we use the set $S_5 = S_4 \cup \{\phi\}$, the most interesting subset as mentioned in the previous section.

**Theorem 4.** $BDP(S_5, k)$ is $\mathcal{NP}$-complete.

*Proof.* It is easy to show that $BDP(S_5, k)$ is in $\mathcal{NP}$. Its $\mathcal{NP}$-hardness is proved by reducing the $\mathcal{NP}$-complete *rectilinear picture compression problem* (RPC in short) [11] to $BDP(S_5, k)$. The RPC problem asks whether given an $m \times m$ matrix $M$ of 0's and 1's and a positive integer $q$, there exists a collection of $q$ or fewer rectangles that cover precisely those entries in $M$ that are 1's. That is, we have to show that for a given $m \times m$ matrix $M$ we can construct a pattern $P$ such that $P$ can be drawn by a drawing procedure of length $k$ or shorter if and only if there exists a collection of $q$ or fewer rectangles that cover precisely those entries in $M$ that are 1's.

First of all, the $m \times m$ matrix $M$ is modified to $(m+2) \times (m+2)$ matrix $M'$ by padding one row of $(m + 2)$ 0's on the top row, one row of $(m + 2)$ 0's under the bottom row, one column of $(m+2)$ 0's in the leftmost, and one column of $(m+2)$ 0's in the rightmost. Namely, the new matrix $M'$ is obtained by surrounding the original matrix $M$ with 0's.

$$M = \begin{pmatrix} 0 & 1 & 1 & 1 \\ 0 & 1 & 0 & 1 \\ 1 & 1 & 1 & 1 \\ 1 & 0 & 0 & 0 \end{pmatrix}, \quad M' = \begin{pmatrix} 0 & 0 & 0 & 0 & 0 & 0 \\ 0 & 0 & 1 & 1 & 1 & 0 \\ 0 & 0 & 1 & 0 & 1 & 0 \\ 0 & 1 & 1 & 1 & 1 & 0 \\ 0 & 1 & 0 & 0 & 0 & 0 \\ 0 & 0 & 0 & 0 & 0 & 0 \end{pmatrix}$$

We next prepare a two-dimensional grid of $(m + 2) \times 3$ rows and $(m + 2) \times 3$ columns, and place black borders on all gridlines except for its outline. Then, if the entry at the $i$th row and $j$th column of $M'$ is 1, then we obtain borders by placing white lines (or deleting the black borders drawn above) on all the outside and inside black borders of nine cells, $c(3i - 2, 3j - 2)$, $c(3i - 2, 3j - 1)$, $c(3i-2, 3j)$, $c(3i-1, 3j-2)$, $c(3i-1, 3j-1)$, $c(3i-1, 3j)$, $c(3i, 3j-2)$, $c(3i, 3j-1)$, $c(3i, 3j)$ for every $1 \le i, j \le m + 2$.

Finally, by surrounding the above grid with $(3m + 7) \times 2$ horizontal and $(3m + 7) \times 2$ vertical black borders of length one, called *scraps*, we obtain our reduced pattern $P$ from the instance of RPC. Figure 8 illustrates $P$, that has $(3m + 10) \times (3m + 10)$ cells.

As shown later in Lemmas 1 and 2, the reduced pattern $P$ has a feasible drawing sequence of length $k = q + 6m + 18$ or shorter for the pattern $P$ if and only if all 1's in $M$ are covered by a collection of $q$ or fewer rectangles. This completes the proof.                                                                                            □

**Lemma 1.** *The pattern $P$ has a feasible drawing sequence of length $k = q + 6m + 18$ or shorter for the pattern $P$ if all 1's in $M$ are covered by a collection of $q$ or fewer rectangles.*

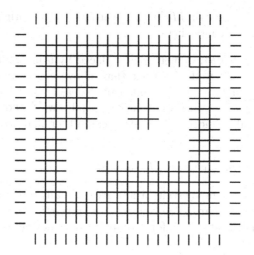

**Fig. 8.** Pattern $P$

*Proof.* We can actually give a drawing sequence of $q + 6m + 18$ steps as follows:
(i) We place $3m + 7$ horizontal black borders of length $3m + 10$, i.e., all of them go across from the left-end to the right-end, by using the first $3m + 7$ steps. (ii) Also, $3m + 7$ vertical borders of length $3m + 10$ are placed in the next $3m + 7$ steps. (iii) By using the $\phi$ style, all black borders of four e-cells of $c(2,2) : c(2, 3m + 9)$, $c(3m + 9, 2) : c(3m + 9, 3m + 9)$, $c(2,2) : c(3m + 9, 2)$, and $c(3m + 9, 2) : c(3m + 9, 3m + 9)$ are completely deleted. (iv) Finally, according to the rectangle covering of RPC, we delete the black borders again by using the $\phi$ style in at most $q$ steps.    □

**Lemma 2.** *The pattern $P$ has a feasible drawing sequence of length $k = q + 6m + 18$ or shorter only if all $1$'s in $M$ are covered by a collection of $q$ or fewer rectangles.*

*Proof.* Suppose that the pattern $P$ can be drawn in at most $k = q + 6m + 18$ steps. Our first claim is that out of this $k = q + 6m + 18$ steps we need $6m + 18$ steps only to draw the $(3m + 7) \times 4$ scraps and the borders corresponding to the $0$'s in $M'$ padded to the original matrix $M$ in its surrounding area. (Since we have so many scraps and at most two scraps are drawn in a single step, one can see easily that the procedures (i) through (iii) in the proof of the previous lemma is the only one way to draw this portion of the pattern.) Moreover, after drawing those scraps and the padded ones in this number of steps, all the gridlines of the central part of the figure must be black. (This is obvious if we have no choice other than using the procedures (i) through (iii).)

So, we now have to complete the drawing with the remaining $q$ steps. Obviously we have to use the $\phi$ style for all those steps to make the "holes" in the central part, but that can be simulated by the same number of rectangles which cover all the 1's of the matrix $M$. Thus the answer the the original RPC problem is also Yes, which is a contradiction.                                                    □

Let $S_5'$ denote $\{\ell, r, t, b, \theta\}$, namely $\phi$ is replaced by $\theta$ in $S_5$. Then the proof of $\mathcal{NP}$-hardness for BDP$(S_5', k)$ is easier than above, since we can simulate the RPC problem almost directly. Also one can see easily that BDP is $\mathcal{NP}$-hard if its style set includes $\{\ell, r, t, b\}$ and $\theta$ or $\phi$ (actually we do not need all the four basic styles). Probably another interesting case other than $S_5$ is the case of the full set.

**Theorem 5.** *BDP$(S_{\text{Excel}}, k)$ is $\mathcal{NP}$-complete.*

*Proof.* Since we can use many styles, there are too many possibilities for drawing, which makes our proof hard. The basic idea is to restrict the original pattern of the RPC problem so that to use, for example, $\phi$ does not help for the simulation. Details are omitted.                                                    □

Now it is natural to consider approximation algorithms or heuristic algorithms for BDP. Among the several intractable cases, the first one to be considered is $S_5'$, because an approximation algorithm for BDP$(S_5')$ might be a prototype for other cases. (The reason will be mentioned later.)

Consider a pattern as an input for BDP$(S_5')$ illustrated in Figure 9-(1). A cell surrounded by black borders is called a *black-cell*; otherwise *gray-cell*. For example, $c(1,2)$ and $c(2,2)$ are black-cells, and $c(1,1)$ and $c(1,3)$ are gray-cells. A sequence of consecutive vertically aligned black-cells bounded by gray-cells on the top and the bottom constitutes a *strip*. See Figure 9-(2); the pattern has 11 strips, $c(3,1)$, $c(1,2) : c(6,2)$, $c(2,3) : c(3,3)$, and so on. Two strips $c(i_1, k_1) : c(j_1, k_1)$ and $c(i_2, k_2) : c(j_2, k_2)$ are said to be *independent* if $i_1 \neq i_2$ or $j_1 \neq j_2$ holds. For example, six stripes $c(1,2) : c(6,2)$, $c(2,3) : c(3,3)$, $c(6,3) : c(7,3)$, $c(3,4)$, $c(4,5) : c(5,5)$, $c(2,6) : c(6,6)$ are *mutually independent*. For each strip $c(i,k) : c(j,k)$, we define its *associated rectangle* to be the unique rectangle that covers this strip, and extends as far as possible to the left and to the right. As shown in Figure 9-(3), there are six rectangles associated with mutually independent six strips.

The basic idea of our approximation algorithm $\mathtt{ALG}(S_5')$ for BDP$(S_5')$ is quite simple: First we draw by the $\ell$- or $r$-style (resp., $t$- or $b$-style) every vertical (resp., horizontal) black line segment which passes through some pair of consecutive horizontally (resp. vertically) aligned gray-cells. For example, a vertical line segment $[(1,8),(7,8)]$ passes through two gray-cells $c(3,8)$ and $c(3,9)$ and thus it is drawn by the $\ell$-style. Similarly, we draw a horizontal line segment $[(1,0),(1,7)]$ by the $t$-style since it passes through two gray-cells, say, $c(1,5) : c(2,5)$. Notice that these draws are indispensable, because other styles cannot draw the line segments. Then only black-cells are left. To draw the black-cells, it is better to use the $\theta$-style. Since drawing the black-cells by the $\theta$-style is essentially the same as the RPC problem, we run a similar procedure introduced in [9] as a subroutine, which has the best approximation factor of $O(\sqrt{\log n})$. Here is a description of $\mathtt{ALG}(S_5')$:

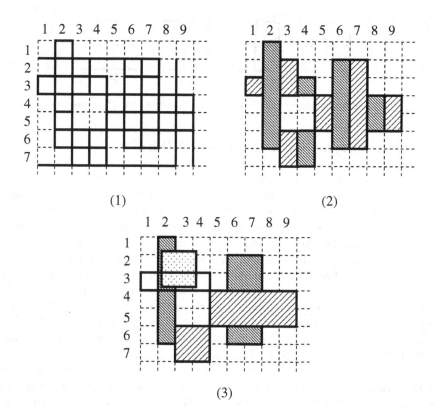

**Fig. 9.** (1) black-cells, gray-cells, (2) strips, and (3) associated rectangles

**Algorithm ALG($S_5'$)**

**Step 1.** Draw every vertical (resp. horizontal) black line segment which passes through some pair of consecutive horizontally (resp. vertically) aligned gray cells by using $\ell$- or $r$-style (resp. $t$- or $b$-style) in column-first order (resp. row-first order).

**Step 2.** Find rectangles associated with mutually independent strips, and draw each of the associated rectangles by using the $\theta$-style.

**Theorem 6.** *Algorithm* ALG($S_5'$) *achieves an approximation ratio of* $O(\sqrt{\log n})$ *for BDP($S_5'$). (Details of the proof are omitted.)* □

If we can use $\phi$ as well, then what we should do first is to look for "holes" for which using $\phi$ helps. Then we once "fill" those holes and apply the above greedy algorithm. After that those holes are dug again by using $\phi$. Unfortunately we have no idea about its approximation factor, the analysis of which appears hard.

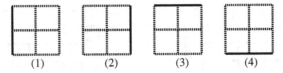

**Fig. 10.** Black-White styles: (1) $\ell^{bw}$-style   (2) $r^{bw}$-style   (3) $t^{bw}$-style   (4) $b^{bw}$-style

## 4   Border Styles with Black and White

Recall that all Excel styles, except for the $\phi$-style, have no white segments and then do not update black borders to white ones. Only the $\phi$-style deletes black borders we have drawn previously or updates black lines back to white ones. As shown in the previous section, this deletion capability gives us efficient drawing sequences. In this section we consider styles which include both Black and White at the same time, as illustrated in Figure 10. In the case of the $\ell^{bw}$-style in Figure 10-(1), all three horizontal line segments 1, 2, and 3 (from top to bottom) are mapped to $W$, and three vertical ones $a$, $b$, and $c$ are mapped to $B$, $W$, and $W$, respectively. The $r^{bw}$-style maps 1, 2, 3, $a$, $b$, and $c$ into $W$, $W$, $W$, $W$, $W$, and $B$, respectively. The $t^{bw}$ and $b^{bw}$-styles are similar. Let $S_4^{bw}$ be $\{\ell^{bw}, r^{bw}, t^{bw}, b^{bw}\}$.

In this section, we assume that the given pattern does not include the gridlines of the boundary of spreadsheets. The reason is as follows: For example, there are no cells above the top horizontal gridline of the sheet itself. Therefore, any border on this gridline cannot be drawn by the $b^{bw}$ style. However, all other horizontal borders can be drawn by that style. One can see that the above assumption excludes such a trivial incompleteness of the style set.

As shown in a moment, $S_4^{bw}$ is sometimes very efficient, which indicates some possibilities of improving the style set of Excel.

**Proposition 1.** *There is a pattern for which $S_4$ needs $\Omega(n^2)$ steps, but $S_4^{bw}$ $O(n)$ steps.*

*Proof.* Figure 11 illustrates one of such patterns. $S_4$ needs $\Omega(n^2)$ since there are $\Omega(n^2)$ vertical segments. Here are rough ideas for $S_4^{bw}$: We first place all vertical $n$-length borders in $O(n)$ steps. Then, with cutting them, we place around $\frac{2n}{3}$ horizontal borders by using the $t^{bw}$ and the $b^{bw}$ styles in $O(n)$ steps. Finally, two outer vertical borders are added.                                                          □

For several patterns, $S_4^{bw}$ is more efficient than $S_4$ but, unfortunately, there is a large class of patterns for which $S_4^{bw}$ has no feasible drawing sequences (other than the trivial ones mentioned at the beginning of this section). Here are some definitions: As shown before, in the case of $S_4^{bw}$, the order of the drawing sequence is very critical and strongly affected by the pattern's layout. In order to discuss the drawing-order of borders, we associate a pattern with an undirected graph, defined below. In the following, we apply each border style in $S_4^{bw}$ only on unit cell, which simplifies the explanation. Actually, this restriction may affect the number of drawing steps, but not the (in)completeness of $S_4^{bw}$.

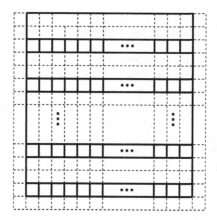

**Fig. 11.** Proof of Proposition 1

For a given pattern, we say (distinct) unit cells are *neighbors* to each other if they share a border. For example, cell $c(x, y)$ and $c(x + 1, y)$ are neighbors if the former has a bottom border, or equivalently the latter a top one. Given a pattern $P$ of borders $\{b_1, b_2, \cdots, b_m\}$, a *neighborhood graph* $G(P)$ is defined by a graph of node set $V(P)$ and edge set $E(P)$, where

$$V(P) = \{u_{i,j} \mid 1 \leq i, j \leq n\} \quad \text{and,}$$
$$E(P) = \{(u_{i,j}, u_{k,\ell}) \mid c(i, j) \text{ and } c(k, \ell) \text{ are neighbors}\}.$$

Note that each node $u_{i,j}$ corresponds to unit cell $c(i, j)$. As for the pattern and its corresponding neighborhood graph, we have the following result:

**Theorem 7.** *Pattern $P$ has no feasible drawing sequences on $S_4^{bw}$ if and only if its neighborhood graph $G(P)$ contains a cycle.*

*Proof.* ($\Rightarrow$) Suppose that $G(P)$ does not have a cycle, i.e., $G(P)$ is a tree. Take an arbitrary node as root $r$, then find paths from $r$ to its leaves. According to the paths, we add direction information $(u_1 \to u_2)$ to each edge $e = (u_1, u_2)$, which means that $u_1$ is a tail node and $u_2$ a head one. If, for example, the edge between $u_{1,2}$ and $u_{2,2}$ has direction $(u_{1,2} \to u_{2,2})$, then the operation $(c(2, 2) : c(2, 2), t^{bw})$ is executed. Due to the orientation, in-degree of each node is at most 1, which implies that a border of each node (or cell) once drawn will not be erased. Therefore, according to the tree-orientation, we can find at least one drawing sequence that can draw $P$.

($\Leftarrow$) We just give a sketch. We show that $G(P)$ containing a cycle cannot be drawn by $S_4^{bw}$ by contradiction. Suppose that $G(P)$ can be drawn by $S_4^{bw}$. This means that $S_4^{bw}$ has a finite drawing sequence of styles for a pattern corresponding to a simple cycle $C$, because drawing sequences that are noncontiguous for the cycle always leave some borders undrawn. Note that if we apply one of the $bw$-type border styles, then one border is added but at the same time three other ones are deleted. Hence, the node corresponding to the cell where the last style of the drawing sequence is placed must be a leaf, which is a contradiction.     □

From the above theorem, every subset of $S_4^{bw}$ is also not complete. Since we have a simple characterization of the drawability and the incompleteness means the number of patterns which can be drawn is small, one might think that, for example, BDP$(S_4^{bw}, k)$ becomes tractable. However, it still remains $\mathcal{NP}$-complete even for BDP$(\{r^{bw}, t^{bw}\}, k)$, although we omit the details due to the space limitation.

**Theorem 8.** *BDP$(S, k)$ is $\mathcal{NP}$-complete for any of the following $S$: $\{r^{bw}, t^{bw}\}$, $\{r^{bw}, b^{bw}\}$, $\{\ell^{bw}, t^{bw}\}$, $\{\ell^{bw}, b^{bw}\}$, $S_4^{bw} \backslash \{\ell^{bw}\}$, $S_4^{bw} \backslash \{r^{bw}\}$, $S_4^{bw} \backslash \{t^{bw}\}$, $S_4^{bw} \backslash \{b^{bw}\}$ and $S_4^{bw}$.*     □

# References

1. Berman, P., DasGupta, B.: Complexities of efficient solutions of rectilinear polygon cover problems. Algorithmica 17(4), 331–356 (1997)
2. Booth, K.S., Lueker, G.S.: Testing for the consecutive ones property, interval graphs, and graph planarity using PQ-tree algorithms. J. of Computer and System Science 13, 335–379 (1976)
3. Culberson, J.C., Reckhow, R.A.: Covering polygons is hard. J. of Algorithms 17, 2–44 (1994)
4. Edmonds, J., Gryz, J., Liang, D., Miller, R.J.: Mining for empty rectangles in large data sets. Theoretical Computer Science 296, 435–452 (2003)
5. Garey, M.R., Johnson, D.S.: Computers and Intractability - A Guide to the Theory of $\mathcal{NP}$-Completeness. W.H.Freeman & Co., New York (1979)
6. Greenberg, D.S., Istrail, S.C.: Physical mapping by STS hybridization: Algorithmic strategies and the challenge of software evaluation. J. Computational Biology 2, 219–274 (1995)
7. Gudmundsson, J., Levcopoulos, C.: Close approximations of minimum rectangular coverings. J. of Combinatorial Optimization 3(4), 437–452 (1999)
8. Hegedüs, A.: Algorithms for covering polygons by rectangles. Computer Aided Geometric Design 14, 257–260 (1982)
9. Anil Kumar, V.S., Ramesh, H.: Covering Rectilinear Polygons with Axis-Parallel Rectangles. SIAM Journal on Computing 32(6), 1509–1541 (2003)
10. Magwene, P.M., Lizardi, P., Kim, J.: Reconstructing the temporal ordering of biological samples using microarray data. Bioinformatics 19(7), 842–850 (2003)
11. Masek, W.J.: Some NP-complete set covering problems. MIT, Cambridge, MA, unpublished manuscript. Referenced in [5] (1978)
12. Reuhman, D.: A short route to tgif pictures.
    ftp://bourbon.usc.edu/pub/tgif/contrib/tgifintro/ (2002)
13. Suri, S.: Polygons. In: Suri, S., Goodman, J.E., O'Rourke, J. (eds.) Handbook of Discrete and Computational Geometry, CRC Press, Boca Raton (1997)
14. Xfig User Manual (ver. 3.2.4),
    http://www.xfig.org/usrman/frm_main_menus.html

# The Ferry Cover Problem

Michael Lampis and Valia Mitsou

School of Electrical & Computer Engineering,
National Technical University of Athens, Greece
mlampis@cs.ntua.gr, valia@corelab.ntua.gr

**Abstract.** In the classical wolf-goat-cabbage puzzle, a ferry boat man must ferry three items across a river using a boat that has room for only one, without leaving two incompatible items on the same bank alone. In this paper we define and study a family of optimization problems called FERRY problems, which may be viewed as generalizations of this familiar puzzle.

In all FERRY problems we are given a set of items and a graph with edges connecting items that must not be left together unattended. We present the FERRY COVER problem (FC), where the objective is to determine the minimum required boat size and demonstrate a close connection with VERTEX COVER which leads to hardness and approximation results. We also completely solve the problem on trees. Then we focus on a variation of the same problem with the added constraint that only 1 round-trip is allowed ($FC_1$). We present a reduction from MAX-NAE-{3}-SAT which shows that this problem is NP-hard and APX-hard. We also provide an approximation algorithm for trees with a factor asymptotically equal to $\frac{4}{3}$. Finally, we generalize the above problem to define $FC_m$, where at most $m$ round-trips are allowed, and $MFT_k$, which is the problem of minimizing the number of round-trips when the boat capacity is $k$. We present some preliminary lemmata for both, which provide bounds on the value of the optimal solution, and relate them to FC.

**Keywords:** approximation algorithms, graph algorithms, vertex cover, transportation problems, wolf-goat-cabbage puzzle.

## 1  Introduction

The first time algorithmic transportation problems appeared in western literature is probably in the form of Alcuin's four "River Crossing Problems" in the book *Propositiones ad acuendos iuvenes*. Alcuin of York, who lived in the 8th century A.D. was one of the leading scholars of his time and a royal advisor in Charlemagne's court. One of Alcuin's problems was the following:

*A man has to take a wolf, a goat and a bunch of cabbages across a river, but the only boat he can find has only enough room for him and one item. How can he safely transport everything to the other side, without the wolf eating the goat or the goat eating the cabbages?*

This amusing problem is a very good example of a constraint satisfaction problem in operations research, and, quite surprisingly for a problem whose solution

P. Crescenzi, G. Prencipe, and G. Pucci (Eds.): FUN 2007, LNCS 4475, pp. 227–239, 2007.

is trivial, it demonstrates many of the difficulties which are usually met when trying to solve much larger and more complicated transportation problems ([2]).

In this paper we study generalizations of Alcuin's problem which we call FERRY problems. In these problems, which belong to a wide family of transportation problems, the goal is to ferry a set of items across a river, while making sure that items that remain unattended on the same bank are safe from each other. The relations between items are described by an incompatibility graph, and the objective varies from minimizing the size of the boat needed to minimizing the number of trips.

There are many reasons which make the study of FERRY problems interesting and worthwhile. First, as they derive from a classical puzzle, they are amusing and entertaining, while at the same time having algorithmic depth. This makes them very valuable as a teaching tool because puzzles are very attractive to students. Several other applications of these concepts are possible. For example in cryptography, the items may represent parts of a key and the incompatibilities may indicate parts that could be combined by an adversary to gain some information. A player wishes to transfer a key to someone else, without allowing him to gain any information before the whole transaction is complete.

The rest of this paper is structured as follows: basic definitions and preliminary notions are given in Section 2. In Section 3 we study the FERRY COVER problem without constraints on the number of trips and present hardness and approximation results, as well as results for several graph topologies. Section 4 consists of an analysis of the TRIP-CONSTRAINED FERRY COVER problem with the maximum number of trips being three, i.e. only one round-trip allowed. We present a reduction from MAX-NAE-{3}-SAT which leads to hardness results for this variation. In Section 5 we analyze the general TRIP-CONSTRAINED FERRY COVER and MIN FERRY TRIPS problems presenting several lemmata that provide bounds on the value of the optimal solution and relate them to FC. Finally, conclusions and directions to further work are given in Section 6.

## 2   Definitions – Preliminaries

The rules of the FERRY games can be roughly described as follows: we are given a set of $n$ items, some of which are incompatible with each other. These incompatibilities are described by a graph with vertices representing items, and edges connecting incompatible items. We need to take all $n$ items across a river using a boat of fixed capacity $k$ without at any point leaving two incompatible items on the same bank when the boat is not there. We seek to minimize the boat size in conjunction with the number of required trips to transfer all items.

Let us now formally define the FERRY problems we will focus on. To do this we need to define the concept of a *legal configuration*. Given an incompatibility graph $G(V, E)$, a *legal configuration* is a triple $(V_L, V_R, b)$, $V_L \cup V_R = V, V_L \cap V_R = \emptyset, b \in \{L, R\}$ s.t. if $b = L$ then $V_R$ induces an independent set on $G$ else $V_L$ induces an independent set on $G$. Informally, this means that when the boat is on one bank all items on the opposite bank must be compatible.

Given a boat capacity $k$ a *legal left-to-right trip* is a pair of legal configurations $((V_{L_1}, V_{R_1}, L), (V_{L_2}, V_{R_2}, R))$ s.t. $V_{L_2} \subseteq V_{L_1}$ and $|V_{L_1}| - |V_{L_2}| \leq k$. Similarly a *right-to-left trip* is a pair of legal configurations $((V_{L_1}, V_{R_1}, R), (V_{L_2}, V_{R_2}, L))$ s.t. $V_{R_2} \subseteq V_{R_1}$ and $|V_{R_1}| - |V_{R_2}| \leq k$. A *ferry plan* is a sequence of legal configurations starting with $(V, \emptyset, L)$ and ending with $(\emptyset, V, R)$ s.t. successive configurations constitute left-to-right or right-to-left trips. We will informally refer to a succession of a left-to-right and a right-to-left trip as a round-trip.

**Definition 1.** *The* FERRY COVER *(FC) problem is, given an incompatibility graph $G$, compute the minimum required boat size $k$ s.t. there is a ferry plan for $G$.*

We will denote by $\text{OPT}_{\text{FC}}(G)$ the optimal solution to the FERRY COVER problem for a graph $G$.

We can also define the following interesting variation of FC.

**Definition 2.** *The* TRIP-CONSTRAINED FERRY COVER *problem is, given a graph $G$ and an integer trip constraint $m$ compute the minimum boat size $k$ s.t. there is a ferry plan for $G$ consisting of at most $2m + 2$ configurations, i.e. at most $2m + 1$ trips, or equivalently $m$ round-trips plus the final trip.*

We will denote by $\text{OPT}_{\text{FC}_m}(G)$ the optimal solution of TRIP-CONSTRAINED FERRY COVER for a graph $G$ given a constraint on trips $m$.

The problem of minimizing the number of trips when the boat capacity is fixed can be defined as follows:

**Definition 3.** *The* MIN FERRY TRIPS *problem is, given a graph $G$ and a boat size $k$ determine the number of round-trips of the shortest possible ferry-plan for $G$ with capacity $k$.*

We will denote by $\text{OPT}_{\text{MFT}_k}(G)$ the optimal solution of MIN FERRY TRIPS for a graph $G$ given a boat capacity $k$. It should be noted that for some values of $k$ there is no valid ferry-plan. In these cases we define $\text{OPT}_{\text{MFT}_k} = \infty$.

For the sake of completeness let us also give the definition of the well-studied NP-hard VERTEX COVER and MAX-NAE-{3}-SAT problems ([3]).

**Definition 4.** *The* VERTEX COVER *problem is, given a graph $G(V, E)$ find a minimum cardinality subset $V'$ of $V$ s.t. all edges in $E$ have at least one endpoint in $V'$ (such subsets are called vertex covers of $G$).*

We denote by $\text{OPT}_{\text{VC}}(G)$ the cardinality of a minimum vertex cover of $G$.

**Definition 5.** *The* MAX-NAE-{3}-SAT *problem is, given a CNF formula where each clause contains exactly 3 literals, find the maximum number of clauses that can be satisfied simultaneously by any truth assignment. In the context of* MAX-NAE-{3}-SAT, *we say that a clause is satisfied when it contains two literals with different values.*

Finally, let us give the definition of the $H$-COLORING problem, which will be useful in the study of $\text{FC}_1$.

**Definition 6.** *For a fixed graph $H(V_H, E_H)$ possibly with loops but without multiple edges, the $H$-COLORING problem is the following: given a graph $G(V_G, E_G)$, find a homomorphism $\theta$ from $G$ to $H$, i.e. a map $\theta : V_G \rightarrow V_H$ with the property that $(u, v) \in E_G \Rightarrow (\theta(u), \theta(v)) \in E_H$.*

The above problem was defined in [4]. Informally, we will refer to the vertices of $H$ as colors.

## 3   The Ferry Cover Problem

In this section we present several results for the FERRY COVER problem which indicate that it is very closely connected to VERTEX COVER. We will show that FERRY COVER is NP-hard and that it has a constant factor approximation.

**Lemma 1.** *For any graph $G$, $\mathrm{OPT_{VC}}(G) \leq \mathrm{OPT_{FC}}(G) \leq \mathrm{OPT_{VC}}(G) + 1$.*

*Proof.* For the first inequality note that if we have boat capacity $k$ and $\mathrm{OPT_{VC}}(G) > k$, then no trip is possible because any selection of $k$ vertices to be transported on the initial trip fails to leave an independent set on the left bank.

For the second inequality, if we have boat capacity $\mathrm{OPT_{VC}} + 1$ then we can use the following ferry plan: load the boat with an optimal vertex cover and keep it on the boat for all the trips. Use the extra space to ferry the remaining independent set vertex by vertex to the other bank. Unload the vertex cover together with the last vertex of the independent set.                    □

**Theorem 1.** *There are constants $\epsilon_F, n_0 > 0$ s.t. there is no $(1+\epsilon_F)$-approximation algorithm for FERRY COVER with instance size greater than $n_0$ vertices unless P=NP.*

*Proof.* It is known that there is a constant $\epsilon_S > 0$ such that there is no $(1-\epsilon_S)$-approximation for MAX-3SAT unless P=NP([1]) and that there is a gap preserving reduction from MAX-3SAT to VERTEX COVER. We will show that there is also a gap-preserving reduction from MAX-3SAT to FERRY COVER.

The gap-preserving reduction to VERTEX COVER in [3] and [6] implies that there is a constant $\epsilon_V > 0$ s.t. for any 3CNF formula $\phi$ with $m$ clauses we produce a graph $G(V, E)$ s.t.

$$\mathrm{OPT_{MAX-3SAT}}(\phi) = m \Rightarrow \mathrm{OPT_{VC}}(G) \leq \frac{2}{3}|V|$$

$$\mathrm{OPT_{MAX-3SAT}}(\phi) < (1 - \epsilon_S)m \Rightarrow \mathrm{OPT_{VC}}(G) > (1 + \epsilon_V)\frac{2}{3}|V|$$

In the first case it follows from Lemma 1 that

$$\mathrm{OPT_{VC}}(G) \leq \frac{2}{3}|V| \Rightarrow \mathrm{OPT_{FC}}(G) \leq \frac{2}{3}|V| + 1.$$

In the second case,

$$\mathrm{OPT_{VC}}(G) > (1 + \epsilon_V)\frac{2}{3}|V| \Rightarrow \mathrm{OPT_{FC}}(G) > (1 + \epsilon_V - \frac{1 + \epsilon_V}{\frac{2}{3}|V| + 1})(\frac{2}{3}|V| + 1).$$

For $|V| > \frac{3}{2}\frac{1}{\epsilon_v}$ there is a constant $\epsilon_F > 0$ s.t. $\epsilon_V - \frac{1+\epsilon_V}{\frac{2}{3}|V|+1} > \epsilon_F$. Setting $n_0 = \lceil \frac{3}{2}\frac{1}{\epsilon_v} \rceil$ completes the proof. $\qquad\square$

**Corollary 1.** FERRY COVER *is NP-hard*

*Proof.* It follows from Theorem 1 that an algorithm which exactly solves large enough instances of FERRY COVER in polynomial time, and therefore achieves an approximation ratio better than $(1 + \epsilon_F)$, implies that $P=NP$. $\qquad\square$

It should be noted that the constant $\epsilon_F$ in Theorem 1 is much smaller than $\epsilon_V$. However, this is a consequence of using the smallest possible value for $n_0$. Using larger values would lead to a proof of hardness of approximation results asymptotically equivalent to those we know for VERTEX COVER. This is hardly surprising, since Lemma 1 indicates that the two problems have almost equal optimum values. Lemma 1 also leads to the following approximation result for FERRY COVER.

**Theorem 2.** *A $\rho$-approximation algorithm for* VERTEX COVER *implies a $(\rho + \frac{1}{OPT_{FC}})$-approximation algorithm for* FERRY COVER.

*Proof.* Consider the following algorithm: use the $\rho$-approximation algorithm for VERTEX COVER to obtain a vertex cover of cardinality $SOL_{VC}$, then set boat capacity equal to $SOL_{FC} = SOL_{VC} + 1$. This provides a feasible solution since loading the boat with the approximate vertex cover leaves enough room to transport the remaining independent set one by one as in Lemma 1. Observe that $SOL_{FC} = SOL_{VC} + 1 \le \rho OPT_{VC} + 1 \le \rho OPT_{FC} + 1$ (the first inequality from the approximation guarantee and the second from Lemma 1). $\qquad\square$

We now present some examples for specific graph topologies.

*Example 1.* If $G$ is a clique, i.e. $G = K_n$, then $OPT_{FC}(G) = OPT_{VC}(G) = n-1$.

*Example 2.* If $G$ is a ring, i.e. $G = C_n$ then $OPT_{FC}(G) = OPT_{VC}(G) = \lceil \frac{n}{2} \rceil$.

*Example 3.* Consider a graph $G(V, E)$, $|V| \ge n + 3$ s.t. $G$ contains a clique $K_n$ and the remaining vertices form an independent set. In addition every vertex outside the clique is connected with every vertex of the clique. For example see Figure 1.

We will show that $OPT_{FC}(G) = OPT_{VC}(G) + 1$. Assume that $OPT_{FC}(G) = OPT_{VC}(G)$. The optimal vertex cover of $G$ is the set of vertices of $K_n$. A ferry plan for $G$ should begin by transferring the clique to the opposite bank and then leaving a vertex there. On return the only choice is to load a vertex from the independent set, because leaving any number of vertices from the clique is impossible. On arrival to the destination bank we are forced to unload the vertex from the independent set and reload the vertex from the clique. We are now at a deadlock, because none of the vertices on the boat can be unloaded on the left bank.

The graph $G$ described in this example is a generalization of the star, where the central vertex is replaced by a clique. The star is the simplest topology where $OPT_{FC}(G) = OPT_{VC}(G) + 1$.

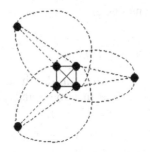

**Fig. 1.** An example of the graph described in Example 3

The following theorem, together with the observation of Example 3 about stars, completely solves the FERRY COVER problem on trees.

**Theorem 3.** *If $G$ is a tree and* $\text{OPT}_{\text{VC}}(G) \geq 2 \Rightarrow \text{OPT}_{\text{FC}}(G) = \text{OPT}_{\text{VC}}(G)$.

*Proof.* Let $v_1$, $v_2$ be two vertices of an optimal vertex cover of $G$. Then $v_1$ and $v_2$ have at most one common neighbor, because if they had two then $G$ would contain a cycle. We denote by $u$ the common neighbor of $v_1$ and $v_2$, if such a vertex exists.

Then a ferry plan for $G$ is the following: load the vertex cover in the boat and unload $v_1$ in the opposite bank. Then transfer all the neighbors of $v_2$ vertex by vertex, leaving vertex $u$ last to be ferried. When $u$ is the only remaining neighbor of $v_2$ on the left bank, unload $v_2$ and load $u$ on the boat. On arrival to the destination bank unload $u$ and load $v_1$. The remaining vertices of the independent set are now transported one by one to the destination bank and finally $v_2$ is loaded on the boat on the last trip and transported across together with the rest of the vertex cover.                                                              □

*Remark 1.* If $\text{OPT}_{\text{VC}}(G)$ for a tree $G$ is 1 (i.e. the tree is a star) then $\text{OPT}_{\text{FC}}(G) = 2$ unless the star has no more than 2 leaves, in which case $\text{OPT}_{\text{FC}}(G) = 1$.

**Corollary 2.** *The* FERRY COVER *problem can be solved in polynomial time in trees.*

*Proof.* The VERTEX COVER problem can be solved in polynomial time in trees. Theorem 3 and Remark 1 imply that determining $\text{OPT}_{\text{VC}}$ is equivalent to determining $\text{OPT}_{\text{FC}}$.

## 4    The Trip-Constrained Ferry Cover Problem with Trip Constraint 1

An interesting variation of FC is the TRIP-CONSTRAINED FERRY COVER problem where there is a limit on the number of trips allowed. In this section we study

TRIP-CONSTRAINED FERRY COVER in the case of a very tight trip constraint, i.e. when only one round-trip is allowed (recall that we denote this variation by $FC_1$). We will show that even in this case the problem is *NP*-hard, by obtaining a reduction from MAX-NAE-{3}-SAT. Our reduction is gap-preserving, and therefore we will also show that $FC_1$ is *APX*-hard.

We will use the $H$-COLORING problem to obtain an equivalent definition for $FC_1$.

**Lemma 2.** *A ferry plan of a graph $G$ for $FC_1$ is equivalent to an $F_1$-coloring of graph $G$, where $F_1$ is the graph of Figure 2.*

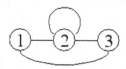

**Fig. 2.** Graph $F_1$ of Lemma 2

*Proof.* Given a ferry plan we can define the following homomorphism $\theta$ from $G$ to $F_1$:

- $\theta(u) = 1$, for all vertices $u$ of $G$ remaining on the boat only during the first trip,
- $\theta(u) = 2$, for all vertices $u$ of $G$ remaining on the boat throughout the execution of the plan,
- $\theta(u) = 3$, for all vertices $u$ of $G$ remaining on the boat only during the final trip.

Given an $F_1$-coloring we can devise a ferry plan from the above in the obvious way. □

**Corollary 3.** *For any graph $G(V, E)$ $OPT_{FC_1}(G) = \min\{|V_2|+\max\{|V_1|, |V_3|\}\}$, where the minimum is taken among all proper $F_1$-colorings of $G$ and $V_1, V_2, V_3$ are the subsets of $V$ that have taken the colors 1, 2 and 3 respectively.*

*Proof.* From Lemma 2 we obtain a ferry plan for $FC_1$: load the subsets $V_1$ and $V_2$ in the first trip and unload the subset $V_1$ in the opposite bank while keeping $V_2$ on the boat. Then return to the first bank and load $V_3$ together with $V_2$ and transport them to the destination bank.

This implies that the boat should have room for $V_2$ together with the larger of the sets $V_1$ and $V_3$. □

From now on we will refer to the value $|V_2| + \max\{|V_1|, |V_3|\}$ as the *cost* of an $F_1$-coloring. Thus, $FC_1$ can be reformulated as the problem of finding the minimum cost over all possible $F_1$-colorings. This reformulation leads to the following theorem:

**Theorem 4.** $FC_1$ *is NP-hard. Furthermore, there is a constant* $\epsilon_F > 0$ *s.t. there is no polynomial-time* $(1 + \epsilon_F)$*-approximation algorithm for* $FC_1$*, unless P=NP.*

*Proof.* We present a gap-preserving reduction from MAX-NAE-{3}-SAT. Our first step in the reduction is, given a formula $\phi$ with $m$ clauses, to construct a formula $\phi'$ with $2m$ clauses by adding to $\phi$ for every clause $(l_1 \vee l_2 \vee l_3)$ the clause $(\overline{l_1} \vee \overline{l_2} \vee \overline{l_3})$. Observe that if a formula contains the clause $(l_1 \vee l_2 \vee l_3)$, we can add the clause $(\overline{l_1} \vee \overline{l_2} \vee \overline{l_3})$ without affecting the formula's satisfiability, since a truth assignment satisfies the first clause (in the NAESAT sense) iff it satisfies both. Note that this also has no effect on the ratio of satisfied over unsatisfied clauses for any truth assignment. In addition, for any $i$, literals $l_i$ and $\overline{l_i}$ appear in $\phi'$ the same number of times. Note that, since this is the version of NAESAT where every clause has exactly three literals, the sum of the numbers of appearances of all variables in $\phi'$ is equal to $6m$.

Next, we construct a graph $G$ from $\phi'$. Every variable $x_i$ must appear an even number of times in $\phi'$, half of them as $x_i$ and half as $\neg x_i$. Let $2f_i$ denote the total number of appearances of the variable $x_i$. Then, for every variable $x_i$ we construct a complete bipartite graph $K_{f_i, f_i}$. One half of the bipartite graph represents the appearances of the literal $x_i$ and the other half the appearances of the literal $\neg x_i$.

For every clause $(l_1 \vee l_2 \vee l_3)$, we construct a triangle. We connect each vertex of the triangle to a vertex of the bipartite graph that corresponds to its literal, and has not already been connected to a triangle vertex. This is possible, since the vertices in the bipartite graphs that correspond to a literal $l_i$ are as many as the appearances of the literal $l_i$ in $\phi'$, and therefore as many as the vertices of triangles that correspond to $l_i$. This completes the construction, and we now have a graph where every vertex of a triangle has degree 3 and every vertex of a $K_{f_i, f_i}$ has degree $f_i + 1$.

Suppose that our original MAX-NAE-{3}-SAT formula $\phi$ had $m$ clauses, and we are given a truth assignment which satisfies $t$ of them. Let us produce an $F_1$-coloring of $G$ with cost $8m - t$. The given truth assignment satisfies $2t$ of the $2m$ clauses of $\phi'$. Assign colors 1 and 3 to the vertices of the bipartite graphs, depending on the truth value assigned to the corresponding literal (1 for false and 3 for true). Every triangle corresponding to a satisfied clause can be colored using all three colors, by assigning 1 to a true literal, 3 to a false literal and 2 to the remaining literal. Triangles corresponding to clauses with all literals true are colored with two vertices receiving 2 and one receiving 1. Similarly, triangles corresponding to clauses with all literals false are colored with two vertices receiving 2 and one receiving 3. Note that, due to the construction of $\phi'$, the number of clauses with all literals true, is the same as the number of clauses with all literals false. Therefore, $|V_1| = |V_3| = \sum_i f_i + 2t + \frac{2m - 2t}{2} = 4m + t$, while $|V_2| = 2t + 2(2m - 2t) = 4m - 2t$ making the total cost of our coloring equal to $8m - t$.

Conversely, suppose we are given an $F_1$-coloring of $G$ with cost at most $8m - t$, we will produce a truth assignment that satisfies at least $2t$ clauses of $\phi'$ and therefore at least $t$ clauses of $\phi$. We will first show that this can be done when the color 2 is not used for the vertices of the bipartite graphs, and then show that any coloring which does not meet this requirement can be transformed to one of at most equal cost that does.

If color 2 is not used in the bipartite graphs, then the cost for these vertices is $\sum_i f_i = 3m$. Therefore, the cost for the $2m$ triangles is at most $5m - t$. No triangle can have cost less than 2, therefore there are at most $m - t$ triangles with cost 3, or equivalently at least $m + t$ triangles of cost 2. Suppose that no triangle uses color 2 three times (if not, pick one of its vertices arbitrarily and color it with 1 or 3, without increasing the total cost). Also, wlog suppose that $|V_3| \geq |V_1|$ (if not, colors 1 and 3 can be swapped without altering the cost).

Now, triangles can be divided in the following categories:

1. Triangles that use color 2 once. These triangles also use colors 1 and 3 once and their cost is 2.
2. Triangles that use color 2 twice and color 1 once. These triangles have a cost of 2.
3. Triangles that use color 2 twice and color 3 once. The cost of these triangles is 3.

Suppose that the first category has $k$ triangles (these correspond to clauses that will be satisfied by the produced truth assignment). Now, $|V_3| \leq \sum_i f_i + m - t + k$, but $|V_3| \geq |V_1| \geq \sum_i f_i + m + t$, thus, $m + t \leq m - t + k \Rightarrow k \geq 2t$. Produce a truth assignment according to the coloring of the bipartite graphs ($1 \rightarrow$ false and $3 \rightarrow$ true). The assignment described above satisfies at least $k$ clauses.

If color 2 is used in the bipartite graphs, we distinguish between two separate cases: first, suppose that the same side of a bipartite graph does not contain both colors 1 and 3. In other words, one side is colored with 1 and 2, and the other with 2 and 3. On the first side, pick a vertex with color 2. If its only neighbor from a triangle has received colors 2 or 3, change its color to 1. If its neighbor has received color 1 exchange their colors. Repeat, until no vertices on that side have color 2 and proceed similarly for the other side, thus eliminating color 2 from the bipartite graphs without increasing the total cost.

Finally, suppose that the same side of a bipartite graph contains both colors 1 and 3 (let $A$ denote the set of vertices of this side). Then, the other side (the set of its vertices is denoted by $B$) must contain only color 2. We will reduce this case to the previous one. Let $A_1$ be the subset of $A$ consisting of vertices colored with 1 and $A_3$ the subset of vertices colored with 3 ($|A_1| + |A_3| \leq |A|$). Let $B_1$ be the subset of $B$ consisting of vertices connected with triangle vertices colored with 2 or 3, and let $B_3$ be the subset of $B$ consisting of vertices connected with triangle vertices colored with 2 or 1 ($|B_1| + |B_3| \geq |B|$). Since $|A| = |B|$ then $|A_1| \leq |B_1|$ or $|A_3| \leq |B_3|$. If $|A_1| \leq |B_1|$ then assign color 2 to all vertices of $A_1$ and color 1 to all vertices of $B_1$ (this does not increase the total cost), thus eliminating color 1 from side $A$. If $|A_3| \leq |B_3|$ similarly assign color 2 to the vertices of $A_3$ and color 3 to the vertices of $B_3$.

The above reduction shows that given a MAX-NAE-{3}-SAT formula $\phi$ with $m$ clauses we can construct a graph $G$ s.t.

$$\text{OPT}_{\text{MAX-NAE-}\{3\}\text{-SAT}}(\phi) = m \Rightarrow \text{OPT}_{\text{FC}_1}(G) = 7m$$
$$\text{OPT}_{\text{MAX-NAE-}\{3\}\text{-SAT}}(\phi) < (1 - \epsilon)m \Rightarrow \text{OPT}_{\text{FC}_1}(G) > (1 + \epsilon_F)7m$$

where $\epsilon_F = \frac{\epsilon}{7}$. In other words we have constructed a gap-preserving reduction from MAX-NAE-{3}-SAT to FC$_1$, by making use of the reformulation with $H$-colorings. Well-known hardness results for MAX-NAE-{3}-SAT (see for example [5]) complete the proof of this theorem.    $\square$

**Theorem 5.** *There is a $\frac{3}{2}$-approximation algorithm for* FC$_1$ *on trees.*

*Proof.* First observe that $\text{OPT}_{\text{FC}_1}(G) \geq \frac{n}{2}$, since the boat only arrives to the destination bank twice, and therefore it must be able to carry at least half of the vertices of $G$. Next, it can be shown that $\text{OPT}_{\text{VC}}(G) \leq \frac{n}{2}$, since on trees there is always an independent set of size at least $\frac{n}{2}$. This can be trivially shown, since trees are bipartite graphs.

A ferry plan for a tree is the following: compute an optimal vertex cover (its size is at most $\frac{n}{2}$) and place all its vertices on the boat. Fill the boat with enough of the remaining vertices so that it contains $\lceil \frac{n}{2} \rceil$ vertices. Move to the other side, compute an optimal vertex cover of the graph induced on the original graph by the vertices on the boat (its size is at most $\frac{\lceil \frac{n}{2} \rceil}{2}$) and keep only those vertices on the boat. Return to transfer the remaining vertices to the destination bank. Clearly, a boat capacity of at most $\frac{\lceil \frac{n}{2} \rceil}{2} + \lfloor \frac{n}{2} \rfloor \leq \frac{3n}{4}$ is sufficient to execute this plan, and this is at most $\frac{3}{2}$ times the optimal.    $\square$

The ideas of the previous theorem can be further refined to produce the following result:

**Theorem 6.** *There is an approximation algorithm for* FC$_1$ *on trees with approximation guarantee asymptotically equal to $\frac{4}{3}$.*

*Proof.* Suppose now that instead of transporting $\frac{n}{2}$ vertices on the first trip we wish to transport $\frac{n}{k}$ vertices for some $k > 1$. Upon arrival to the destination bank we unload at least half of them and return with at most $\frac{n}{2k}$ vertices. Now we need to take all the remaining vertices to the other side.

This plan requires a boat capacity of $\max\{\frac{n}{k}, \frac{n}{2k} + n - \frac{n}{k}\}$. It is not hard to see that this is minimized for $k = \frac{3}{2}$. Thus, by taking two thirds of the vertices on the initial trip we devise a ferry plan that requires a capacity of $\frac{2n}{3}$ vertices. Clearly, this is at most $\frac{4}{3}$ times the optimal.

Unfortunately, the preceding analysis requires that $n$ is a multiple of 3. If this is not the case we would be required to take $\lceil \frac{2n}{3} \rceil \leq \frac{2n}{3} + 1$ vertices. This results to an approximation ratio bounded by $\frac{4}{3} + \frac{2}{n}$ which tends to $\frac{4}{3}$ as $n$ tends to infinity.    $\square$

# 5   The Trip-Constrained Ferry Cover and Min Ferry Trips Problems

In this section we study TRIP-CONSTRAINED FERRY COVER for general values of the trip constraint and present several lemmata which provide bounds on the optimal solution and relate $FC_m$ to FC. We extend the reasoning behind those lemmata to prove a set of similar results for MIN FERRY TRIPS.

First note that a very loose constraint on the number of trips makes the problem equivalent to the FERRY COVER problem.

**Lemma 3.** *For any graph* $G(V, E)$, $|V| = n$, $OPT_{FC_{2^n-1}}(G) = OPT_{FC}(G)$.

*Proof.* Any solution to $FC_{2^n-1}(G)$ allows a ferry plan with at most $2^{n+1}$ configurations. There are at most $2^n$ partitions of the vertices of $G$ into two sets, therefore there are at most $2^{n+1}$ possible legal configurations. No optimal ferry plan repeats the same configuration twice, since the configurations found between two successive appearances of the same configuration in a ferry plan can be omitted to produce a shorter plan. Therefore, any optimal ferry plan for the unconstrained version has at most $2^{n+1}$ configurations and can be realized within the limits of the trip constraint.                                            □

Loosening the trip constraint can only improve the value of the optimal solution.

**Lemma 4.** *For any graph* $G$ *and any integer* $i \geq 0$, $OPT_{FC_i}(G) \geq OPT_{FC_{i+1}}(G)$.

*Proof.* Observe that a ferry plan with trip constraint $i$ can also be executed with trip constraint $i + 1$.                                            □

A different lower bound is given by the following Lemma.

**Lemma 5.** *For any graph* $G(V, E)$, $OPT_{FC_m}(G) \geq \frac{|V|}{m+1}$

*Proof.* Observe that a trip constraint of $m$ implies that for any ferry plan the boat will arrive at the destination bank at most $m + 1$ times. Therefore, at least one of them it must carry at least $\frac{|V|}{m+1}$ vertices.                                            □

**Corollary 4.** *There is an* $(m + 1)$-*approximation algorithm for* $FC_m$.

*Proof.* A boat of capacity $|V|$ can trivially solve the problem. From Lemma 5 it follows that this solution is at most $m + 1$ times the optimal.                                            □

Setting the trip constraint greater or equal to the number of vertices makes the constrained version of the problem similar to the unconstrained version.

**Lemma 6.** *For any graph* $G(V, E)$, *with* $|V| = n$, $OPT_{VC}(G) \leq OPT_{FC_n}(G) \leq OPT_{VC}(G) + 1$.

*Proof.* For the first inequality, a boat capacity smaller than the minimum vertex cover allows no trips. For the second inequality it suffices to observe that the ferry plan of Lemma 1 can be realized within the trip constraint.                                            □

**Corollary 5.** *Determining* $OPT_{FC_m}$ *is NP-hard for all* $m \geq n$. *Furthermore, there are constants* $\epsilon_F, n_0 > 0$ *s.t. there is no* $(1 + \epsilon_F)$-*approximation algorithm for* $OPT_{FC_m}$ *with instance size greater than* $n_0$ *vertices unless P=NP.*

*Proof.* By using Lemmata 6 and 4 we can show that $OPT_{VC}(G) \leq OPT_{FC_m} \leq OPT_{VC}(G) + 1$. The rest of the proof is similar to that of Theorem 1.     □

It is unknown whether there are graphs where $OPT_{FC_n}(G) > OPT_{FC}(G)$. We conjecture that there is a threshold $f(n)$ s.t. for any graph $G$, $OPT_{FC_{f(n)}}(G) = OPT_{FC}(G)$ and that $f(n)$ is much closer to $n$ than $2^n - 1$ which was proven in Lemma 3.

Following similar reasoning as in Lemmata 3 - 6 we reach the following results for $MFT_k$:

**Lemma 7.** *For any graph* $G(V, E)$, $|V| = n$

1. *If* $k = OPT_{VC}(G)$ *then* $OPT_{MFT_k}(G) \leq 2^n - 1$ *or* $OPT_{MFT_k}(G) = \infty$
2. *For any integer* $k$, $OPT_{MFT_k}(G) \geq OPT_{MFT_{k+1}}(G)$
3. $OPT_{MFT_k} \geq \frac{n}{k} - 1$
4. *If* $k \geq OPT_{VC}(G) + 1$ *then* $OPT_{MFT_k}(G) \leq n$

*Proof.* Similar to proofs of Lemmata 3,4,5,6 respectively.

$MFT_k$ is at least as hard as FC since determining the optimal number of trips involves deciding whether a ferry-plan is possible with the given boat capacity, which is exactly the decision version of FC. However, it would be interesting to investigate whether $MFT_k$ remains NP-hard even for $k \geq OPT_{VC} + 1$, in which case there is always a valid ferry plan. We conjecture that the problem remains NP-hard in that case.

## 6     Conclusions and Further Work

In this paper we have investigated the algorithmic complexity of several variations of FERRY problems. For the unconstrained FERRY COVER problem we have presented results that show it is very closely related to VERTEX COVER, which is a consequence of the fact that the optimal values for the two problems are almost equal.

For $FC_1$ we have presented hardness results, but the question of how the problem can be efficiently approximated is open. It would be interesting to see an approximation algorithm which achieves a ratio better than 2, which can be achieved trivially by setting boat size $n$.

For the TRIP-CONSTRAINED FERRY COVER and MIN FERRY TRIPS problems, we have presented several lemmata that point out their relation to FC. We believe that these variations are more interesting because they appear to be less related to VERTEX COVER than FC. It remains an open problem to determine at which value of the trip constraint $FC_m$ becomes equivalent to FC (however, an upper bound on this value is $2^n - 1$, as shown in Lemma 3). This is a particularly interesting question since so far it remains open whether FC is in NPO, or there exist graphs where every optimal ferry plan is of exponential length. However, we believe it is highly unlikely that FC is not in *NPO*. Finally, it would be interesting to investigate whether there are values of $m$ for which $FC_m$ can be solved in polynomial time, but we believe that hardness results similar to those for $m = 1$ and $m \geq n$ hold for all values of $m$.

**Acknowledgements.** This paper is based on work done at the Computation and Reasoning Lab at NTUA. The authors would like to thank Aris Pagourtzis and Georgia Kaouri for the original inspiration of the algorithms exercise from which this paper began, Stathis Zachos for useful discussion and the anonymous referees for their comments which helped to clarify the presentation of some of our results.

# References

1. Arora, S., Lund, C., Motwani, R., Sudan, M., Szegedy, M.: Proof verification and intractability of approximation problems. In: Proc. 33rd IEEE annual Symposium on Foundations of Computer Science (FOCS), pp. 13–22 (1992)
2. Borndörfer, R., Grötschel, M., Löbel, A.: Alcuin's transportation problems and integer programming (1995)
3. Garey, M.R., Johnson, D.S.: Computers and Intractability: A guide to the theory of NP-completeness. W. H. Freeman and Co., New York, NY (1979)
4. Hell, P., Nešetřil, J.: On the complexity of h-coloring. J. Comb. Theory Ser. B. 48(1), 92–110 (1990)
5. Petrank, E.: The hardness of approximation: Gap location. In: Computational Complexity, vol. 4, Springer, Heidelberg (1994)
6. Vazirani, V.V.: Approximation Algorithms. Springer, Heidelberg (2001)

# Web Marshals Fighting Curly Link Farms*

Fabrizio Luccio and Linda Pagli

Dipartimento di Informatica, Università di Pisa
{luccio,pagli}@di.unipi.it

**Abstract.** *Graphides cincinnatae* (also known as circulant graphs) $Ci_n(L)$ of $n$ vertices are studied here as link farms in the Web, built automatically by a spammer to promote visibility of a page $T$. These graphs are *k-consecutive*, denoted by $Ci_{n,k}$, if each vertex $v_i$ is connected to $v_{i+j}$ and $v_{i-j}$ with $j = 1, 2, ..., k$. *Graphides cirratae* are cincinnatae with some irregularities. We discuss how to fight this phenomenon with a set of *Web marshals*, that is autonomous agents that visit the farm for cutting the links to $T$. The farm reacts reconstructing the links through majority voting among its pages. We prove upper and lower bounds on the number of marshals, and of link hops, needed to dismantle the farm. We consider both synchronous and asynchronous operations.

**Regular Keywords:** Circulant graph, Web graph, Page-rank, Web spam, Link farm, Autonomous agent, Majority rule.

**Elegant Keywords:** Graphis cincinnata, Graphis cirrata, Curly graph, Web marshal.

## 1 Background

A known definition (aside perhaps for the wording) is the following:

**Definition 1.** *A* graphis cincinnata[1] $Ci_n(\mathcal{L})$ *is a graph of $n$ vertices $v_0, v_1, ...,$ $v_{n-1}$ in which $v_i$ is connected to $v_{i+j}$ and $v_{i-j}$ for each $j$ in a list $\mathcal{L}$ (indices are taken modulo $n$).*

In particular we say that a graphis cincinnata $Ci_n(\mathcal{L})$ is *k-consecutive*, denoted by $Ci_{n,k}$, if $\mathcal{L}$ is the list of consecutive integers $1, 2, ..., k$. $Ci_n(1, 2, 3)$ and $Ci_n(1, 4)$ are shown in Figure 1. The former is 3-consecutive, also denoted as $Ci_{n,3}$.

Graphides cincinnatae emerge from pure graph theory to the Web, to assume a role in the so called "visibility bubble". Reference is generally made to Google's page-ranking, although all search engines suffer similar attacks. The main technique for boosting page visibility is well known (e.g., see [5,6]). A spammer creates a *link farm* (also called *spam farm*) $F$ of $n$ boosting pages pointing to a *target page* $T$ whose ranking is to be artificially increased. Then the spammer

---

* As to now, no foreseeing institution has sponsored this research.
[1] Plural: "Grafides cincinnatae", literally "curly graphs". Dully called "circulant graphs" in the literature [2,10].

P. Crescenzi, G. Prencipe, and G. Pucci (Eds.): FUN 2007, LNCS 4475, pp. 240–248, 2007.
© Springer-Verlag Berlin Heidelberg 2007

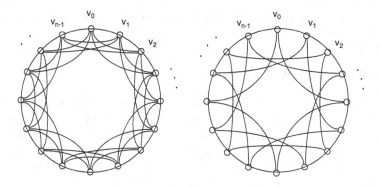

**Fig. 1.** Graphides cincinnatae $Ci_n(1,2,3) = Ci_{n,3}$ and $Ci_n(1,4)$

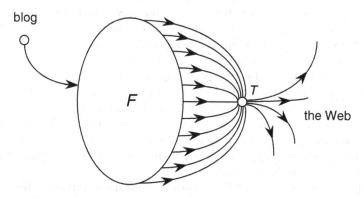

**Fig. 2.** Schematic of a link farm $F$ promoting a target page $T$

provides links to $F$ from regular pages outside $F$, for example by posting a hyperlink to $F$ in a public blog or in a Web page open to be written in. Such a structure is sketched in Figure 2.

Conditions for enhancing the efficiency of a link farm have been indicated in [7], and then thoroughly studied in [3] and in [1], respectively from a theoretical and an experimental point of view. In particular, in the latter work link farms are detected by a robot (or crawler) as densely connected subgraphs, where each node is related to the rest of the graph only for incoming links from some blog, and for the out link to the target page $T$ (for simplicity we assume that only one page is targeted). Since $F$ contains a high number of pages and is built automatically, it seems reasonable to assume that it has a regular structure. In fact we shall assume that $F$ is a graphis cincinnata or some of its variations.

We propose that, after a link spam $F$ has been detected by a crawler, a group of autonomous agents called *Web marshals* be sent to dismantle $F$ with minimum amount of changes done in its pages. On its side the spammer has provided a recovery mechanism against the marshals, based on automatic information exchange among the pages of $F$. The fighting is carrying on, and its terms will now be explained.

## 2   Recovering by Proximity

Assume that $m$ marshals $M_0, M_1, ..., M_{m-1}$ are sent all together to a vertex of $F$, from which they start moving along the links. Before the attack takes place all the vertices of $F$ are *harmful*. When a marshal reaches a harmful vertex $v$, it makes it *harmless* by changing one bit in the link from $v$ to the target page $T$. Periodically the vertices of $F$ interrogate their adjacent peers to check the value of the link to $T$, and "repair" it automatically by taking as correct value the one stored in the *majority* (i.e., more than one half) of its neighbors. So a vertex $v$ made harmless might become harmful again, or vice-versa. Our algorithms, however, insures that, after $v$ has been made harmless it will be unable to recover. That is, after the attacking marshal has gone, $v$ is never surrounded by a majority of harmful neighbors.

This acting model is a combination of the ones based on *mobile agents* [4], and on *dynamos* [9], as already been proposed in [8] for repairing faulty networks. Clearly it can be modified at will. Our problem is finding the minimum number of agents needed to dismantle a link farm, together with the attacking algorithm. Of minor importance is the time (or the number of link hops) needed by the marshals, in view of the fact that they will work asynchronously. Still, for simplicity we will examine synchronous functioning first, and then show how such a limitation can be removed.

Let the harmful vertices of $F$ be *white*; the harmless vertices still containing a marshal be *black*; and the harmless vertices previously visited by a marshal,

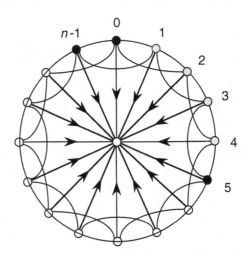

**Fig. 3.** $Ci_{n,2}$ promoting the ranking of one page, and the three marshals. Edges without an arrow represent links in two directions.

and presently without one, be *gray* (see Figure 3). As the link farms considered here are graphides cincinnatae, all their vertices have the same degree $\delta$ and the majority of their neighbors in $F$ is equal to $\delta/2 + 1$ (recall that only the edges among vertices of $F$ are considered, and that $\delta$ is even). For a vertex $v$, let $\sigma(v)$ denote the number of its harmless neighbours (black or gray) at any given moment. Our model must obey to the following basic lemma, that is easily derived from a result of [8] noting that, if the stated conditions were violated, $v$ would become harmful again.

**Lemma 1.** *Let a vertex $v$ contain one marshal $M$:*

1. *if $\sigma(v) < \delta/2$, $M$ cannot move from $v$;*
2. *if $\sigma(v) = \delta/2$, $M$ can only move to a white neighbour of $v$;*
3. *if $\sigma(v) > \delta/2$, $M$ can move to any neighbour of $v$.*

## 3  Fighting Link Farms

The three basic parameters here are: $n$ = number of vertices in $F$; $m$ = number of marshals; and $k$ for $Ci_{n,k}$, or in general $k$ = greatest element in $\mathcal{L}$ for $Ci_n(\mathcal{L})$. Note that $\delta = 2k$ for $Ci_{n,k}$, or in general $\delta = 2|\mathcal{L}|$ for $Ci_n(\mathcal{L})$. With the vertex numbering of Definition 1, we recognize a *main cycle* $v_0, v_1, v_2, ..., v_{n-1}$ in $Ci_n(\mathcal{L})$ (e.g., see Figure 1). We assume that the marshals can discriminate among the links of each page of $F$, in particular, they can recognize the links of the main cycle.

We start by presenting a simple algorithm for dismantling a link farm $F$ in the form of a $k$-consecutive graphis cincinnata $Ci_{n,k}$, with $m = k + 1$ marshals. We prove that this value is minimum, i.e., less marshals cannot do the job, and that the total number of link hops is also minimum if the marshals work synchronously. Note that it is not required that the marshals know the value of $n$, so they can work with fixed memory size, but they must be able to recognize each other upon meeting in a vertex.

**Algorithm 1.** *Dismantling $Ci_{n,k}$ with $m = k + 1$ marshals $M_0, M_1, ..., M_{m-1}$.*

**start** with the marshals in $v_0$;

$c = \lfloor m/2 \rfloor$;

**for** $i \in \{1, .., c\}$ **move** $M_i$ to $v_{n-i}$ along the link $(v_0, v_{n-i})$;

**for** $i \in \{c + 1, .., m - 1\}$ **move** $M_i$ to $v_{i-c}$ along the link $(v_0, v_{i-c})$;

**repeat** (**move** $M_{m-1}$ one step forward along the main cycle)

    **until** ($M_{m-1}$ meets $M_c$).

For example three marshals can dismantle $Ci_{n,2}$ as shown in Figure 3, where two of them are maintained in vertices $n - 1$ and $0$, and the third one travels along the main cycle. In fact, starting from vertex 1 the third marshal can move forward since it always has two harmless neighbors in $F$, that is the number required by point 2 of Lemma 1. We have:

**Theorem 1.** *A link farm $F$ in the form of $Ci_{n,k}$ can be dismantled by $m = k+1$ synchronous marshals with $n$ link hops.*

**Proof.** Use algorithm 1. Correctness of the algorithm derives from Lemma 1 point 2, as marshal $M_m$ can move forward always maintaining $\delta/2 = k$ harmless neighbors in $F$. The number of hops is the sum of the $k$ initial moves of marshals $M_0,..., M_{m-1}$ (two **for** cycles of the algorithm), plus the next $n - k$ moves of $M_{m-1}$ along the main cycle. The assumption of synchronicity is required because marshal $M_{m-1}$ could be moving faster than the others, thus moving along the main cycle before $k$ of its neighbors have become harmless.     *Q.E.D.*

We now prove the the upper bounds of Theorem 1 are tight. We have:

**Theorem 2.** *To dismantle a link farm $F$ in the form of $Ci_{n,k}$ at least $m = k+1$ synchronous marshals, and $n$ link hops, are needed.*

**Proof.** 1. Number of marshals. By contradiction, assume that $m' \leq k$ marshals are used. Since all the vertices of $F$ must be visited, gray vertices will inevitably appear at some point. Let $v$ be the gray vertex appearing first, that is, all the other vertices are black or white at this point. $v$ can now have at most $m'$ harmless neighbors, not a majority as required. Hence $v$ would become white again, against the attacking rules. (In fact, the marshal that left from $v$ could not have done it by Lemma 1, because $v$ had at most $m' - 1 < k$ harmless neighbors).

2. Number of hops. At least $n - 1$ moves are needed to reach the vertices $v_1,..., v_{n-1}$ from $v_0$, and an additional move is needed for a final meeting of two marshals in the same vertex, to ensure termination.     *Q.E.D.*

We now explain how Algorithm 1 can be simply modified for dismantling a general $Ci_n(\mathcal{L})$ using $m = k + 1$ marshals, where $k$ is now the greatest element in $\mathcal{L}$. This is shown in Figure 4 where four marshals $M_0$ to $M_3$ are used for attacking $Ci_n(1,3)$. The marshals are initially moved to the vertices $n - 2, n - 1$, $0, 1$ along the main cycle, as direct links between $v_0$ and the other vertices may not exist in this case. This is done with an obvious modification of the two **for** cycles of the algorithm. At this point each marshal occupies a vertex $v$ with $\sigma(v) = 2 = \delta/2$ harmless neighbors, then $M_{m-1}$ can move along the main cycle as in Algorithm 1 (the first move is shown in Figure 4) because it always reaches vertices $v$ with $\sigma(v) = 2$.

It can be easily seen that the total number $h$ of link hops required by the two **for** cycles of the algorithm is now given by:

$$h = \begin{cases} k^2/4 + k/2 & \text{for } k \text{ even} \\ k^2/4 + k/2 + 1/4 & \text{for } k \text{ odd} \end{cases} \quad (1)$$

Note that $h$ is integer. We then have:

**Theorem 3.** *A link farm $F$ in the form of $Ci_n(\mathcal{L})$ can be dismantled by $m = k + 1$ synchronous marshals with $n - k + h$ link hops, where $k$ is the greatest element in $\mathcal{L}$, and $h$ is given by equation (1).*

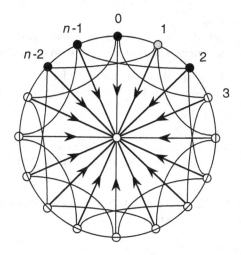

**Fig. 4.** $Ci_n(1,3)$ dismantled by four marshals

However, we are unable to establish whether these bounds are tight. Noting that $\delta = 2|\mathcal{L}|$ we can only state:

**Theorem 4.** *To dismantle a link farm $F$ in the form of $Ci_n(\mathcal{L})$ at least $m = k + 1$ synchronous marshals and $n$ link hops are needed, with $k = |\mathcal{L}|+1$.*

## 4   Asynchronous Marshals

As already stated, we expect the marshals to work asynchronously. Therefore Algorithm 1 and its modification for $Ci_n(\mathcal{L})$ may not work properly. To overcome this difficulty we send an additional marshal $M_m$ that travels along the graph to inform the other marshals of their mutual positions. As before let $m = k+1$, where $k$ is the greatest element in $\mathcal{L}$ (then the marshals will be $k + 2$). To this end we propose the new Algorithm 2 below.

It can be easily seen that the total number $h$ of link hops required to place the marshals $M_1,...,\ M_{m-2}$ into the initial vertices $v_{n-c},...,\ v_{m-2+c}$, including the hops of $M_m$, is now given by:

$$h = \begin{cases} 3(k^2/4 - k/2) & \text{for } k \text{ even} \\ 3(k^2/4 - k/2 + 1/4) & \text{for } k \text{ odd} \end{cases} \tag{2}$$

**Algorithm 2.** *Dismantling $Ci_n(\mathcal{L})$ with $m + 1$ marshals $M_0, M_1, ..., M_m$.*

**start** with the marshals in $v_0$;

$c = \lfloor m/2 \rfloor$;

**move** $M_1$ and $M_m$ to $v_{n-1}$ along the main cycle;
 **when** the two marshals meet at $v_{n-1}$ **move** $M_m$ back to $v_0$;

**for** $i = 2$ **to** $c$ **do**

**when** $M_m$ returns to $v_0$
  **move** $M_i$ and $M_m$ to $v_{n-i}$ along the main cycle;
  **when** the two marshals meet at $v_{n-i}$ **move** $M_m$ back to $v_0$;

**move** $M_{c+1}$ and $M_m$ to $v_1$ along the main cycle;
  **when** the two marshals meet at $v_1$ **move** $M_m$ back to $v_0$;

**for** $i = c + 2$ **to** $m - 2$ **do**
  **when** $M_m$ returns to $v_0$
   **move** $M_i$ and $M_m$ to $v_{i-c}$ along the main cycle;
   **when** the two marshals meet at $v_{i-c}$ **move** $M_m$ back to $v_0$;

**when** $M_m$ returns to $v_0$
  **repeat** (**move** $M_{m-1}$ one step forward along the main cycle)
  **until** ($M_{m-1}$ meets $M_c$).

Referring to Algorithm 2 we can state:

**Theorem 5.** *A link farm $F$ in the form of $Ci_n(\mathcal{L})$ can be dismantled by $m = k + 2$ asynchronous marshals with $n - c + h$ link hops, where $k$ is the greatest element in $\mathcal{L}$, $c = \lfloor (k + 1)/2 \rfloor$, and $h$ is given by equation (2).*

However, we are unable to establish whether these bounds are tight. Theorem 4 still applies, changing asynchronous for synchronous.

## 5  Concluding with Other Graphides

Graphides cinncinnatae are called *cirratae*[2] if some irregularities arise in their structure. Since the farm $F$ is built automatically by the spammer, we assume that only some systematic deviations from the standard form of $Ci_n(\mathcal{L})$ may occur. Assuming that the main cycle $v_0, v_1, ..., v_{n-1}$ is maintained, and that the graph has a repetitive structure, we consider two cases of graphides cirratae:

1. The graph presents alternation of links of different "length" connecting vertices of the main cycle, as in graph $C_1$ of Figure 5 where links of length 2 and 4 alternate. In terms of graph theory, these are *regular* but not circulant graphs [2,10].
2. The graph has vertices of different degree, as graph $C_2$ of Figure 5. These are *not* regular graphs.

In the graphs of type 1 the vertex degree $\delta$ is constant, and we let $k = max(j - i) \bmod n$, where $j, i$ are vertex indices such that the link $(v_j, v_i)$ exists (i.e., $k$ is the length of the "longest" link outside the main cycle). In graph $C_1$ of Figure 5 we have $\delta = 4$ and $k = 4$.

In the graphs of type 2 we let $\delta = $ maximum vertex degree, and $k$ defined as for type 1. In graph $C_2$ of Figure 5 we have $\delta = 6$ and $k = 4$.

---

[2] Literally, "with natural curls".

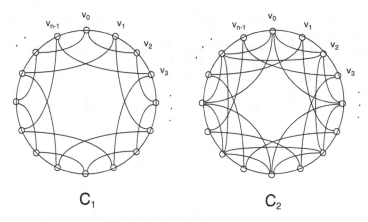

**Fig. 5.** Two graphides cirratae

It can be easily verified that Algorithm 1 (extended for handling $Ci_n(\mathcal{L})$ as indicated), and Algorithm 2, work properly also for graphides cirratae of type 1 and 2, if the start vertex $v_0$ is chosen as the source of a "longest" link. In our example five synchronous marshals, or six asynchronous marshals, can dismantle both $C_1$ and $C_2$. We have:

**Theorem 6.** *A link farm $F$ in the form of a graphis cirrata of type 1 or 2 can be dismantled by $k+1$ synchronous marshals or $k+2$ asynchronous marshals.*

To conclude, we are aware that the model presented here is too constrained to be realistic. In fact we regard it as a way to initiate studying the dismantlement of link farms by autonomous agents. Further work is needed for extending our algorithms to different proximity majority rules, and to more general graphs.

# References

1. Becchetti, L., et al.: Link-Based Characterization and Detection of Web Spam. In: Proc. AIRWeb'06, Seattle (2006)
2. Buckley, F., Harary, F.: Distance in Graphs. Addison-Wesley, Redwood City,CA (1990)
3. Du, Y., Shi, Y., Zhao, X.: Using Spam Farm to Boost Page Rank, Manuscript under publication (2006)
4. Flocchini, P., Huang, M.J., Luccio, F.L.: Contiguous Search in the Hypercube for Capturing an Intruder. In: Proc. 19th IEEE International Parallel and Distributed Processing Symp. Denver pp. 62–71 (2005)
5. Gori, M., Witten, I.: The Bubble of Web Visibility. Communications of the ACM 48(3), 115–117 (2005)
6. Gyöngyi, Z., Garcia-Molina, H.: Web Spam Taxonomy. In: Proc. AIRWeb'05, Chiba (2005)

7. Gyöngyi, Z., Garcia-Molina, H.: Link Spam Alliances. In: Proc. VLDB'05, Trond-heim pp. 517–528 (2005)
8. Luccio, F., Pagli, L., Santoro, N.: Network Decontamination with Local Immuniza-tion. In: Proc. APDCM'06, Rhodes (2006)
9. Peleg, D.: Size Bounds for Dynamic Monopolies. In: Proc. SIROCCO 97, Ascona, pp. 151–161 (1997)
10. West, D.B.: Introduction to Graph Theory. Prentice-Hall, Englewood Cliffs (2000)

# Intruder Capture in Sierpiński Graphs

Flaminia L. Luccio

Dipartimento di Informatica, Università Ca' Foscari Venezia, Venice, Italy
luccio@unive.it
http://www.dsi.unive.it/~luccio

**Abstract.** In this paper we consider the problem of capturing an intruder in a networked environment. The intruder is defined as a mobile entity that moves arbitrarily fast inside the network and escapes from a team of software agents. The agents have to collaborate and coordinate their moves in order to isolate the intruder. They move asynchronously and they know the network topology they are in is a particular fractal graph, the Sierpiński graph $SG_n$.

We first derive lower bounds on the minimum number of agents, number of moves and time steps required to capture the intruder. We then consider two models: one in which agents have a capability, of "seeing" the state of their neighbors; the second one in which the actions of the agents are leaded by a coordinator. One of our goals is to continue a previous study on what is the impact of visibility on complexity: we have found that in this topology the visibility assumption allows us to reach an optimal bound on the number of agents required for the cleaning strategy. On the other hand, the second strategy relies only on local computations but requires an extra agent and a higher (by a constant) complexity in terms of time and number of moves.

**Keywords:** Mobile agents, intruder capture, Sierpiński graphs.

## 1 Introduction

In this paper we consider a networked environment in which a set of software agents move along the network links from a node to a neighboring one. The *intruder* is a mobile (possibly malicious) entity that moves from node to node, escaping from the other agents.

The problem we consider consists of devising a strategy that allows a team of agents, initially located on the same node, the *homebase*, and unaware of the position of the intruder, to search for it and to surround it (so that it does not have a free link to escape through). We assume a worst case scenario in which the intruder is a very powerful entity, that can move arbitrarily fast inside the network and can permanently "see" the position of all the other agents, thus avoiding them as long as possible.

This problem can be easily reformulated in terms of another known problem called the *decontamination (or cleaning)* problem, in which a network, initially *contaminated* has to be *decontaminated*. More formally, initially all nodes are *contaminated*, except for the ones where agents are located, which are *guarded*. A node becomes *clean* whenever an agent passes through it an all its neighboring nodes are either clean or guarded; on the other hand, the intruder contaminates the node and edges it traverses. In the

P. Crescenzi, G. Prencipe, and G. Pucci (Eds.): FUN 2007, LNCS 4475, pp. 249–261, 2007.

*node-decontamination problem* at the end of the computation all the nodes of the network have to be clean, analogously, in the *edge-decontamination problem* at the end all the nodes and edges have to be clean. In this case the intruder can be seen as a virus that spreads inside a system, and that has to be isolated and recovered from.

In this paper we are interested in solving the node-decontamination problem and we are interested in complexity issues, i.e., devising efficient strategies, where efficiency is measured in terms of the size of the agent team, the number of moves agents have to perform and the time.

## 1.1   Related Work

The problem of capturing an intruder, firstly introduced by Breish [4] and Parson [13], is well known in the literature with the name of *graph search* or *decontamination problem* ([5,10,11,12]) and considers a system of tunnels, initially *contaminated*, that have to be *decontaminated* by a set of searchers. The tunnels are the edges of the graph.

There are several variants of the graph search problem, and in most of them the actions that can be performed consist in placing and removing searchers at any time and on any node, i.e., searchers may "jump" across the network. In the *node and edge search problems* the searchers have to lead the network to a state in which all the edges are simultaneously decontaminated. In the node search problem an edge is decontaminated whenever two searchers are placed on the edge extremes, in the edge search problem, the edge has also to be traversed, thus in this case this action is also allowed. The strategy has also to minimize the number of searchers involved, i.e., to consume the minimum amount of resources. However, determining whether the smallest number of searchers for which a search strategy exists (the so called *search number*) is at most $k$, for any integer $k$, is an $NP$-hard problem in most topologies.

Note that, graph search, intruder detection, and decontamination are equivalent problems.

In this paper we consider another variant of the problem, called the *contiguous, monotone, node search problem*, where agents may only move contiguously inside the network (i.e., may not be removed), nodes may not be recontaminated and finally, the set of clean nodes at any instant of time has always to define a connected subnetwork.

This problem, first formulated in [2], is still interesting, as the authors prove that also in this variant finding if the optimal number of agents required to capture the intruder with a contiguous strategy is at most $k$, for any integer $k$, is an $NP$-complete problem for arbitrary graphs, moreover the known techniques for node and edge search, do not generally apply.

Some specific topologies have been investigated as trees, chordal rings and tori, meshes and hypercubes [2,6,7,8], and in arbitrary networks some heuristics have been proposed [9] and a move-exponential solution, that passes through a temporary recontamination of the network (but provides a final contiguous and monotone strategy) has been given in [3].

## 1.2   Our Results

In this paper we study the problem of capturing an intruder in a nontrivial family of fractal graphs, called Sierpiński graphs [14], that have been lately vastly investigated (see e.g.,[1,15]). We have chosen these graphs, because they have a very nice fractal representation. Moreover, they are interesting from an applicative point of view, as a small variation of them has been lately investigated as it simultaneously exhibits fractality and small world effect, the latter being a feature of real interconnection networks [1].

We first prove a lower bound on the number of agents, number of moves and of time steps required to solve the contiguous monotone search problem in finite Sierpiński graphs. We then propose two different strategies: the first one in which agents have the capability of "seeing" the state of their neighbors, thus can move autonomously in a fully distributed way. The second one, in which actions are leaded by a coordinator. We show that the first strategy is optimal in terms of number of agents used, whereas the second one requires an extra agent; both their time and move complexities are distant for a logarithmic factor from the optimal. The second strategy will also require a higher (by a constant) number of moves and times steps compared to the first strategy.

## 2   Definitions and Basic Properties

### 2.1   The Sierpiński Graphs

We consider a family of nontrivial fractal graphs, the *Sierpiński graphs* $SG_n$. These graphs are obtained by $n$ out of an infinite number of iterations of the procedure that builds a fractal, the *Sierpiński gasket* $S_n$ (also called the *Sierpiński sieve* or the *Sierpiński triangle*), first presented by Sierpiński in 1915 [14]. The procedure, illustrated in Figure 1, is as follows:

a)                          b)                          c)

**Fig. 1.** $S_1$, $S_2$ and $S_3$

---

**Construction of a Sierpiński gasket.**
1. Build an equilateral triangle (Figure 1 a));
2. choose the middle point of each of the three sides of the triangle and build a new equilateral triangle $T'$ (white triangle in Figure 1 b));
3. repeat step 2 with the three triangles obtained by the construction of $T'$ (the black triangles in Figure 1 b).

---

Figure 1 c) is the output of the procedure after three steps and represents $S_3$. $S_1$ and $S_2$ are represented in Figure 1 a) and b), respectively. In general, at step $i$ of the procedure we obtain $S_i$, and the actual fractal is obtained after an infinite number of steps.

A nice property of the Sierpiński fractal is that it can be visualized from Pascal's triangle by coloring black all odd numbers, and white all even numbers.

The *Sierpiński graph* $SG_n$ is constructed after $n$ iterations of the procedure that builds $S_n$, by considering as a vertex set $V_n$, the intersecting points of the line segments of $S_n$, and as the edge set $E_n$ the line segments connecting two intersecting points. An example of $SG_4$ and $SG_5$ is illustrated in Figure 2.

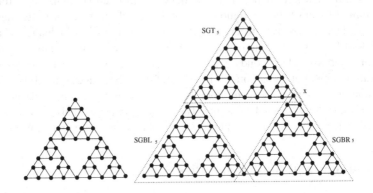

**Fig. 2.** The Sierpiński graph $SG_4$ on the left, $SG_5$ on the right

By construction, $SG_n$, for $n > 1$, is composed of three copies of $SG_{n-1}$, one on the top, one on the bottom left and one in the bottom right. We will denote them by $SGT_n$, $SGBL_n$, and $SGBR_n$, respectively (see Figure 2 right for $SGT_5$).

The number of vertices $V_n$ and edges $E_n$ of $SG_n$ are given by:

**Proposition 1.** *[2]* $SG_n$ *has* $|V_n| = \frac{3^n+3}{2}$ *vertices and* $E_n = 3^n$ *edges.*

We also need the following property:

*Property 1.* The vertices of the Sierpiński graph $SG_n$ have all degree 4 but the three corner vertices that have degree 2.

PROOF. By induction. $S_1$ has three corner vertices all of degree 2. Assume $SG_{n-1}$ has all the vertices of degree 4 but the corner vertices that have degree 2. $SG_n$ may alternatively be built glueing together three copies of $SG_{n-1}$ that is $SGT_n$, $SGBL_n$ and $SGBR_n$. $SGT_n$ and $SGBL_n$ will share a common corner, that is one of their old corners will not be a corner anymore and will now have degree 4. The same will hold for $SGBL_n$ and $SGBR_n$ that will share a corner and for $SGT_n$ and $SGBR_n$ that will share another corner. All the other vertices will maintain their original degree that by inductive hypothesis is either 4 or 2 (the vertices with degree 2 will be the new corners). The claim follows.  ∎

## 2.2   The Model

Consider an undirected (respectively directed) graph $G = (V, E)$. We say that if $\{x, y\} \in E$ (respectively $(x, y) \in E$) then $x$ and $y$ are neighbors and that $\{x, y\}$ (respectively $(x, y)$) is an edge outgoing from $x$ and ingoing into $y$. The degree (in-degree) of a vertex (node) is the number of ingoing and outgoing (for the out-degree) edges (arcs).

In $SG_n$ operates a team of autonomous mobile agents, that can move from a node to a neighboring one, have computing capabilities, computational storage ($O(\log n)$ bits suffice for all our algorithms), and obey the same set of behavioral rules. The intruder escapes from the team, whereas agents in the team have to collaborate in order to capture the intruder. Initially, all agents are located at the same node, the homebase and have distinct Identities (Ids for short), otherwise agents are all identical.

The network and the agents are *asynchronous*, i.e., all actions agents perform (i.e., computing, moving), take a finite but unpredictable amount of time. In this setting efficiency is measured in terms of the number of agents to be involved, the traffic (i.e., number of moves the agents have to perform), and the time (or steps).[1]

The cleaning strategy has to be contiguous, i.e., agents start the procedure at a specific vertex, the homebase, and expand the cleaning around the homebase, forming a connected subnetwork, until the whole network is clean. Moreover, it has to be monotone, that is, whenever a vertex has been cleaned, it will always remain clean.

## 3 The Strategy

In this section we first compute a lower bound on the number of agents, number of moves and time steps needed to clean the Sierpiński graph $SG_n$. We then propose two different strategies for the capture of the intruder in this graph, the first in which agents have the capability of "seeing" the state of their neighbors, the second in which agents are leaded in the cleaning by a coordinator.

### 3.1 Lower Bound

We first state a lower bound, that holds for all models, on the number of agents needed to clean the Sierpiński graph $SG_n$.

**Theorem 1.** *To solve the contiguous monotone search problem in $SG_n$ at least $n + 1$ agents are required. Using $n + 1$ agents, the number of time steps is at least $\frac{3^n+1}{2(n+1)} + 1$ and the number of moves is at least $\frac{3^n+3}{2}$.*

PROOF. We prove the bound on the number of agents by induction. For $n = 1$ in $SG_1$ at least two searchers are required as a unique searcher, while leaving a vertex would not be able to protect it from recontamination. Thus, for the base case of the induction the proof holds. Let us now assume by inductive hypothesis that in $SG_{n-1}$ at least $n$ searchers are required to capture the intruder. Let us now prove that at least $n + 1$ are required to capture the intruder in $SG_n$. Assume by contradiction that $n$ agents are enough. Any possible cleaning strategy has either to clean one of the three $SG_{n-1}$ subgraphs of $SG_n$ and then move to the others, or may start on two or three of them together. In this latter case, however, less than $n$ agents would be used for the cleaning of $SG_{n-1}$, thus a contradiction on the inductive hypothesis. Assume now the first case. W.l.o.g., by inductive hypothesis $n$ agents are used to clean $SGT_n$. These

---

[1] As the system is asynchronous, we will measure ideal time, i.e., we assume - for the purpose of time complexity only - that it takes one unit of time for an agent to traverse a link.

$n$ agents could be first used to clean, e.g., $SGBL_n$, and then $SGBR_n$. However, no matter which strategy would be used, if no agent would be placed on the bottom corner of $SGT_n$ (vertex $x$ of Figure 2 left), recontamination of $SGT_n$ could occur. Thus an agent has to be placed in this corner to protect $SGT_n$. At this point $n - 1$ searchers would not be enough to clean $SGBL_n$, thus a contradiction.

Let us now recall that we are interested in monotone and contiguous procedures, thus that all agents start from the same vertex. To compute the number of time steps let us first observe that the $n + 1$ agents will first be placed in a vertex and then will be able to clean at most $n + 1$ vertices at a time, thus the entire cleaning will require at least $\frac{\frac{3^n+3}{2}-1}{(n+1)} + 1 = \frac{3^n+1}{2(n+1)} + 1$ time steps. Moreover, all vertices will have to be visited and cleaned, thus at least $\frac{3^n+3}{2}$ moves will have to be executed.     ∎

## 3.2   Cleaning with Visibility

In this section we propose a cleaning strategy where agents have "visibility", i.e., they are located at a node and can "see" whether their neighboring vertices are clean or guarded or contaminated. Thus decisions rely only on some local knowledge. This visibility capability could be easily achieved if the agents have also communication power and send a message (e.g., a single bit) to their neighboring nodes after cleaning a node or guarding a node [6,7,8].

Formally, we assume what follows:

**Definition 1.** *[6,7,8] An agent located at node $x$ has a* visibility *capability if it can see the state of its neighbors $N(x)$.*

This property will allow us to define a strategy that requires an optimal number of agents.

**Cleaning strategy.** The main idea of this strategy applied to $SG_n$, is that all the required $n + 1$ agents, starting from the same homebase $H$, move on, and recursive clean one subgraph $SG_{n-1}$ at the time. They are identical and autonomous, and they all follow the same local rule. In order to obtain this recursive procedure we first build a special directed subgraph of $SG_n$ called $Sub_n$: this graph has a recursive structure. Agents will only move along the arcs of $Sub_n$. We store on the vertices of $Sub_n$ a value that ranges from 0 to 2 and that associates to at most one of the 2 outgoing arcs of $Sub_n$ the number of agents that have to be sent along that arc. The remaining agents will be sent along the other arc. An agent on a node $x$ moves on an outgoing directed arc of $Sub_n$ to a node $y$ given that: 1) all the neighbors of $x$ not reachable with a directed arc from $x$ in $Sub_n$ are clean or guarded; 2) if there is another contaminated neighbor $z$ of $x$ reachable from $x$ on $Sub_n$ then there is another available agent at $x$. That is, the agent moves as long as $SG_n$ is not recontaminated.

Let us now formally define the strategy starting from the one for $SG_1$ and $SG_2$ that is illustrated in Figure 3 a) and b), respectively. $H$ is the homebase from which all the agents start the procedure. White vertices have been cleaned, black vertices are contaminated. Thick arrows indicate the arcs of $Sub_1$ and $Sub_2$ which in this special case are binary trees (the dotted arrow in $Sub_2$ will be used in the general

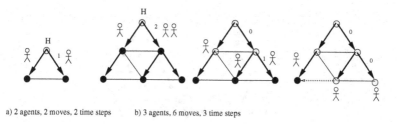

a) 2 agents, 2 moves, 2 time steps        b) 3 agents, 6 moves, 3 time steps

**Fig. 3.** Capture of an intruder with visibility in $SG_1$ and $SG_2$

construction of $Sub_n$). The values on the arcs indicate the number of agents that have to be sent along that arc. This number will decrease up to 0, in this case the remaining agents will be sent along the other arc. For simplicity we have represented a synchronous execution of the cleaning procedure.

$Sub_n$, for $n > 2$ is built through the recursive procedure $SB_n$, for $n \geq 2$, which returns the tuple $(Sub_n, v_1, v_2, v_3, e)$ where $v_1, v_2, v_3$ are the corners of $Sub_n$ and $e = (x, y)$ is the only arc ingoing into the bottom right corner $v_3$. $\bar{e}$ represents arc $(y, x)$.

**Definition 2.** *The graph $Sub_n = (V', E')$ is a directed subgraph of $SG_n$ recursively built as follows:*

**Base Case,** $i = 2$  $SB_2 = (Sub_2, v_1', v_2', v_3', (v_4', v_3'))$ *(see Figure 4).*
  *Nodes of $Sub_2$ are "fresh", i.e., every call of $SB_2$ produces a graph such as $Sub_2$, with new node names.*
**Recursive Case,** $i > 2$  *Let*
$$(G', v_1', v_2', v_3', e') = SB_{i-1},$$
$$(G'', v_1'', v_2'', v_3'', e'') = SB_{i-1},$$
$$(G''', v_1''', v_2''', v_3''', e''') = SB_{i-1}.$$
*in*
$$SB_i = (G' \bigcup G''[v_2'/v_1''] \bigcup G'''[v_2''/v_1''', v_3'/v_3''', \bar{e}'''/e'''], v_1', v_2'', v_3'', e''),$$
*where $e''$ is the only arc ingoing into $v_3''$, $v/v'$ is an operation that renominates vertex $v'$ in $v$ and $e/e'$ replaces arc $e'$ with arc $e$.*

An example of how to construct $Sub_3$ from $Sub_2$, and an example of $Sub_5$ are shown in Figure 4. Note that the construction of $SG_n$ has a very simple recursive structure. Moreover, the values associated to some of the arcs require only $O(1)$ bits of memory, as we store only values $0, 1, 2, \emptyset$ (an example of how to associate these values is given in Figure 4), and there is at most one of these values stored on the node they are outgoing from. This follows from the construction of subgraph $Sub_n$, and also from the following property:

*Property 2.* All the vertices of $Sub_n$ have at most 2 ingoing and at most 2 outgoing arcs.

PROOF. $Sub_n$ is a subgraph of $SG_n$ that by Property 1 has vertices of degree 4 or 2. We now have to prove that the nodes of $Sub_n$ have in and out-degree at most 2. We prove it by induction. This is implied by the construction of $Sub_n$: for $n = 1, 2, 3$ the

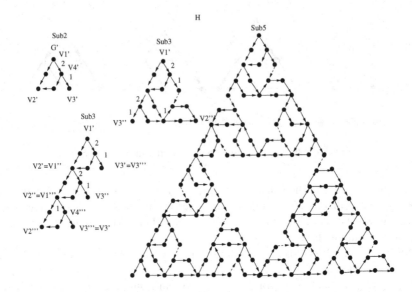

**Fig. 4.** The construction of subgraph $Sub_n$

property trivially holds. Assume it is true for $Sub_{n-1}$, $n-1 \geq 2$. Also assume $n$ is odd. $Sub_n$ is composed of three copies of $Sub_{n-1}$, i.e., $G', G'', G'''$ with corners respectively, $v_1', v_2', v_3', v_1'', v_2'', v_3''$, and $v_1''', v_2''', v_3'''$, that are such that: $v_2'$ and $v_1''$ are merged and respectively have in-degree 2 and out-degree 2; $v_2''$ and $v_1'''$ are merged and respectively have in-degree 2 and out-degree 2; and finally, $v_3'$ and $v_3'''$ are merged and respectively have in-degree 1 and in-degree 1 (that will become an out-degree). Thus the property is maintained. The proof for the case $n$ even is analogous. ∎

We will now present the procedure for the capture of the intruder with visibility in $SG_n$. The cleaning will be carried out along the arcs of $Sub_n$. Let us call $C_{SG_n}(x)$ (respectively $C_{Sub_n}(x)$), the set of contaminated neighbors of $x$ not connected (respectively, connected) by a directed arc from $x$ in $Sub_n$. Note that, by construction, $|C_{SG_n}(x)| \leq 2$

---

**Algorithm.** CAPTURE WITH VISIBILITY.

– Start from the homebase $H$;
– On a node $x$, **if** ($x$ has out-degree $= 0$) **then** terminate **else**
  1. **while** ($|C_{SG_n}(x)| > 0$) **or** ($|C_{Sub_n}(x)| \geq |A(x)|$) **do** wait at $x$.
     /* there are contaminated neighbors not reachable from $x$ in $Sub_n$ or there are more contaminated neighbors of $x$ in $Sub_n$ than available agents */

  2. choose (coherently with the other agents in $A(x)$) a neighbor $y$ of $x$ in $Sub_n$.
     /* that is, if $v(x,y) > 0$, then at most $v(x,y)$ agents choose arc $(x,y)$, the remaining choose the other arc */
     **if** $v(x,y) > 0$ **then** $v(x,y) := v(x,y) - 1$;
     move along $(x,y)$, decontaminate and guard $y$.

(the trivial proof is omitted), and $|C_{Sub_n}(x)| \le 2$). Call $A(x)$ the set of available agents at node $x$. Finally, call $v_{(x,y)}$ the value associated to arc $(x, y)$ at $x$. This value can be $\{0, 1, 2\}$ or $\emptyset$. If $v_{(x,y)} \in \{1, 2\}$ this arc may be chosen, else the other arc has to be chosen (i.e., enough agents have traversed it).

Let us now define the cleaning procedure for an agent $a$ at node $x$. The procedure starts from the homebase $H$.

Figure 5 shows the order in which the nodes get cleaned with our strategy (a synchronous execution with all agents starting together from $H$). The number close to every agent represents the size of the agent set, whereas the number close to an arc represents its value. Note that, at step 2 only agents on the right may move as the one on the left have a contaminated neighbor not reachable with an arc of $Sub_n$ from the node their are guarding. At step 4 the agent on the right waits up to step 9 (not shown) to move.

Note that, with this strategy nodes are not cleaned sequentially; several nodes, in fact, could be cleaned independently.

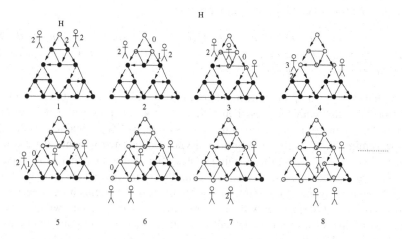

**Fig. 5.** An almost complete execution of Algorithm CAPTURE WITH VISIBILITY on a Sierpiński graph $SG_3$

**Correctness and complexity analysis.** We now prove the correctness of Algorithm CAPTURE WITH VISIBILITY and we compute the number of agents required by the procedure.

**Theorem 2.** *The cleaning process of Algorithm* CAPTURE WITH VISIBILITY *on a Sierpiński graph* $SG_n$ *decontaminates all nodes using* $n + 1$ *agents. During the execution clean nodes cannot be recontaminated.*

PROOF. By induction on $n$. From Figure 3 and 5 we have that the algorithm is correct for $SG_1$, $SG_2$ and $SG_3$ and that respectively 2, 3, 4 agents are used by the procedure. Moreover we want to check if all agents but one reach one bottom corner of $SG_n$, and all the others reach the other bottom corner. This is also true for $SG_1$, $SG_2$ and $SG_3$ (by the value assignment that has been given). The base of the induction follows. Let us now assume all the above properties are true for $SG_{n-1}$. $SG_n$ will be composed of three copies $SGT_n$, $SGBL_n$, $SGBR_n$, of $SG_{n-1}$. On each of them the algorithm will correctly work as long as $n$ agents start the procedure from the a corner. If $n$ is odd (respectively, even), then $n + 1$ agents are sent and by the assignment of values, the arc connecting the bottom left (respectively, right) corner of $SGT_n$, let us call it $bl$ ($br$), will have associated value 1, and only one agent will arrive at $bl$ ($br$). Moreover, by construction, $bl$ ($br$) will have only one outgoing arc that connects it to $SGBL_n$ ($SGBR_n$), thus the agent will stop in $bl$ ($br$) up to when the cleaning will arrive at $SGBL_n$ ($SGBR_n$). All the remaining $n$ agents will arrive at the bottom right (left) corner of $SGT_n$, let us call it $br$ ($bl$). Thus, all the $n$ agents will be able to clean $SGBR_n$ ($SGBL_n$). Also note that no recontamination may occur as $SGT_n$ is protected from recontamination by the agent on $bl$ ($br$) as the $n$ agents will first correctly (by the inductive step) clean $SGBR_n$ ($SGBL_n$) and then move to $SGBL_n$ ($SGBR_n$). Thus the claim is true.     ∎

Note that, $n + 1$ is an optimal number of agents. We now calculate the total number of moves and the time needed for the entire process.

**Theorem 3.** *The total number of moves needed to perform the cleaning of the Sier-piński graph $SG_n$ with Algorithm* CAPTURE WITH VISIBILITY *is $O(n \cdot 3^n)$. The time complexity is $O(3^n)$ time units.*

PROOF. We start by computing the number of moves and the time required for the procedure on $SG_3$, first with 4 agents, then with $n + 1$. By construction this will have to be repeated $3^{n-3}$ times (once for each of the $3^{n-3}$ subgraphs $SG_3$ contained inside $SG_n$). We now remind that we are working in an asynchronous environment, so we consider the ideal time complexity for the cleaning strategy (i.e., we assume that it takes one unit of time for an agent to traverse an edge). The number of moves and of steps necessary to clean $SG_3$ with 4 agents is constant. Thus trivially, if $n + 1$ agents have to execute the algorithm on $SG_3$, $O(n)$ moves will be required (the extra $n - 3$ agents will also execute a constant number of moves) and trivially 11 time steps (i.e., $O(1)$) are required. This has to be repeated for $3^{n-3}$ times thus the total number of moves needed to perform the cleaning of the Sierpiński graph $SG_n$ with Algorithm CAPTURE WITH VISIBILITY is $O(n \cdot 3^n)$. The time complexity is $O(3^n)$ time units.     ∎

Note that, from Proposition 1 we have that $SG_n$ has $|V_n| = \frac{3^n + 3}{2}$ vertices, thus the number of time steps is linear in the number of vertices and the number of moves has an extra logarithmic factor. Moreover they differ by a logarithmic factor from the lower bound.

### 3.3   Cleaning with a Coordinator

In this section we propose a cleaning strategy where actions are leaded by a coordinator. The strategy we propose in very similar to the one of the previous section, however an extra agent will be required to coordinate the moves of the other agents. On the other hand this strategy does not require the extra capability of "visibility" and actions may be taken by solely accessing local information.

**Cleaning strategy.** The main idea of this strategy is that all the agents, starting from the same homebase, move on and clean one subgraph $SG_{n-1}$ at the time, coordinated by an agent (e.g., the one with smallest Id), that acts as a leader. While moving they protect system from recontamination. The moves the agents will execute will be the same of the Algorithm CAPTURE WITH VISIBILITY with the exception of the coordinator that will have to move back and forth to lead the moves of the other agents. As in the other algorithm, the new procedure will work recursively using the subgraph $Sub_n$. Thus it will only be necessary to show the coordinated moves inside a $Sub_3$.

Figure 6 shows the order in which the nodes are cleaned by the agents leaded by the coordinator. Note that for $SG_1$, 2 agents would be enough as the coordinator would not need to lead both agents to the vertex (they will eventually arrive there). However we have shown a strategy with 3 agents as will will use this as a subroutine for the cleaning of $SG_2$.

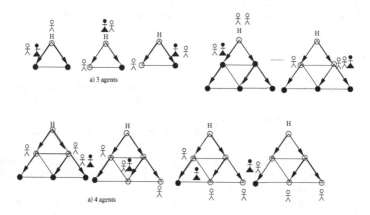

**Fig. 6.** Algorithm COORDINATED CAPTURE on a Sierpiński graph $SG_2$ and $SG_3$. One agent (the lady in black) is the coordinator.

The procedure for $SG_3$ is the same of the one used in Algorithm CAPTURE WITH VISIBILITY with the difference that here moves may not rely on local decisions, but the coordinator has to decide where and when agents have to move, and thus has to simulate the decisions of the previous strategy by leading the agents. It is trivial to prove that this generalizes to a correct strategy for $SG_n$ with the extra expense of an additional agent (the coordinator), extra moves (the one executed by the coordinator, whereas the other agents execute the same move), and extra time steps (the extra time spent for the movements of the coordinator). However both time and move complexities will globally differ by a constant factor from the one of the previous strategy.

## 4   Conclusion

In this paper we have considered the problem of capturing an intruder in a Sierpiński graph $SG_n$. We have considered two different models: one where agents have a visibility capability, i.e., can see the state of their neighbors, the other in which agents are leaded by a coordinator and execute actions by accessing only local information.

Our goal was to continue a study on the impact that these additional assumptions have on the efficiency of the solution process for the intruder capture problem in general networks. From our observations Visibility seems to be a crucial assumption for the reduction of all the complexities.

An open problem is to show if the algorithms in both models also reach an optimal complexity in the number of moves and time steps and if the second model may require only $n + 1$ agents. Another interesting issue would be to study the problem in a higher dimensional fractal graph, i.e., the Sierpinski tetrahedron.

## Acknowledgement

I would like to thank Riccardo Focardi for all his helpful suggestions.

## References

1. Barrière, L., Comellas, F., Dalfó, C.: Fractality and the small-world effect in Sierpiński graphs. Journal of Physics A: Mathematical and General 39, 11739–11753 (2006)
2. Barrière, L., Flocchini, P., Fraigniaud, P., Santoro, N.: Capture of an intruder by mobile agents. In: Proceedings of the 14-th ACM Symposium on Parallel Algorithms and Architectures (SPAA), Winnipeg, Manitoba, Canada, pp. 200–209 (2002)
3. Blin, L., Fraigniaud, P., Nisse, N., Vial, S.: Distributed Chasing of Network Intruders by Mobile Agents. In: Flocchini, P., Gąsieniec, L. (eds.) SIROCCO 2006. LNCS, vol. 4056, pp. 70–84. Springer, Heidelberg (2006)
4. Breish, R.: An intuitive approach to speleotopology. Southwestern cavers VI(5), 28–72 (1967)
5. Ellis, J.A., Sudborough, I.H., Turner, J.S.: The vertex separation and search number of a graph. Information and Computation 113, 50–79 (1994)
6. Flocchini, P., Huang, M.J., Luccio, F.L.: Decontamination of chordal rings and tori. In: Proceedings of the 8th Workshop on Advances in Parallel and Distributed Computational Models (APDCM), Rodi, Greece, 2006. Extended version to appear in International Journal of Foundations of Computer Science (IJFCS)
7. Flocchini, P., Huang, M.J., Luccio, F.L.: Contiguous Search in the Hypercube for Capturing an Intruder. In: Proc. 19th IEEE International Parallel and Distributed Processing Symposium (IPDPS) (2005)
8. Flocchini, P., Luccio, F.L., Song, L.X.: Size Optimal Strategies for Capturing an Intruder in Mesh Networks. In: Proceedings of the International Conference on Communications in Computing (CIC), Las Vegas, USA, pp. 200–206 (2005)
9. Flocchini, P., Nayak, A., Shulz, A.: Cleaning an arbitrary regular network with mobile agents. In: Proceedings of the 2nd International Conference on Distributed Computing & Internet Technology (ICDCIT), Bhubaneswar, India, pp. 132–142 (2005)
10. Kirousis, L.M., Papadimitriou, C.H.: Searching and pebbling. Theoretical Computer Science 47, 205–218 (1986)
11. Lapaugh, A.: Recontamination does not help to search a graph. Journal of the ACM 40(2), 224–245 (1993)

12. Megiddo, N., Hakimi, S., Garey, M., Johnson, D., Papadimitriou, C.: The complexity of searching a graph. Journal of the ACM 35(1), 18–44 (1988)
13. Parson, T.: Pursuit-evasion problem on a graph. In: Theory and applications of graphs. Lecture Notes in Mathematics, pp. 426–441. Springer, Heidelberg (1976)
14. Sierpiński, W.: Sur une courbe dont tout point est une point de ramification. C.R. Acad. Sci. Paris 160, 302–305 (1915)
15. Teufl, E., Wagner, S.: The number of spanning trees of finite Sierpiński graphs. In: Proceedings of the 4th Colloquium on Mathematics and Computer Science, DMTCS, 411–414 (2006)

# On the Complexity of the Traffic Grooming Problem in Optical Networks
## (Extended Abstract)

Mordechai Shalom[1,3], Walter Unger[2], and Shmuel Zaks[1,*]

[1] Department of Computer Science, Technion, Haifa, Israel
{cmshalom,zaks}@cs.technion.ac.il
[2] Lehrstuhl für Informatik I RWTH Aachen Ahornstraße 55 52056 Aachen, Germany
quax@i1.informatik.rwth-aachen.de
[3] TelHai Academic College, Upper Galilee, 12210, Israel
cmshalom@telhai.ac.il

**Abstract.** A central problem in optical networks is to assign wavelengths to a given set of lightpaths, so that at most $g$ of them that share a physical link get the same wavelength ($g$ is the *grooming factor*). The switching cost for each wavelength is the number of distinct endpoints of lightpaths of that wavelength, and the goal is to minimize the total switching cost. We prove NP-completeness results for the problem of minimizing the switching costs in path networks. First we prove that the problem is NP-complete in the strong sense, when all demands are either 0 or 1, the routing is single-hop, and the number of wavelengths is unbounded. Next we prove that the problem is NP-complete for any fixed $g \geq 2$, and when the number of wavelengths is bounded. These results improve upon existing results regarding the complexity of the traffic grooming problem for ring and path networks.

**Keywords:** Wavelength Assignment, Wavelength Division Multiplexing(WDM), Optical Networks, Add-Drop Multiplexer(ADM).

## 1 Background

Optical wavelength-division multiplexing (WDM) is today the most promising technology, that enables us to deal with the enormous growth of traffic in communication networks, like the Internet. A communication between a pair of nodes is done via a *lightpath*, which is assigned a certain wavelength. In graph-theoretic terms, a lightpath is a simple path in a graph, with a color assigned to it. Most of the studies in optical networks dealt with the issue of assigning wavelengths to lightpaths, so that every two lightpaths sharing a common edge get different wavelengths.

When the various parameters comprising the switching mechanism in these networks became clearer, the focus of studies shifted, and today a large portion

* This research was partly supported by the EU Project "Graphs and Algorithms in Communication Networks (GRAAL)" - COST Action TIST 293.

P. Crescenzi, G. Prencipe, and G. Pucci (Eds.): FUN 2007, LNCS 4475, pp. 262–271, 2007.

of the studies concentrate with the total hardware cost. The key point here is that each lightpath uses an Add-Drop Multiplexer (ADM) at each of its two endpoints. If two lightpaths sharing a common endpoint are assigned the same wavelength, then they can use the same ADM. An ADM may be shared by at most two lightpaths. The total cost considered is the total number of ADMs. Lightpaths sharing ADM's in a common endpoint can be thought of as con-catenated, so that they form longer paths or cycles. These paths/cycles do not use any edge $e \in E$ twice, for otherwise they cannot use the same wavelength which is a necessary condition to share ADM's. In graph-theoretic terms, this can viewed as assigning colors to given paths so that no two paths that get the same color have any edge in common. Each path uses two ADM's, one at each endpoint, and at most two paths of the same color can share an ADM at their common endpoint. The goal is to minimize the total number of ADMs.

Moreover, in studying the hardware cost, the issue of *grooming* became central. This problem stems from the fact that the network usually supports traffic that is at rates which are lower than the full wavelength capacity, and therefore the network operator has to be able to put together (= groom) low-capacity demands into the high capacity fibers. In graph-theoretic terms, this can be viewed as assigning colors to given paths so that at most $g$ of them ($g$ being the *grooming factor*) can share one edge. Each path uses two ADM's, one at each endpoint, and in case $g$ paths of the same wavelength enter through the same edge to one node, they can all use the same ADM (thus saving $g - 1$ ADMs). The goal is to minimize the total number of ADMs. Note that the above coloring problem is simply the case of $g = 1$.

When considering these problems, some parameters play important role:

1. Lightpaths: we either deal with the case where the lightpaths are given, or with the case where we are given a set of pairs of nodes, and the problem is to design a set of lightpaths that satisfies the communication between these pairs.
2. Number of colors: this can be either unbounded, or we might be given a bound on the number of colors, and we have to find the best solution satis-fying this bound.
3. Mode of communication: The communication between pairs of nodes is using either a single lightpath (the *single-hop* case) or a concatenation of lightpaths (the *multi-hop* case). In the multi-hop case the assumption is that the traffic between a given pair of nodes can be routed through a sequence of lightpaths of possibly different wavelengths.
4. Communication links: can be directed or undirected, and the underlying graph is either directed or undirected. In case of a directed graph, the con-dition applies to directed edges.
5. Routing: the traffic between a given pair of nodes can be routed either through different routes (*splittable* case) or it is constrained to be routed through a unique route (*non-splittable* case). In this case the assumption is that each re-quirement is an integer, and the splitting can be only to integral components.

We show that the grooming problem is NP-complete in the strong sense, for graphs of path topology, even in the case where all demands are either 0 or 1, the routing is single-hop, and the number of colors is unbounded. Next we prove that the problem is NP-complete for graphs of path topology, for any fixed $g \geq 2$, and when the number of colors is bounded. These results improve the complexity analysis of the traffic grooming problem for the fundamental ring and path networks. In particular, they extend the result of [EMZ02], where it was shown that the problem in NP-complete for a ring network for $g = 1$, and the result of [CM00] (and others - see Section 2), where a weak NP-completeness proof is given for a general $g$ and general demands. We present the results in the case where the lightpaths are given, the mode of communication is single-hop, the links are directed, and the routing is non-splittable. We refer to other possibilities throughout the discussion and in the summary section.

The paper is organized as follows: We start with presenting the formal model in Section 3, continue with showing the two NP-complete results in Sections 4 and 5, respectively, and end with a summary and open problems in Section 6.

## 2   Previous Work

The problem of minimizing the number of ADMs for the case $g = 1$ was introduced in [GLS98] for ring topology. The problem was shown to be NP-complete for ring networks in [EMZ02]. The discussion in [EMZ02] easily implies that minimizing the number of ADMs, with $g = 1$, for a ring network, where all edge loads are equal, is NP-hard. In both cases, between each pair of nodes the number of requests is arbitrary. These reductions follow immediately from the NP-completeness of the problem of coloring of circular arc graphs. As is often the case, showing NP-completeness of a problem for a path network turns out to be more difficult than doing so for the case of a ring network (this is due to the fact that coloring of interval graphs is polynomial whereas coloring of circular arc graphs is NP-complete). We note that minimizing the number of ADMs for a path network is trivial for this case of $g = 1$.

An approximation algorithm for the ring topology, for the case of $g = 1$, with approximation ratio of $3/2$, was presented in [CW02], and was improved in [SZ04, EL04] to $10/7 + \epsilon$ and $10/7$, respectively. For a general topology [EMZ02] describes an algorithm with approximation ratio of $8/5$. The same problem was studied in [CFW02], and an algorithm with approximation ratio $3/2 + \epsilon$ was presented.

The notion of traffic grooming ($g > 1$) was introduced in [GRS98] for the ring topology. The problem was shown to be NP-complete in [CM00] for a directed ring network, and a general $g$, and where all the lightpaths connecting the nodes to a single node, termed *egress node*. The authors discuss the single-hop case, the reduction is done from the Bin Packing problem (see [GJ79]), and the complexity of the problem stems from the fact that the number of lightpaths that connect a node to the egress node is arbitrary. Actually, the result of [CM00] holds also for a path topology, since in the reduction there is no lightpath connecting the egress node to others. In unidirectional ring networks the problem is equivalent to partitioning the requests into sets of size at most $g$, so that the sum of the nodes induced by each set is minimum. This problem is termed the SONET edge partition problem in [GHLO03], which proves that the problem is NP-complete for any $g \geq 3$, and gives an $O(\sqrt{g})$-approximation algorithm for it. A different version of the traffic grooming problem is presented in [DHR06, H02], where the authors discuss the multi-hop case, and use a cost function in which each ADM is counted not only once - like in our case - but a number of times that is equal to the number of lightpaths that use it. The authors show an NP-complete result by using a reduction from the Knapsack problem (see [GJ79]); this result is presented in more detail in [H02]. In this case the network topology is a directed path, and the complexity of the problem stems - as in [CM00] - from the fact that the number of lightpaths between pairs of nodes is arbitrary. The NP-complete results of [CM00, DHR06, H02] are all in the weak sense (see [GJ79]).

The uniform all-to-all traffic case, in which there is the same demand between each pair of nodes, is studied in [CM00, BC03] for various values of $g$; an optimal construction for the uniform all-to-all problem, for the case $g = 2$ in a path network, was given in [BBC05]. A review on traffic grooming problems can be found in [ZM03]. A $\log g$ approximation for ring networks was presented in [FMSZ05].

## 3    Formal Model

### 3.1    Problem Definition

An instance of the problem consists of an undirected graph $G = (V, E)$, a set of paths $P = \{p_1, p_2, \cdots, p_N\}$, and a *grooming factor* $g$. We have to assign colors to the paths in $P$ such that at most $g$ paths of the same color can share an edge. Formally, a coloring function $w : P \to \{1, ..., W\}$ is said to be $g$-*feasible* if for every edge $e \in E$ and every color $\lambda \in \{1, ..., W\}$

$$|\{p \in P | p \text{ includes } e, w(p) = \lambda\}| \leq g$$

. A 1-feasible function is termed feasible.

Every colored path $p \in P$ needs one ADM at each of its endpoint nodes. Two lightpaths $p$ and $p'$, with a common endpoint $v$, and such that $w(p) = w(p')$, can share an ADM at $v$. An ADM at a node $v$ can serve paths of two incident edges, and at most $g$ lightpaths of the same color from each such edge. Our aim is to minimize the number of ADMs. Note that one can always assign $N$ distinct colors to the given lightpaths, and then the number of ADMs will be $2N$. On the other hand, the number of ADMs will always be at least $N/g$. This implies that the approximation ratio of any algorithm will be at most $2g$.

Given a graph, a set of paths, and a grooming factor $g$, our aim is thus to find a $g$-feasible coloring function that minimizes the total number of ADMs.

We consider the following decision problem:

THE TRAFFIC GROOMING PROBLEM:

**Input:** A graph $G = (V, E)$, a set $P$ of paths in $G$, integers $g, T, K > 0$.
**Question:** Is there a $g$-feasible coloring $w$, that uses at most $T$ ADMs and at most $K$ colors?

Note that as $P$ is a set, a path can appear at most once in $P$, as opposed to a multiset in which a path can appear more than once, i.e. there is an integer demand associated with a path.

In the case where the graph is a path or a ring, and $w$ is a $g$-feasible coloring function, let $ADM_\lambda^w$ be the number of distinct end-nodes of paths assigned color $\lambda$ by $w$. The goal is to find a feasible $g$-coloring $w$ that minimizes the value $ADM^w = \sum_\lambda ADM_\lambda^w$.

## 4    Path Network, Any Given Grooming Factor

In this section we prove the strong NP-completeness of the grooming problem, as follows:

**Lemma 1.** *The grooming problem for a path, is NP-complete in the strong sense even when the number of colors is unbounded.*

*Proof.* For the proof we use the following problem:

THE 3-PARTITION PROBLEM (see [GJ79]):

**Input:** A set $\mathcal{A} = \{a_1, a_2, \cdots, a_{3m}\}$ and a bound $B > 0$, such that $B/4 < a_i < B/2$ for every $1 \le i \le 3m$, and $\sum_{i=1}^{3m} a_i = mB$.
**Question:** Can the elements of $\mathcal{A}$ be partitioned into triplets $(a_{j_1}, a_{j_2}, a_{j_3})$, $j = 1, \cdots, m$ such that $a_{j_1} + a_{j_2} + a_{j_3} = B$ for every $1 \le j \le m$?

This problem is $NP$-complete in the strong sense, i.e., $NP$-complete even if $B$ is bounded by a polynomial in $m$. Given such an instance $\alpha$ of 3-partition, we construct an instance $\alpha'$ of the grooming problem, as follows (note that the reduction is polynomial, in view of the fact that $B$ is bounded by a polynomial): $G = (V, E)$ is a directed path with $mB + 3m + 1$ nodes, where

$$V = V_0 \cup V_1 \cup \{\tilde{v}\},$$
$$V_0 = \{v_{i,j} \mid 1 \le i \le 3m, 1 \le j \le a_i\}, V_1 = \{v_i \mid 1 \le i \le 3m\},$$
$$E = \{(v_i, v_{i,1}) \mid 1 \le i \le 3m\} \cup \{(v_{i,j}, v_{i,j+1}) \mid 1 \le i \le 3m, 1 \le j < a_j\} \cup$$
$$\{(v_{i,a_i}, v_{i+1}) \mid 1 \le i < 3m\} \cup \{(v_{3m,a_{3m}}, \tilde{v})\},$$
$$P = \{(v_i, v_{i,j}) \mid 1 \le i \le 3m, 1 \le j \le a_j\} \cup \{(v_{i,j}, \tilde{v}) \mid 1 \le i \le 3m, 1 \le j \le a_j\},$$
$$g = B, T = m(B + 4).$$

We now show that there is a solution to the instance $\alpha$ of the 3-partition problem if and only if there is a solution to the instance $\alpha'$ of the grooming problem.

Assume we are given a solution to the 3-partition problem $(a_{j_1}, a_{j_2}, a_{j_3})$, $j = 1, \cdots, m$ such that $a_{j_1} + a_{j_2} + a_{j_3} = B$ for every $j$. Then, for each of the $m$ triplets $(a_{j_1}, a_{j_2}, a_{j_3})$, we can assign the same color to all the lightpaths

$$\{(v_{j_1}, v_{j_1,j}) \mid 1 \leq j \leq a_{j_1}\} \cup \{(v_{j_2}, v_{j_2,j}) \mid 1 \leq j \leq a_{j_2}\} \cup$$
$$\{(v_{j_3}, v_{j_3,j}) \mid 1 \leq j \leq a_{j_3}\} \cup \{(v_{j_1,j}, \tilde{v}) \mid 1 \leq j \leq a_{j_1}\} \cup$$
$$\{(v_{j_2,j}, \tilde{v}) \mid 1 \leq j \leq a_{j_2}\} \cup \{(v_{j_3,j}, \tilde{v}) \mid 1 \leq j \leq a_{j_3}\}.$$

These lightpaths require one ADM at each of the nodes in

$$\{v_{j_1}, v_{j_2}, v_{j_3}, \tilde{v}\} \cup \{v_{j_1,j} \mid, 1 \leq j \leq a_{j_1}\} \cup \{v_{j_2,j} \mid 1 \leq j \leq a_{j_2}\} \cup \{v_{j_3,j} \mid 1 \leq j \leq a_{j_3}\}.$$

This is true since each of the nodes except for $\tilde{v}$ has at most $B$ incident lightpaths, and $\tilde{v}$ has exactly $B$ incident lightpaths (since $a_{j_1} + a_{j_2} + a_{j_3} = B$). This amounts to $a_{j_1} + a_{j_2} + a_{j_3} + 4 = B + 4$ ADMs. Summing for all the $m$ triplets we thus get a solution for the instance of the grooming problem which uses $T = m(B + 4)$ ADMs.

Conversely, assume we are given a solution to the grooming problem that uses at most $T = (B + 4)m$ ADMs. We clearly need an ADM at each of the nodes in $V - \{\tilde{v}\}$; this amounts to $|V - \{\tilde{v}\}| = |V_0| + |V_1| = \sum_{i=1}^{3m} a_i + 3m = m(B + 3)$ ADMs. Since the total number of ADMs cannot exceed $T = m(B+4)$, it follows that the number of ADMs in $\tilde{v}$ is at most $m$. On the other hand, since the number of lightpaths with endpoint at $\tilde{v}$ is $\sum_{i=1}^{3m} a_i = mB$, and since the grooming factor is $B$, we need at least $m$ ADMs at $\tilde{v}$.

We thus conclude that the given solution to the grooming problem must use exactly $m$ ADMs at $\tilde{v}$, and exactly one ADM at each of the nodes in $V_0$ and $V_1$. This implies that, for every $1 \leq j \leq 3m$, all the $a_j$ lightpaths from the node $v_j$ to the nodes $v_{j,k}$, $1 \leq k \leq a_j$, have the same color, and that this same color must also be used by all the $a_j$ lightpaths from the nodes $v_{j,k}$, $1 \leq k \leq a_j$, to $\tilde{v}$.

Therefore we have $m$ ADMs at $\tilde{v}$, and each of them must be the endpoint of exactly $B$ lightpaths. Since all the $a_j$ lightpaths from the nodes $v_{j,k}$, $1 \leq k \leq a_j$, to $\tilde{v}$ have the same color, it follows that the numbers $a_i$ have to be partitioned into subsets, each summing up to $B$. Now, since $B/4 < a_i < B/2$ for every $i$, it follows that each of these sets must contain exactly three elements. In other words, we conclude that the numbers $a_i$ can be partitioned into triplets $(a_{j_1}, a_{j_2}, a_{j_3})$ such that $a_{j_1} + a_{j_2} + a_{j_3} = B$. This establishes a solution to the given instance of the 3-partition problem. $\qquad \square$

## 5    Path Network, Fixed Grooming Factor

We now show the NP-completeness for any fixed value of $g$. We show the construction for $g = 2$, and in the next lemma show how to extend it to any fixed value of $g > 1$. Note that for $g = 1$, the greedy path coloring algorithm to minimize the number of colors solves this problem too, optimally.

**Lemma 2.** *The grooming problem for a path is NP-complete for $g = 2$, for bounded number of colors.*

**Sketch of proof:** For the proof we use the following problem:
K-COLORING OF CIRCULAR ARC GRAPH (see [GJ79]):

**Input:** A cycle, whose nodes $\mathcal{N} = \{1, 2, \cdots, n\}$ are numbered clockwise, a set of arcs $\mathcal{A} = \{(a_i, b_i) | 1 \leq i \leq t\}, a_i, b_i \in \mathcal{N}$ (the arc $(a_i, b_i)$ is clockwise from $a_i$ to $b_i$), such that each of the edges $\{(1,2), (2,3), \cdots, (n-1, n), (n, 1)\}$ is covered by exactly $k$ arcs in $\mathcal{A}$.

**Question:** is there a coloring function $f : \mathcal{A} \rightarrow \{1, 2, \cdots, k\}$, such that $f(a) \neq f(b)$ if the arcs $a$ and $b$ intersect.

Given an instance $\alpha$ of the k-coloring of circular arc graph problem we construct an instance $\alpha'$ of the grooming problem.

Rather than formally present the construction, we show it by an example. Assume the circle is as depicted in Figure 1; namely, it is of size $n = 5$, and its nodes are 1,2,3,4,5. The arcs are $a = (2,5), b = (5,2), c = (4,1), d = (1,3)$ and $e = (3,4)$. The load on each edge is $k = 2$. In the first step we cut the paths that go through the edge $(5,1)$; namely, we cut the paths $b$ and $c$. We construct a path, whose vertices are, from right to left, $c', b', 1, 2, 3, 4, 5, b", c"$. The lightpaths are the original arcs $a$, $d$ and $e$ that were not cut, the lightpaths $b_1, b_2, b_3$ (that correspond to the arc $b$ that was cut), and the lightpaths $c_1, c_2, c_3$ (that correspond to the arc $c$ that was cut). This is depicted in Figure 2.

In the next step we duplicate each of the vertices 1,2,3,4 and 5, and change the paths so that they do not have any endpoint in common, except for the pairs $b_1, b_3$ (that meet at $b"$), $b_2, b_3$ (that meet at $b'$), $c_1, c_3$ (that meet at $c"$), and $c_2, c_3$ (that meet at $c'$). For example, the two paths $c_2$ and $d$ met at vertex 1 (Figure 2), so they are separated so that one ends at 1 and the other at 1' (the choice is arbitrary). The resulting path consists of 5 more vertices, and is depicted in Figure 3.

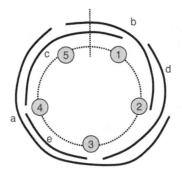

**Fig. 1.** A given instance $\alpha$ of the k-circular arc coloring problem

In general, given an instance of the k-circular arc graph problem $\alpha$, that consists of a total of $m$ arcs, we construct a corresponding instance of the grooming problem on a path of $2m + 2k$ nodes and with $m + 2k$ lightpaths (in our example $m = 5$ and $k = 2$), the bound $T$ on the number of ADMs is equal to the number of nodes, namely $T = 2m + 2k$, and the bound on the number of colors is $k$.

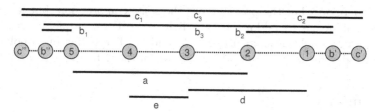

**Fig. 2.** The first step in the construction of a path

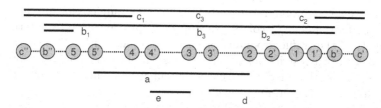

**Fig. 3.** The second step in the construction - the instance $\alpha'$ of the traffic grooming problem

If there is a solution to the circular arc graph coloring problem that uses $k$ colors, then we can assign the same colors to the lightpaths that do not go through the edge $(n, 1)$, and the three paths that correspond to a path that does go through $(n, 1)$ get the same color as the original path. In our example, the lightpaths $a, d, e$ in Figure 3 will get the same color as they got in the coloring of the instance of Figure 3, the lightpaths $b_1, b_2, b_3$ will get the same color that was assigned to $b$, and $c_1, c_2, c_3$ will get the same color that was assigned to $c$. The inner lightpaths use a total of $2(m - k)$ ADMs (since all of their endpoints are distinct), and the other lightpaths use a total of $4k$ ADMs, and thus we got a coloring with grooming of $g = 2$ that uses a total of $T = 2m + 2k$ ADMs and $k$ colors.

Conversely, assume that we have a coloring, with grooming of $g = 2$, that uses at most $T = 2m + 2k$ ADMs and $k$ colors. Since each node is an endpoint of at least one lightpath, this means that each of the nodes must use exactly one ADM. This means that the three lightpaths that correspond to an arc that was 'cut' must get the same color. In our example, the three lightpaths $b_1, b_2, b_3$ must be colored with the same color, and so do the lightpaths $c_1, c_2, c_3$. Moreover, these $k$ lightpaths get $k$ distinct colors, due to the grooming $g = 2$. Since on each edge each of the $k$ colors is used once by the 'long' lightpaths, and since the capacity is $2k$, two intersecting inner lightpaths must get different colors. We thus get a coloring of the arcs in $\alpha$ with exactly $k$ colors.                          □

In a similar manner, we can show that

**Lemma 3.** *The grooming problem for a path is NP-complete for any fixed $g \geq 2$, for bounded number of colors.*

**Sketch of proof:** For the proof we use the same construction, but for each of the arcs that we cut we add $g - 1$ 'long' lightpaths. For example, for case $g = 3$, we will add two lightpaths identical to $b_3$ and $c_3$. The reduction is polynomial for any fixed value of $g$. We skip here the rest of the details.     □

## 6   Summary

In this paper we proved the NP-completeness of the traffic grooming problem on a path. We proved that the problem is NP-complete in the strong sense when the number of colors is unbounded, and that it is NP-complete for any fixed $g \geq 2$ for a bounded number of colors. Natural open problems are to determine the complexity of the traffic grooming problem for a path with unbounded number of colors, for other restrictions on the various parameters and special topologies, for cost measures that involve also other switching components except the ADMs, and for dynamic (e.g. on-line) scenarios.

Regarding the notes at the end of Section 1:

1. Our proofs apply also to the case where we are only given the pairs of nodes to connect; to this end we can use the same reduction, since, given the pairs to be connected, the routing is unique.
2. Number of colors: the first reduction applies for any number of colors, whereas the second reduction assumed a bound on the number of colors.
3. Our proofs also apply for the multi-hop case. For this it suffices to note that the choice of parameters in the first reduction imply that at each of the nodes, except for $\tilde{v}$, we could use only one ADM. This implies that all the lightpath starting at $v_i$, for any $i$, must have the same color; in addition, this implies that we cannot use multi-hop routing and save in the number of ADMs, since we cannot change a color in any intermediate node (since changing a color implies the use of an additional ADM). This comment can be shown to apply also to the second case.
4. Our proofs apply clearly to undirected graphs.
5. Our proofs apply to rings.
6. Our proofs apply to the non-splittable case, since in our instance no splitting is possible.

## References

[BBC05]   Bermond, J.-C., Braud, L., Coudert, D.: Traffic grooming on the path. In: 12th Colloqium on Structural Information and Communication Complexity, Le Mont Saint-Michel, France (May 2005)

[BC03]    Bermond, J.-C., Coudert, D.: Traffic grooming in unidirectional WDM ring networks using design theory. In: IEEE ICC, Anchorage, Alaska (May 2003)

[CFW02]   Călinescu, G., Frieder, O., Wan, P.-J.: Minimizing electronic line terminals for automatic ring protection in general WDM optical networks. IEEE Journal of Selected Area on Communications 20(1), 183–189 (2002)

[CM00]    Chiu, A.L., Modiano, E.H.: Traffic grooming algorithms for reducing elec-
          tronic multiplexing costs in WDM ring networks. Journal of Lightwave
          Technology 18(1), 2–12 (2000)
[CW02]    Călinescu, G., Wan, P.-J.: Traffic partition in WDM/SONET rings to mini-
          mize SONET ADMs. Journal of Combinatorial Optimization 6(4), 425–453
          (2002)
[DHR06]   Dutta, R., Huang, S., Rouskas, G.N.: Traffic grooming in path, star and
          tree networks: Complexity, bounds, and algorithms. IEEE Journal on Se-
          lected Areas in Communications 24(4), 66–82 (2006)
[EL04]    Epstein, L., Levin, A.: Better bounds for minimizing SONET ADMs. In:
          2nd Workshop on Approximation and Online Algorithms, Bergen, Norway
          (September 2004)
[EMZ02]   Eilam, T., Moran, S., Zaks, S.: Lightpath arrangement in survivable rings
          to minimize the switching cost. IEEE Journal of Selected Area on Com-
          munications 20(1), 172–182 (2002)
[FMSZ05]  Flammini, M., Moscardelli, L., Shalom, M., Zaks, S.: Approximating the
          traffic grooming problem. In: Deng, X., Du, D.-Z. (eds.) ISAAC 2005.
          LNCS, vol. 3827, Springer, Heidelberg (2005)
[GHLO03]  Goldschmidt, O., Hochbaum, D.S., Levin, A., Olinick, E.V.: The sonet
          edge-partition problem. NETWORKS 41(1), 32–37 (2003)
[GJ79]    Garey, M., Johnson, D.S.: Computers and Intractability, A Guide to the
          Theory of NP-Completeness. Freeman, New York (1979)
[GLS98]   Gerstel, O., Lin, P., Sasaki, G.: Wavelength assignment in a WDM ring to
          minimize cost of embedded SONET rings. In: INFOCOM'98, Seventeenth
          Annual Joint Conference of the IEEE Computer and Communications So-
          cieties, pp. 69–77 (1998)
[GRS98]   Gerstel, O., Ramaswami, R., Sasaki, G.: Cost effective traffic grooming in
          WDM rings. In: INFOCOM'98, Seventeenth Annual Joint Conference of
          the IEEE Computer and Communications Societies (1998)
[H02]     Huang, S.: Traffic Grooming in Wavelength Routed Path Networks. PhD
          thesis, North Carolina State University (2002)
[SZ04]    Shalom, M., Zaks, S.: A $10/7 + \epsilon$ approximation scheme for minimiz-
          ing the number of ADMs in SONET rings. In: First Annual Interna-
          tional Conference on Broadband Networks, San-José, California, USA, pp.
          254–262, October (2004)
[ZM03]    Zhu, K., Mukherjee, B.: A review of traffic grooming in WDM optical
          networks: Architecture and challenges. Optical Networks Magazine 4(2),
          55–64 (2003)

# Author Index

Printing: Mercedes-Druck, Berlin
Binding: Stein+Lehmann, Berlin

# Lecture Notes in Computer Science

For information about Vols. 1–4411

please contact your bookseller or Springer

Vol. 4476: V. Gorodetsky, C. Zhang, V.A. Skormin, L. Cao (Eds.), Autonomous Intelligent Systems: Multi-Agents and Data Mining. XIII, 323 pages. 2007. (Sublibrary LNAI).

Vol. 4475: P. Crescenzi, G. Prencipe, G. Pucci (Eds.), Fun with Algorithms. X, 273 pages. 2007.

Vol. 4472: M. Haindl, J. Kittler, F. Roli (Eds.), Multiple Classifier Systems. XI, 524 pages. 2007.

Vol. 4471: P. Cesar, K. Chorianopoulos, J.F. Jensen (Eds.), Interactive TV: a Shared Experience. XIII, 236 pages. 2007.

Vol. 4470: Q. Wang, D. Pfahl, D.M. Raffo (Eds.), Software Process Dynamics and Agility. XI, 346 pages. 2007.

Vol. 4465: T. Chahed, B. Tuffin (Eds.), Network Control and Optimization. XIII, 305 pages. 2007.

Vol. 4464: E. Dawson, D.S. Wong (Eds.), Information Security Practice and Experience. XIII, 361 pages. 2007.

Vol. 4463: I. Măndoiu, A. Zelikovsky (Eds.), Bioinformatics Research and Applications. XV, 653 pages. 2007. (Sublibrary LNBI).

Vol. 4462: D. Sauveron, K. Markantonakis, A. Bilas, J.-J. Quisquater (Eds.), Information Security Theory and Practices. XII, 255 pages. 2007.

Vol. 4459: C. Cérin, K.-C. Li (Eds.), Advances in Grid and Pervasive Computing. XVI, 759 pages. 2007.

Vol. 4453: T. Speed, H. Huang (Eds.), Research in Computational Molecular Biology. XVI, 550 pages. 2007. (Sublibrary LNBI).

Vol. 4452: M. Fasli, O. Shehory (Eds.), Agent-Mediated Electronic Commerce. VIII, 249 pages. 2007. (Sublibrary LNAI).

Vol. 4451: T.S. Huang, A. Nijholt, M. Pantic, A. Pentland (Eds.), Artifical Intelligence for Human Computing. XVI, 359 pages. 2007. (Sublibrary LNAI).

Vol. 4450: T. Okamoto, X. Wang (Eds.), Public Key Cryptography – PKC 2007. XIII, 491 pages. 2007.

Vol. 4448: M. Giacobini et al. (Ed.), Applications of Evolutionary Computing. XXIII, 755 pages. 2007.

Vol. 4447: E. Marchiori, J.H. Moore, J.C. Rajapakse (Eds.), Evolutionary Computation,Machine Learning and Data Mining in Bioinformatics. XI, 302 pages. 2007.

Vol. 4446: C. Cotta, J. van Hemert (Eds.), Evolutionary Computation in Combinatorial Optimization. XII, 241 pages. 2007.

Vol. 4445: M. Ebner, M. O'Neill, A. Ekárt, L. Vanneschi, A.I. Esparcia-Alcázar (Eds.), Genetic Programming. XI, 382 pages. 2007.

Vol. 4444: T. Reps, M. Sagiv, J. Bauer (Eds.), Program Analysis and Compilation, Theory and Practice. X, 361 pages. 2007.

Vol. 4443: R. Kotagiri, P.R. Krishna, M. Mohania, E. Nantajeewarawat (Eds.), Advances in Databases: Concepts, Systems and Applications. XXI, 1126 pages. 2007.

Vol. 4440: B. Liblit, Cooperative Bug Isolation. XV, 101 pages. 2007.

Vol. 4439: W. Abramowicz (Ed.), Business Information Systems. XV, 654 pages. 2007.

Vol. 4438: L. Maicher, A. Sigel, L.M. Garshol (Eds.), Leveraging the Semantics of Topic Maps. X, 257 pages. 2007. (Sublibrary LNAI).

Vol. 4433: E. Şahin, W.M. Spears, A.F.T. Winfield (Eds.), Swarm Robotics. XII, 221 pages. 2007.

Vol. 4432: B. Beliczynski, A. Dzielinski, M. Iwanowski, B. Ribeiro (Eds.), Adaptive and Natural Computing Algorithms, Part II. XXVI, 761 pages. 2007.

Vol. 4431: B. Beliczynski, A. Dzielinski, M. Iwanowski, B. Ribeiro (Eds.), Adaptive and Natural Computing Algorithms, Part I. XXV, 851 pages. 2007.

Vol. 4430: C.C. Yang, D. Zeng, M. Chau, K. Chang, Q. Yang, X. Cheng, J. Wang, F.-Y. Wang, H. Chen (Eds.), Intelligence and Security Informatics. XII, 330 pages. 2007.

Vol. 4429: R. Lu, J.H. Siekmann, C. Ullrich (Eds.), Cognitive Systems. X, 161 pages. 2007. (Sublibrary LNAI).

Vol. 4427: S. Uhlig, K. Papagiannaki, O. Bonaventure (Eds.), Passive and Active Network Measurement. XI, 274 pages. 2007.

Vol. 4426: Z.-H. Zhou, H. Li, Q. Yang (Eds.), Advances in Knowledge Discovery and Data Mining. XXV, 1161 pages. 2007. (Sublibrary LNAI).

Vol. 4425: G. Amati, C. Carpineto, G. Romano (Eds.), Advances in Information Retrieval. XIX, 759 pages. 2007.

Vol. 4424: O. Grumberg, M. Huth (Eds.), Tools and Algorithms for the Construction and Analysis of Systems. XX, 738 pages. 2007.

Vol. 4423: H. Seidl (Ed.), Foundations of Software Science and Computational Structures. XVI, 379 pages. 2007.

Vol. 4422: M.B. Dwyer, A. Lopes (Eds.), Fundamental Approaches to Software Engineering. XV, 440 pages. 2007.

Vol. 4421: R. De Nicola (Ed.), Programming Languages and Systems. XVII, 538 pages. 2007.

Vol. 4420: S. Krishnamurthi, M. Odersky (Eds.), Compiler Construction. XIV, 233 pages. 2007.

Vol. 4419: P.C. Diniz, E. Marques, K. Bertels, M.M. Fernandes, J.M.P. Cardoso (Eds.), Reconfigurable Computing: Architectures, Tools and Applications. XIV, 391 pages. 2007.

Vol. 4418: A. Gagalowicz, W. Philips (Eds.), Computer Vision/Computer Graphics Collaboration Techniques. XV, 620 pages. 2007.

Vol. 4416: A. Bemporad, A. Bicchi, G. Buttazzo (Eds.), Hybrid Systems: Computation and Control. XVII, 797 pages. 2007.

Vol. 4415: P. Lukowicz, L. Thiele, G. Tröster (Eds.), Architecture of Computing Systems - ARCS 2007. X, 297 pages. 2007.

Vol. 4414: S. Hochreiter, R. Wagner (Eds.), Bioinformatics Research and Development. XVI, 482 pages. 2007. (Sublibrary LNBI).

Vol. 4412: F. Stajano, H.J. Kim, J.-S. Chae, S.-D. Kim (Eds.), Ubiquitous Convergence Technology. XI, 302 pages. 2007.